Managing
COMPUTER
NUMERICAL
CONTROL
Operations

Managing
COMPUTER
NUMERICAL
CONTROL
Operations

How to
Get the Most
Out of Your
CNC Machine Tools

BY MIKE LYNCH

Society of Manufacturing Engineers
One SME Drive, P.O. Box 930
Dearborn, Michigan 48121

Library of Congress Catalog Card Number: 95-70370
International Standard Book Number: 0-87263-466-3

Additional copies may be obtained by contacting:

Society of Manufacturing Engineers
Customer Service
One SME Drive
Dearborn, Michigan 48121
1-800-733-4763

SME staff who participated in producing this book:

Donald A. Peterson, Senior Editor
Dorothy M. Wylo, Production Assistant
Rosemary K. Csizmadia, Operations Administrator
Sandra J. Suggs, Editorial Assistant
Judy D. Munro, Manager, Graphic Services

Printed in the United States of America

For everyone who aspires to improve.

Table of Contents

PART III—IMPROVING UTILIZATION THROUGH SETUP AND CYCLE TIME REDUCTION

9 Cycle Time Reduction Techniques 257

Preface

Computer numerical control (CNC) machine tools are among the most popular forms of manufacturing equipment in existence today. What they can do to improve a company's productivity ranges from producing simple workpieces to helping a user accomplish previously impossible tasks. With the range of CNC equipment now available, almost all companies that manufacture *anything* utilize CNC equipment in one form or another.

The impact CNC machine tools have on a company's success, and the diversity they offer, present the CNC user with a wide variety of challenges. The greatest challenge any CNC user faces is utilizing the available CNC equipment to its optimum capacity. While this broad challenge can come in many forms, the company's very survival may depend on how well the challenge is met. It is the primary purpose of this text to address the issue of improving CNC machine tool utilization.

UTILIZATION DEFINED

CNC machine utilization, defined as *the effectiveness level of a company's CNC machine tools,* can be measured in many ways, depending on a company's specific goals and criteria for its own production success. For any two companies, the goals for CNC machine utilization and the criteria for measuring them may vary dramatically and often conflict. This can lead to confusion and controversy among CNC users about how their CNC equipment should be best utilized.

Do not confuse utilization with *application.* To show the difference between utilization and application, compare CNC machine tool utilization to the use of personal computers. Identical personal computers can be purchased by different companies (or different departments within the same company) and applied in totally different ways. One computer may be used for word processing, another for accounting purposes, yet another for computer-aided design and manufacturing (CAD/CAM). The applications for personal computers are virtually limitless. While one application may push a computer closer to its computing limits than another, to say that a computer is being underutilized solely on the basis of its application would be incorrect. The computer used for simple word processing is as important to its user as the one used for complex CAD/CAM work.

The same goes for CNC machine applications. For example, two identical CNC machining centers may be used for totally different applications. One may be used as little more than a glorified drill press, simply drilling and tapping holes in mold bases. The other may be used by an aerospace company to machine complex 3-dimensional aircraft components. While the machine producing aerospace components may be performing more complex machining operations and pushing the machine closer to the limits of its capabilities, the machine simply drilling and tapping holes in a mold base is every bit as important to its owner.

In the computer analogy, any of the computers may be *underutilized* if its user does not apply proper usage techniques. For example, if the computer user does not understand the basics of directory structure and file management, it is likely that this lack of understanding will eventually lead to usage problems. Files may be misplaced or, worse, lost, causing wasted time and duplication of effort. Optimum utilization will not be possible, because of the user's inadequacies.

In the same way, either CNC vertical machining center could be underutilized if its user is not taking advantage of the best usage techniques. If, for example, the user of the machine producing complex aerospace components is unaware of an off-line program verification system that could completely eliminate program verification time at the machine, some setup time will be wastefully expended on program verification. This, of course, means the machine will not be utilized to its fullest capacity. In similar fashion, if the owner of the machine producing mold bases is not using the best available method of program storage and retrieval, program transfer time will suffer, as will the machine's overall utilization.

Note the subtle distinction between machine application and machine utilization. Each company using CNC machines will have different applications that warrant differing and often conflicting usage techniques. Though each company will push the machine to different levels of its capacity with regard to horsepower, spindle speeds, complexity of work, etc., the machine's application by itself has little to do with how well the company is utilizing the machine. *Any* company could be underutilizing any of its CNC machine tools if the best usage method available for the particular application is not being applied. It is the primary focus of this text to present techniques you can use to improve the utilization of your CNC machine tools.

OBJECTIVES

CNC machine tools tend to be so productive that most may run at less than optimum levels and still satisfy their owners. A company could be dramatically underutilizing its CNC equipment and still be thoroughly satisfied with its current level of performance. Couple this with the human tendency to leave well enough alone and you can see why CNC machine tools tend to remain underutilized, even *after* a user recognizes the potential for improvement.

Given the global competition manufacturing companies face today, it should be of paramount importance to ensure that all production machine tools are running at their peak performance levels, regardless of how well they *appear* to be running. We attempt in this text to show you ways to achieve peak utilization for your CNC machine tools, based on your company's own criteria for success. We will do so by presenting countless ideas and suggestions related to understanding and evaluating your own CNC environment, providing CNC training for your personnel, reducing CNC setup and cycle time, and a variety of other important CNC issues.

It is difficult to show every reader how to achieve peak CNC utilization in every CNC environment. The sheer number of CNC machine types, different application techniques, and operating differences among companies make it impossible in this text to give clear and concise instructions for every scenario. For this reason, our secondary objective is to help you adapt what you learn from this text to your own CNC environment. While a given topic or technique being presented may not directly apply in your particular CNC environment, you may gain insight into how improvements can be made by comparing your needs to the needs of others.

Keep in mind that there are many successful methods of applying CNC machine tools. Additionally, there are limitless and sometimes conflicting criteria that determine any single company's optimum CNC machine utilization. What one company considers optimum utilization may be perceived as subpar by another. Suggestions for improvement given here include criteria for judging whether the suggestion is right for you.

Also remember that a solution that overcomes one problem may create another (possibly unforeseen) problem. As you consider implementing any improvement or a change of any kind, you must constantly be on the lookout for potential problems. As we present the potential improvements to your CNC environment, we will attempt to acquaint you with the potential problem areas as well.

THE EVOLUTION OF CNC IN YOUR COMPANY

One of the largest problems associated with attaining optimum CNC machine utilization has to do with what brought your company to its current CNC environment. When your company purchased its first CNC machine tool (how many years ago?), it is likely that absolutely no consideration was given to future needs. A CNC machine was purchased to perform the machining operations required at the time.

It is quite likely that your company's management staff was so thrilled at how well that first CNC machine performed (compared to previous methods) that another was purchased soon after the first. And then another, and another. Your company probably purchased several CNC machine tools before someone realized that you had a *CNC environment,* and that it required organizing. If you ask the people involved with initial CNC machine purchases, they would probably tell you that things would be done much differently if they had to do it over again.

Unfortunately, this scenario is common among CNC users. *Many* CNC users have accumulated a hodgepodge of CNC machine tools that were purchased individually, with little (if any) consideration for future company needs and growth.

We liken the evolution of a company's CNC environment to the evolution of any city. Often, there is no rhyme or reason to the city's initial development.

Someone wants to open a grocery store and simply buys the land and builds the store. Then someone else decides to open a department store. Then a drug store. A liquor store. Sooner or later, someone in the local government determines the need for a *city planner* or *planning committee* to organize the opening of future businesses to suit the needs of the community. The new city planners are commonly faced with the difficult task of implementing restrictions and rules that allow future growth while still accommodating the current hodgepodge of local businesses.

Just as cities have found it necessary to organize their communities with city planners, so have many manufacturing companies found it necessary to organize their CNC environments with *CNC coordinators* and *CNC coordination committees*. And unfortunately, just as city governments have found that substantial problems exist when they finally decide to begin organizing, managers in manufacturing companies find that the initial task of organizing their CNC environments is a difficult one.

A third objective of this text assumes your company has evolved in the manner described. We discuss the problems a company faces when there is little or no continuity among the CNC machine tools a company owns. Using the most common problems, we show ways to implement needed changes in a congruent manner. As stated, when you implement *any* change, it is likely to have repercussions. Where applicable, we show you how you can organize and coordinate implementation of change while still considering the overall needs of your company.

CNC MACHINE TYPES TO BE DISCUSSED

The main topics of this text can be applied to virtually any form of CNC equipment, but the specific examples are related to the two most popular forms, the CNC machining center and the CNC turning center. Almost all companies with CNC equipment have one or both of these highly productive CNC machine types.

Admittedly, there are many other forms of CNC equipment currently being used in manufacturing. When it is required to make a specific point, this text will discuss other types as well. However, since almost all CNC-using companies have one or more machining centers and/or turning centers, they will be used as the example machines to stress most of the points made in this text.

ORGANIZATION OF THIS TEXT

Though this text can be used as a reference guide, allowing you to quickly scan to your subjects of interest, our intention is for you to read it from cover to cover. The presentations constantly build on previous information, using a building-block approach. We also assume that you understand concepts and ideas presented during earlier presentations to avoid duplicating presentations.

The text is divided into four parts, each with three chapters. Chapter One introduces the elements of the CNC environment, ensuring your understanding of core elements, satellite elements, and target elements affecting and affected by your CNC machine tools. Chapter Two describes basic principles related to an important manufacturing philosophy called *value added*. This will give you a good appreciation for what is truly important in your CNC environment. Chapter Three will help you define your company's CNC needs. You will learn the three most basic company types, the main factors that contribute to your company profile, and how the CNC coordinator can organize the CNC environment based on your company's goals.

Part II is dedicated to helping you implement your own in-plant CNC training program. Chapter Four introduces you to key industrial training issues, including the importance of well-trained personnel, the benefits of implementing your own training program, how to find and hire CNC people, and the resources available for help with CNC training. In Chapter Five, we show you how to develop your own in-plant training program: how to limit the scope of CNC training, set up a good learning environment, and select your instructor. In Chapter Six, we offer course content your instructor can use to present courses on CNC machining centers and turning centers. Course outlines are included, as well as detailed instructions for presenting CNC courses based on a proven *key concepts* approach.

Part III shows how to improve CNC machine tool utilization by the two most important ways possible: setup and cycle time reduction. Chapter Seven introduces you to the principles of setup time reduction. We discuss the importance of setup time reduction, define the tasks related to setup time, and describe the three most basic ways to reduce it. Chapter Eight is a lengthy chapter offering many specific techniques and examples that show how setup time can be reduced. Included are techniques that can be used from the time the last workpiece in the current production run is completed to the time the first good workpiece in the next production run is machined. Chapter Nine offers the same kind of detailed information for cycle time reduction techniques. We introduce cycle time reduction principles, and discuss ways to reduce workpiece loading time, reduce program execution time, and minimize cutting tool maintenance time.

Part IV is a collection of ideas aimed at addressing other important CNC issues. In Chapter Ten, we address documentation issues, focusing on why documentation is so important. The chapter also offers suggestions on how documentation should be done. In Chapter Eleven, we present alternatives for program preparation, storage, transfer, and verification devices, while Chapter Twelve discusses service and maintenance issues.

PART

I

UNDERSTANDING YOUR CNC ENVIRONMENT

Throughout this text, we make references to your computer numerical control environment. In Part I, we take a conscious look at your entire CNC environment. Since your CNC environment includes all that affects, or is affected by, your CNC equipment, you may be surprised to find that there are facets that you have not previously considered.

We begin in this manner for three reasons. First, we wish to ensure that all readers understand how large a role CNC equipment plays in the manufacturing companies that use it.

As you will see in Chapter One, there is almost no area of your company that is not in some way touched by these highly productive machines. With a firm understanding of the full scope of your CNC environment, you will be better prepared to make improvements.

Second, you must understand how to judge your company's current level of CNC utilization. While the evaluation of CNC machine utilization is affected by many factors and will change from one company to another, it should be based primarily on how productive your CNC machines are versus how productive they could be. In Chapter Two, we introduce you to a very important manufacturing philosophy to help you better gage your CNC utilization.

Third, you must be able to determine how to implement improvements based on your own company's needs. As we discuss throughout this text, companies using CNC vary dramatically with regard to how CNC is applied, and no two companies will have exactly the same needs. In Chapter Three, we introduce you to the three basic company types to help you better understand what should be important to your own organization.

Chapter One

Elements of the CNC Environment

I n this chapter, we introduce the term *CNC environment* and discuss its basic elements. While our focus will be on manufacturing companies that produce a product, most of the points made will apply to any company that uses CNC equipment. Some of this information will be little more than a review for experienced CNC people. However, we wish to make sure that all readers have a firm understanding of these important points.

CNC ENVIRONMENT DEFINED

The CNC machine tools in your company are influenced by many factors. The methods by which you prepare programs, what kinds of cutting tools you use, how you make setups, what kind of distributive numerical control (DNC) system you use, how aggressive you are with cutting conditions, and the complexity of your work are a few of the many factors that determine how well your CNC machine tools function for your company. Indeed, how well you utilize your CNC machine tools is the sum total of these influences.

Conversely, CNC machine tools influence many company functions. In fact, there is almost no area of your company that is not in some way touched by CNC technology. In some areas of your company, the impact felt from CNC machines is quite dramatic and obvious. In product-producing companies, for example, people who assemble component workpieces have probably come to appreciate the consistency among workpieces produced on CNC machine tools.

In other cases, the impact of CNC machines may be less direct and not as easily felt. Production control people, for example, have likely found that smaller inventories are necessary due to the shorter lead times CNC machine tools allow (especially as compared to previous production methods), since one CNC machine can commonly complete a workpiece that may have previously required several machining operations. Though this is the case, improved production control capability is probably not one of your company's initial reasons for purchasing CNC equipment.

In other, rather unfortunate cases, the impact of the CNC machines may be so obscure that it goes completely unnoticed by a key person in your company. If, for example, design engineers are not aware of all your company's CNC machine tools can do, their ability to come up with the most efficient and competitive workpiece designs will be severely impaired.

Since CNC machine tools both influence and are influenced by the way your company functions, we offer a 2-part definition for the term *CNC environment*. First and foremost, your CNC environment includes *any portion of your company that affects the utilization of your company's CNC equipment*.

Within this first portion of the definition, there are certain highly important elements that are part of almost every CNC environment. These we call *core elements* of the CNC environment. Since core elements are so important, they are relatively obvious and their influences on the CNC environment are easy to visualize. As we discuss them, it is likely that you will easily recognize each one. However, there are other, not-so-obvious elements that affect the CNC environment to a lesser extent, and their influences are not quite so easy to visualize. These we call *satellite elements* of the CNC environment. Since these elements are less obvious, and since the consequences of ignoring satellite elements can be severe, we discuss them in greater detail.

Second, your CNC environment includes *any portion of your company that is in some way affected by the use of your CNC equipment*. We call these *target elements* of the CNC environment.

All too often, so much emphasis is placed on core elements that satellite and target elements are ignored. While core elements do demand the most attention, there are many implications of satellite and target that can lead to gross underutilization of CNC equipment if they are overlooked. Figures 1-1 and 1-2 illustrate the relationship of many core, satellite, and target elements in a typical CNC company.

TYPES OF ELEMENTS

Elements within the CNC environment are made up of machinery, personnel, and actions. *Machinery and mechanical devices* are the machine tools themselves, as well as other machines and accessories that in any way influence (or are influenced by) the utilization of CNC equipment. *Personnel* in your CNC environment include any people who affect (or are affected by) the utilization of your company's CNC equipment. *Actions* are performed by the people in your CNC environment and include any actions that affect (or are affected by) the way your CNC equipment is utilized.

CORE ELEMENTS

Though no two companies will have identical CNC environments, *every company* will have in its CNC environment those elements it considers to be of extreme importance to the utilization of its CNC machine tools. The list of items we offer represents core elements that are common among CNC users. You must be prepared for variations resulting from your own company's application of its CNC equipment.

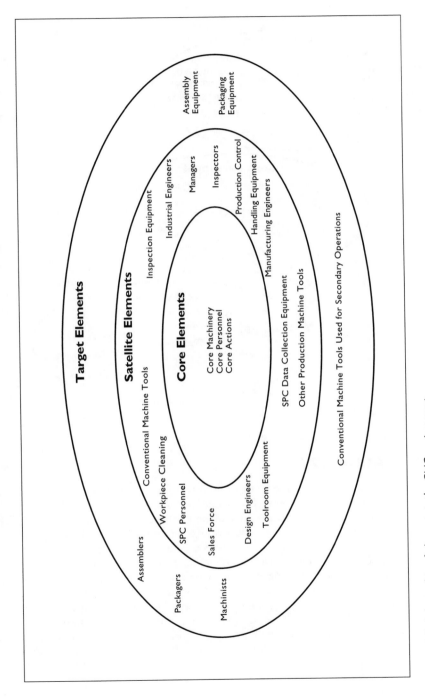

Figure 1-1. Relationship of elements in the CNC environment.

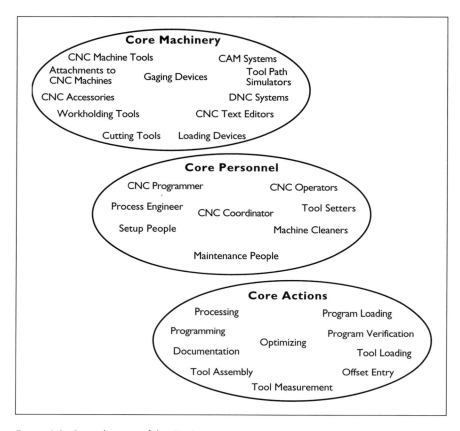

Figure 1-2. Core elements of the CNC environment.

CORE MACHINERY

Your CNC machine tools are the heart of your CNC environment. As CNC machines are purchased, their axis travel specifications must accommodate the range of workpiece sizes your company machines. Their horsepower specifications must be chosen to machine workpieces as efficiently as possible. Their accuracy specifications must reflect the workpiece tolerances your company expects to hold. Their automatic tool changers must hold the number of tools needed to machine the workpieces machined by your company. Compromises may have been made during the machine's purchase because of cost constraints, floor space available, and other considerations. Despite these compromises, generally speaking, *all* qualities of your CNC machine tools, right down to their placement in the shop, must be chosen to match your company's specific needs based on the kind of work your company does. Because they are so very important in that environment, most improvements you consider making in the environment will involve your CNC machine tools.

As important as your CNC machine tools are, they are but one type of mechanical device included in the core of your CNC environment. Here we discuss seven other types of mechanical devices you must consider to be part of your CNC environment.

Machine accessories

Accessories are attachments to the CNC machine tool that enhance its use. They may be purchased from the machine tool builder during the initial machine purchase or bought from an after-market supplier at any time during the machine's life. Turning-center accessories include bar feeders, part catchers, tailstocks, steady rests, live tooling, special workholding devices, tool touch-off probes, postprocess gaging systems, tool-life management systems, and automatic loading devices. Machining center accessories include automatic tool changers, pallet changers, indexers, spindle-speed increasers, right-angle heads, tool-life management systems, and probing systems.

Accessory devices can often spell the difference between the success and failure of a given CNC machine tool. In some cases, machines not producing enough workpieces can be dramatically improved by accessories. A machining center application ailing from poor workpiece loading time may be improved by adding a pallet changer. In other cases, the correct accessory can make seemingly impossible applications possible. Programmable steady rests, for example, can be added to a turning center to allow the machining of extremely long shafts.

Workpiece loading devices

Before any CNC machine can perform its function, a workpiece (or series of workpieces) must be placed into the machine by one means or another. While many times a CNC operator acts as the sole workpiece loader, mechanical devices often can enhance or replace the operator. Mechanical loading equipment ranges from simple devices that help an operator manually load workpieces (like overhead and boom cranes) to fully automatic systems (robots, autoloaders, etc.). Since workpiece loading must often be done while the CNC machine is down, workpiece loading devices can have a dramatic impact on CNC utilization and must be considered as important core elements of your CNC environment.

Other CNC accessories

There are countless devices that can be used to streamline the operation of CNC machine tools on the shop floor. Most are designed to enhance the CNC setup person's ability to make setups or the operator's ability to run the CNC machine. A tool length measuring gage, for example, helps the setup person determine the length of each cutting tool. In similar fashion, edge finders help the setup person determine the location of the program zero point needed for the program. A chuck pressure gage helps the turning center operator to precisely set the clamping force needed to grip thin-wall workpieces. Since most of these devices are inexpensive and can help streamline the operation of CNC equipment, you should be constantly on the lookout for new accessories that can help your CNC people.

Tooling

Tooling has a direct impact on the productivity of CNC machines. This is true of both workholding tools and cutting tools. Good fixturing, for example, can ensure that workpieces are held with the rigidity required for machining. Additionally, the proper use of fixtures can minimize tool changes if multiple workpieces are held. Similarly, cutting tools have a direct relationship to machining time. The better the tooling, the faster machining can take place. Cutting tool maintenance can also be kept to a minimum if the proper cutting tools are used. All of this means that tooling must be considered an essential core element of your CNC environment.

Keep in mind that CNC machine tool technology and cutting tool technology are constantly leapfrogging one another. The machine tool builder designs a CNC machine that can withstand all that today's cutting tools can do. Tomorrow, next week, or next year, a tooling manufacturer will develop a new grade of cutting tool that improves on current technology. It is then the machine tool builder's turn once again to catch up.

Though it can almost be a full-time job, you must strive to keep track of changes in tooling technology. If you have been working with the same cutting tools for the past five years, for example, you can probably realize substantial improvements in productivity simply by implementing state-of-the-art cutting tools.

Tool storage devices

How quickly and easily your setup people can locate tooling can have a large impact on CNC machine productivity. During setup, a great deal of time can be wasted if setup people have to search the shop for needed fixtures and cutting tools. In similar fashion, production will be held up if a CNC operator must search for replacement inserts and cutting tools. For this reason, any devices you use to store, organize, and track tooling must be considered core elements of your CNC environment.

Gages

The functionality of your company's gaging equipment will directly affect the productivity of your CNC environment. Since many companies require their CNC operators to take measurements while a machine is running production, the easier you can make it for your people to use and interpret gaging equipment, the faster they will be able to perform gaging-related tasks, and the less likely it will be that the machine will sit idle while a CNC operator completes the gaging procedures on a workpiece. For this reason, any gages your CNC setup people and operators use must be considered core elements of your CNC environment.

Program preparation, transfer, and verification devices

The need to create, transfer, and verify CNC programs varies from one company to the next. A product-producing company that has only a limited number of different workpieces to machine may not even need a program preparation device. It may simply need a way to transfer completed programs to and from its CNC machine tools. Another company that commonly runs low production quantities and rarely sees the same job a second time constantly creates new programs. Though the device needs of companies vary, the devices themselves have a dramatic impact on the CNC environment and must be considered as core elements.

CORE PERSONNEL

Here we list a few of the people closest to the core of the CNC environment, commonly those who work with CNC on a full-time basis. Companies employ their CNC people in different ways; larger companies tend to isolate functions, minimizing the number of different duties one person performs, while smaller companies tend to have each person perform several duties. Though we list each person as if he or she performs only one function, the same person may actually be performing several of the listed assignments.

CNC coordinator

While not all companies designate a person with the title of *CNC coordinator*, all companies require coordination among the people working in the CNC environment. The goal of the CNC coordinator is to make sure that all people working in the CNC environment understand the needs of the others. As noted throughout this text, dramatic improvements in CNC utilization are possible with something as simple as improved communication among the people in the CNC environment. The more people in the CNC environment, the more important it is that someone be given the responsibility of coordinating the activities of the people involved.

Process engineer

The person who prepares the method by which workpieces are produced is the process engineer. This person must have a good understanding of machining practice and know the capabilities of all machine tools the company owns. Using this knowledge, the process engineer routes the workpiece through the related machine tools.

Programmer

The responsibilities of the programmer vary dramatically from one company to another. In companies that dedicate their CNC equipment to running extremely high production, the need for a programmer is all but eliminated as soon as a few programs are prepared. In companies that run lower production quantities with limited repeat business, a programmer (or group of programmers) can be kept quite busy trying to keep up with the demands of the CNC machine tools. In extreme cases, when a company produces only one or two workpieces per production run, the CNC operator may actually be the person preparing the program, possibly using *shop-floor programming* techniques.

While a programmer's main responsibility is to prepare programs, most companies rarely utilize their programmers solely for this purpose. More likely, the programmer will also be involved with the development of the process, the design of cutting tools and fixtures, and the preparation of any documentation necessary to help other people involved with the CNC operation (setup sheet, tool station designations, process drawings, etc.). The programmer is usually expected to help the setup person verify and optimize new programs. Additionally, the tasks of storing and organizing programs, maintaining program transfer devices, and helping operators and setup people on an ongoing basis commonly fall on the shoulders of the programmer.

Setup people

Almost all CNC machine tools require major changes in setup whenever a new workpiece must be run. At the very least, a new program must be loaded into the CNC control. More likely, workholding devices must be changed, cutting tools must be added and repositioned, and the program zero position must be measured and entered. This commonly requires someone with very good mechanical aptitude who understands the machine tool, programming, fixturing, and cutting tools.

Some companies break up the tasks related to setup even further. One person may perform the workholding setup. This person is commonly called the *setup person*. Another person, called the *tool setter*, may assemble and measure (or even preset) the cutting tools used within the setup.

Machine operators

CNC machine operators are the people who actually run the machines in production. Their responsibilities typically include loading and removing workpieces, activating the machining cycle, checking critical dimensions and adjusting tool offsets, and, in general, doing whatever it takes to keep production moving. Because many companies have several CNC machines and since they are usually run for two or more shifts, machine operators commonly make up the largest percentage of people working in the CNC environment.

Trainers

As it becomes more difficult to find, hire, and keep qualified CNC people, more and more companies are developing their own in-house training programs. Since they will have a profound influence on the capabilities of CNC workers, the people who teach CNC functions for your company must be considered essential core resources of your CNC environment.

Service and maintenance people

One important core resource of the CNC environment that is often overlooked is the person who performs the servicing and preventive maintenance on core machinery. Wise companies implement preventive maintenance programs that ensure maintenance is scheduled at *regular* intervals. This ensures optimum utilization of CNC machines and helps avoid the need for *corrective*, or reactive, maintenance.

Machine cleaners/laborers

Many companies, especially larger ones, employ people whose sole responsibility is to facilitate the use of CNC equipment (as well as other forms of capital equipment) for the people actually using the machine tools, predominantly CNC setup people and operators. These people empty chips, keep the machines clean, get needed tools, and, in general, do the running for setup people and operators.

CORE ACTIVITIES

Many core activities take place in the CNC environment. We offer a few here in the order they are typically performed when developing new CNC programs.

Processing

Most forms of CNC equipment, including machining centers and turning centers, can accommodate several tools. This gives them the ability to perform several machining operations during one cycle. For a given CNC operation, the process includes the order (sequence) by which machining operations are to be performed, the tooling to be used to machine the workpiece, and the cutting conditions for each tool.

In some companies, the process engineer is very focused, specifying the exact order in which machining operations must be done. In other companies, he or she may offer only a vague list of operations that must be completed by each machine tool, leaving it to the programmer to develop the actual step-by-step operations.

Since the process is of paramount importance to the success of each job, the task of processing must be considered one of the most important core activities performed in the CNC environment. Even a poorly written program can be made to work if the process is good. Conversely, a perfectly written program (one that does precisely what the programmer intends) will not make good workpieces if the process is bad.

Programming

The success of any CNC environment is directly influenced by the quality of programming, making programming one of the most important core elements. Programming for CNC machine tools can be done in several ways. If done by a person off line (away from the CNC machine tool), there are two common methods of programming. The first is to prepare the program *manually*, at the same level at which the CNC control will accept it. This method of programming, commonly used for simple work, requires the programmer to perform numerous math calculations and code instructions to the machine into a series of commands.

The second method of CNC programming off line is with the help of a computer. Computer-aided manufacturing (CAM) systems eliminate much of the drudgery from programming and free the programmer from much of the math required to generate manual programs. These systems also tend to be quite graphic, allowing the programmer to see exactly what the CNC program will do on the screen of the computer *before* the program is sent to the CNC machine tool.

For certain applications, programming off line may not be the best method. Some companies expect their CNC machine operators to do the programming. This is commonly the case in many smaller shops and especially when lower production quantities are run. These companies tend to prefer *shop-floor programming* over *off-line programming*. This form of programming requires a special kind of CNC control, commonly called a *conversational control*, which can be programmed quickly and easily. Conversational controls are programmed in much the same way a CAM system is, but at the CNC machine tool itself.

Documentation

CNC machine tools require a great deal of documentation. The better the documentation, the easier it will be for all those who are part of the CNC environment to perform their duties.

Most CNC program documentation will be done by the programmer. Within the program itself, many programmers document basic things including part name, part number, date, revision number, and tools. Separate from the program, additional documentation may be needed. Programmers will typically use forms, drawings, and even computer files to document important points about the program.

A common set of documentation for a program includes a setup sheet telling how to make the setup, a tooling layout specifying how tools are to be assembled and placed into the machine, a drawing or photograph illustrating how the setup is made, and any other information considered useful to the setup person.

Cutting tool assembly

This core activity is usually performed by the tool setter or setup person. Tools must be assembled according to the programmer's instructions, from information contained in the setup sheet documentation.

Cutting tool measurement, loading, and offset entry

CNC machining centers and turning centers must be aligned with the cutting tools being driven by the program. Some kind of measurement is usually taken, and measured values are commonly entered into a tool offset. Machining centers, for example, require tool length measurement (and possibly tool radius measurement) for every tool. The length and radius values then must be entered into corresponding tool offsets. Additionally, the tools must be loaded into the automatic tool changer magazine (or turret) for use by the program. Since the efficiency of these activities can impact the productivity of your CNC machines, they must be considered as core activities of your CNC environment.

Program zero measurement and entry

When preparing to run the program, a setup person must specify to the control by one means or another the distance between the program zero point of the program and the machine's reference point. Though the method for determining this distance varies greatly from one machine to another, it usually involves some measurements on the machine. Once measured values are found, they must be entered into the control.

Program loading

In most applications, programs must reside within the control's memory before they can be activated. While all current-model CNC controls allow programs to be entered through the keyboard and display screen of the control, CNC controls make extremely expensive typewriters. Instead of typing programs at the machine tool, most CNC users utilize some form of program transfer device to load the program (created during the programming step) into the CNC control.

Program verification and optimizing

Finally, the setup person is ready to test-run the program. If the program is new, extreme caution must be exercised. This is especially true if the program has been prepared manually. Once verified, and after successfully running the first workpiece, the program should be optimized. The higher the production quantities, the more important the optimizing step.

Program retrieval

During verification and optimizing, it is likely that several changes will be made to the program residing in the control's memory. This means the program in the control no longer matches the original version of the program. In order to be able to run the program again at some future date, it will be necessary to transfer a copy of the program from the CNC control to a permanent program storage device (computer hard disk, floppy disk, etc.).

Running production

Once a program is efficiently running workpieces, the production run is turned over to the CNC operator. Though actual responsibilities vary from company to company, at the very least, the operator will be responsible for loading and removing workpieces as well as activating the cycle.

Cutting tool and offset maintenance

As the production run continues, and especially for higher workpiece quantities, tools will wear. Though a tool may not be dull enough to be replaced, tool wear commonly causes the size of the workpiece being machined to change, especially with turning-center tooling. A CNC operator may have to adjust a tool's offset several times during its life to compensate for tool wear. Eventually, the cutting tool (or insert) must be replaced, and the offset set back to its original value.

Other core activities

Though we have described only the most basic core activities, there could be many more in your particular CNC environment. They may include:

- Preventive maintenance activities.
- Designing cutting tools for CNC machine tools.
- Designing fixtures and other workholding devices.
- CNC machine cleaning and general upkeep.

- CNC machine warm-up (if required).
- Any activity essential to the use of CNC machine tools.

SATELLITE ELEMENTS OF THE CNC ENVIRONMENT

Most of the core elements mentioned here are quite obvious and easy to understand. You can easily see their relationship to the CNC environment. Satellite elements tend to be a bit more obscure.

While we list the most common satellite elements, there may be more in your own company's CNC environment. Truly, any piece of equipment, any person, or any action that affects CNC utilization in any way, regardless of how small the impact, must be considered part of the CNC environment.

SATELLITE MACHINERY

Here we list machines and devices that are not actually considered part of a company's CNC machinery, yet impact the CNC environment.

Conventional equipment

Most companies have a wide variety of machine tools with which they produce their products. Conventional machine tools like cutoff saws, screw machines, turret lathes, engine lathes, drill presses, milling machines, surface grinders, and all kinds of fabrication equipment work side by side with CNC equipment.

Any conventional machine tool must be considered part of a company's CNC environment if it affects any facet of CNC utilization. There are at least two occasions when this is the case.

The first is when a CNC operator is expected to perform secondary operations on conventional equipment while the CNC machine tool is in cycle, running a workpiece. These secondary operations may be done on the workpiece being machined by the CNC machine tool, or on unrelated workpieces.

During a long machining center cycle, for example, the CNC operator may be expected to perform tapping operations (on a standard drill press with a tapping head) for the workpiece being machined on the CNC machining center. As long as the tapping can be accomplished while the CNC machining center is still in cycle, the tapping is *internal* to the machining of the balance of the workpiece.

In this case, timing is quite important. The CNC operator must have the skill and ability to complete the tapping operation *during* the CNC machining center cycle. If for any reason this cannot be done, optimum CNC utilization will not be possible. Companies considering this method of CNC operator utilization must consider *all* operator capabilities and responsibilities and be on the lookout for conflicts. If cutting tools must be occasionally changed, if offsets must be occasionally updated, if chips must be occasionally cleared from the machine table, or if operators must perform *any* function that keeps them from performing the secondary operations internal to the CNC cycle, CNC utilization will suffer.

Some companies attempt to utilize any free time the operator has during the CNC machining cycle by having the operator perform secondary tasks on unrelated workpieces. In these cases, there is no urgency to the operations being performed. Whatever work the CNC operator can accomplish will be seen as a benefit to the company's overall performance. Operations like deburring, cleaning, tapping, polishing, and honing are commonly used to fill the CNC operator's time.

The second occasion when conventional machinery dramatically affects CNC utilization is when an operation is performed on a workpiece (by a conventional machine tool or process) prior to the CNC operation. In such cases, the CNC equipment is at the mercy of the preliminary operation. Following are some examples of instances when preliminary operations performed by conventional machines and processes can dramatically affect the success of subsequent CNC machining operations.

Processes that produce blanks. In many cases, the CNC operation is the first machining function to occur on a workpiece. The quality of the blank (the material as it comes to the CNC machine tool), regardless of how it is produced, influences the CNC operation. Any variations in blanks will lead to problems during the CNC operation. Processes used to produce blanks that are notorious for variation include casting, forging, and molding. The cold and hot forming of steel bar can also present problems for CNC machines, especially bar-feeding turning centers, when the straightness of the bar determines how fast the bar feeder can rotate without vibration. Unfortunately, blanks produced by these processes are commonly purchased from outside vendors, and your company may not then have total control of these preliminary processes. However, these processes must be considered as part of the CNC environment, especially when variations occur.

Cutoff saws. Cutoff saws are commonly used to cut the blank workpiece to the size it must be prior to a CNC machine's operation. Many companies that perform chucking work on CNC turning centers, for example, use cutoff saws to saw blanks for the turning center operation. The length of the blank is critical to the turning center operation. Any variation in length leads to problems in the CNC operation.

Facing and centering machines. These machines prepare the ends of shafts with a facing and center-drilling operation. Shaft length is critical to the subsequent CNC turning center operation.

Bar preparation machines. Many CNC bar-feeding turning centers require that the bar being machined be previously chamfered to allow easy loading through the bar feeder and spindle of the turning center. Since it has application only with CNC equipment, this device may also be considered a CNC accessory and, as such, part of the core elements.

Screw machines. Screw machines are commonly used as blanking machines for subsequent CNC operations. Many companies use them to simply prepare the raw material for eventual machining on CNC machine tools. Screw machines used for this purpose must be considered part of the CNC environment since they affect CNC utilization.

Any machine used prior to the CNC operation. Truly, any machine that alters the workpiece in any way (including other CNC machine tools) prior to the CNC operation will have an impact on the CNC operation itself and must be considered part of the CNC environment. In many cases, even if tolerances are held in prior operations, any variations among workpieces can wreak havoc with the CNC operation.

Inspection equipment

During our discussion of core elements, we mentioned gages and measuring tools. These are devices the CNC operator uses during the machining cycle to check and maintain tolerances by changing offsets. However, most companies also have more elaborate inspection equipment. Many companies, for example, use coordinate measuring machines (CMMs) to check workpieces. An inspector (usually *not* the CNC operator or setup person) will check a sampling of workpieces being produced by the machines in the shop (including the CNC machines). At the very least, the first workpiece must pass this kind of inspection before production can be started.

Any inspection machinery used to check workpieces machined on CNC equipment must be considered a satellite element of the CNC environment. In many cases, CNC machines are held up while the inspection of the first workpiece is taking place, meaning setup time for CNC machines must include inspection time.

SPC data collection equipment

Increasingly, companies are using statistical process control (SPC) techniques to document and ensure consistency among workpieces. Any equipment used by CNC people for data recording (personal computers, bar code readers, automatic gages, data entry terminals, etc.) must be considered part of the CNC environment.

Other production equipment

Any machine, device, or accessory a CNC person comes into contact with on a regular basis (including the time clock) should be considered part of the CNC environment. If the device takes time to operate, it could detract from the operation of CNC equipment.

Toolroom equipment

While some manufacturing companies have their tooling manufactured by an outside vendor, many maintain toolrooms in which they manufacture tooling themselves. Since the in-house toolrooms in these cases will be supplying the cutting tools, fixtures, and other tooling needed for CNC machine tools, these toolrooms must be considered parts of the CNC environment.

In fact, some of the machine tools used to produce tooling are often CNC machines themselves (in which case they could also be considered core elements of the CNC environment). While most of the points made in this chapter have been directed to *production machinery*, many of them do apply to CNC toolroom machine tools as well. However, since toolroom production quantities are generally quite low, and since the urgency of producing tooling is not normally as great as producing product, toolroom CNC machines present their own special concerns. For now, consider toolroom CNC machines as simply satellite elements of the CNC environment that support production CNC equipment.

SATELLITE PERSONNEL

Certain workers who do not necessarily work directly with CNC equipment have an impact on CNC utilization.

Design engineers

The impact of a design engineer on the product's cost and ease of manufacture outweighs that of *any* other single individual in manufacturing. Though this text deals only with matters of importance to the CNC environment, careful examination and consideration must be given to all facets of design engineering. The entire process of manufacturing begins with design engineering. A clean, thought-out design, developed with manufacturability as a prime concern, leads to easily interpreted workpiece drawings and efficient manufacturing processes.

We cannot stress this enough. If your company employs its own design engineering department, and if your design engineers understand CNC, you will likely find many manufacturing-related improvements in workpiece design that can simplify manufacturing without compromising the quality or functionality of your product. Even seemingly simple changes can lead to cost savings of thousands of dollars.

All too often, for example, design engineers take shortcuts with dimensioning and tolerancing to simplify their *own* work. One classic example of this is dimensioning of bolt hole patterns. Many design engineers dimension bolt hole patterns with the radius of the bolt hole centers, a starting angle, and the number of evenly spaced holes (Figure 1-3).

While this dimensioning technique requires a minimum of dimensions on the print (making it very easy for the design engineer), it forces *everyone else* involved in the manufacturing process (CNC programmer, CNC operator, inspector, etc.) to calculate the centerline coordinates of each hole on the pattern. The only person who does not have to make calculations is the person who should have made them in the first place. Figure 1-4 shows another way of dimensioning bolt hole patterns so that people in the CNC environment do not have to make calculations.

There are many dimensioning techniques that can make life easier for everyone in manufacturing. Datum surface dimensioning techniques, for example, allow all dimensions to be taken from a central location and can minimize (if not eliminate) the need for countless calculations.

In similar fashion, tolerancing techniques can dramatically impact the ease of manufacturing a workpiece. A tolerance of ±0.002 inch (0.051 mm), for example, is easier to read and interpret than +0.001 -0.003 inch (+0.025 mm -0.076 mm). If the mean value of the tolerance band is always specified as the nominal dimension (with an even plus/minus tolerance), the countless calculations made to determine the mean value can be eliminated (along with the related mistakes).

While mean value tolerancing will make life easier for everyone involved in the manufacturing process, it is especially helpful for CNC operators in adjusting offsets. In cases where entry-level CNC operators with very little basic machining practice background are hired, companies should take all necessary steps to minimize the calculations these beginning operators must make.

How process drawings can help. In many companies, the design engineering department is not in house. The design engineering department for a company that manufactures a product, for example, may be located in the corporate headquarters. In contract shops (job shops), the design engineering department is, of course, located in the customer's facility.

When you do not have direct control of the design engineer's methods for dimensioning and tolerancing workpieces, or when you wish to make it as simple as possible for everyone in the CNC environment, *process drawings* can be developed for each operation performed during manufacturing.

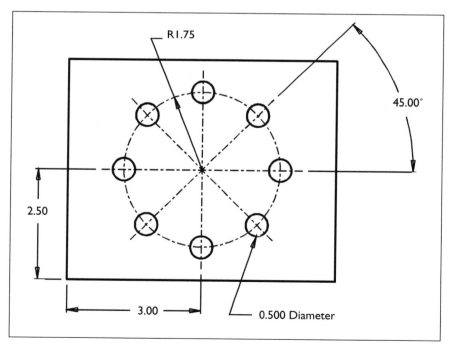

Figure 1-3. Although this method of dimensioning bolt hole patterns has been used for many years, it forces everyone in manufacturing to use trigonometry to calculate the position of each hole.

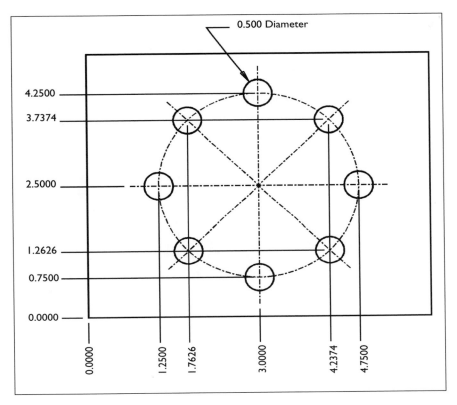

Figure 1-4. This method of dimensioning bolt hole patterns makes the position for each hole very clear to anyone working with the drawing.

A process drawing (commonly made by the process engineer, CNC programmer, or another industrial engineer) is made for each machining operation required in the process. Each is very specific, including only the information related to the operation. Only those dimensions and tolerances related to the operation are shown. They are given in the simplest possible manner, making it easy for everyone involved to understand the machining operations to be performed. Figure 1-5 shows an example of a complete blueprint. Figure 1-6 shows a process drawing made from this blueprint for a specific CNC turning center operation.

Though we have shown only a few simple examples of the design engineers' powerful influence on the CNC environment, we cannot stress enough their importance to the CNC environment. Truly, the manufacturing of a workpiece begins with the design engineer. Since so much information on how a workpiece must be machined is determined from the blueprint (either explicitly or implicitly), the better the design engineer can draw, dimension, and tolerance each workpiece, the easier it will be for everyone in the CNC environment. Design

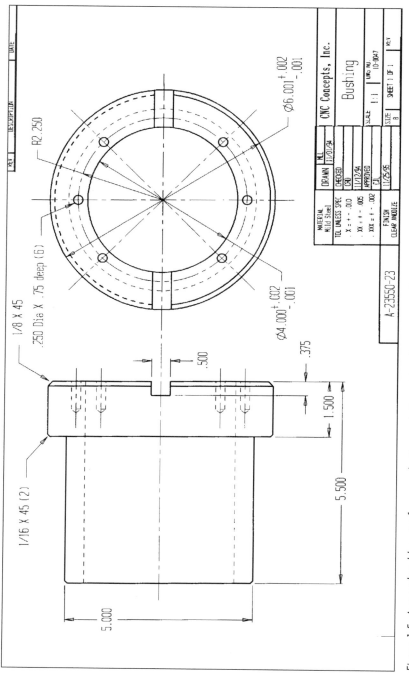

Figure 1-5. A complete blueprint for a production workpiece. Manufacturing people (especially CNC operators) will have difficulty interpreting tolerances.

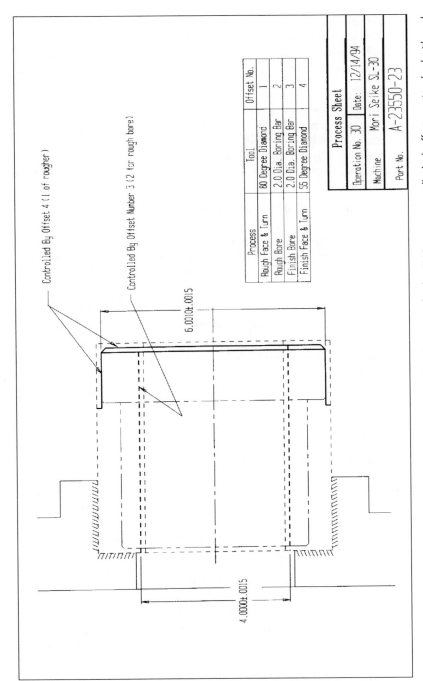

Process	Tool	Offset No.
Rough Face & Turn	80 Degree Diamond	1
Rough Bore	2.0 Dia. Boring Bar	2
Finish Bore	2.0 Dia. Boring Bar	3
Finish Face & Turn	55 Degree Diamond	4

Process Sheet

Operation No. 30	Date: 12/14/94
Machine	Mori Seike SL-30
Part No.	A-23550-23

Controlled By Offset 4 (1 of rougher)

Controlled By Offset Number 3 (2 for rough bore)

6.0010±.0015

4.0000±.0015

Figure 1-6. This process drawing clarifies dimension tolerances and makes it easy for the CNC operator to tell which offsets are involved with each machined surface.

engineers must be made fully aware of the influence they have on the CNC environment. While they do not have to be able to develop CNC programs, they should know the capabilities of your company's CNC equipment and have a good working knowledge of how programs are developed and how CNC machines are run.

Inspectors

Many companies have inspectors check the first workpiece coming off the CNC machine tool in each production run. In most cases, the CNC machine sits idle while the inspector performs the workpiece inspection. If this is the case, the inspector becomes a very important satellite person in your CNC environment. Inspectors must understand the importance of performing first workpiece inspections in a timely manner.

Managers in the CNC environment

While managers and foremen working in the CNC environment may not need the skills required to actually program and operate CNC equipment, at the very least they should be able to communicate intelligently with their staff. Their management skills in scheduling, handling human nature problems, scheduling preventive maintenance, etc. can ensure a successful CNC environment. The more they know about CNC utilization, the more effective they can be in CNC management.

Manufacturing and industrial engineers of all kinds

All industrial engineering personnel, at times, affect the CNC environment. Industrial engineers who perform time and motion studies can often locate and eliminate wasted effort by a CNC operator or setup person. Quality control engineers can minimize the effort of CNC personnel by providing proper measuring devices and other quality assurance equipment. Manufacturing engineers responsible for the procurement of new equipment have a dramatic impact on the future of your CNC environment.

Your company's sales force

Members of the sales force are responsible for selling your company's product and must be considered an important part of your CNC environment. Without them, there would be no CNC environment. Consider how important it is that these people understand the capabilities of your manufacturing facility and, more specifically, what is possible with your CNC equipment. A sales engineer may call on a customer who has a need for something just a little out of the ordinary. With a firm understanding of what is possible with your company's CNC equipment, and as long as your company can be flexible, the sales engineer at least will have the ability to pursue the sale.

This knowledge of your company's CNC capabilities is especially important for contract shops that use outside sales representatives. As you develop your relationship with your sales agents, you must confirm that they truly understand your company's capabilities. And considering how fast this industry changes, you must be sure to keep them abreast of changes within your company.

People who can help from outside your company

We devote an entire chapter (in Part II) to information resources. These include local tooling salespeople, machine tool salespeople, CNC applications engineers, and instructors in local technical schools. Anyone who can help with your CNC equipment utilization is considered part of your CNC environment.

SATELLITE ACTIVITIES

There are many activities not considered in the core of your CNC environment that still substantially affect CNC machine tool utilization. Keep in mind that any activity that affects CNC utilization must be considered part of your CNC environment. Many of these are simply basic factory functions, but in light of how much they affect your CNC equipment performance, they must be considered part of your CNC environment.

Work-in-progress handling

Most companies attach some form of routing sheet to each job as it goes through the shop. Commonly prepared by the process engineer, the routing sheet tells manufacturing people what machining operations must be performed. Depending on the size of your shop and complexity of your processes, it is possible that each lot of workpieces must be moved great distances (possibly several times) in the course of performing all manufacturing operations.

How well your company has designed its work-in-progress handling systems is important to your CNC environment. Ideally, the next lot of workpieces to be machined by a given CNC machine tool should be delivered before the CNC machine has finished its current job. It is likely that the setup person will need one or more workpiece blanks before setup can be made. Precious production time can be wasted if the CNC machine sits idle while an operator or setup person waits for workpiece blanks.

Workpiece cleaning, deburring, polishing, etc.

These are operations that, if performed prior to the CNC machining, can affect consistency among workpieces. Any variations in workpiece blank consistency can have a negative impact on the CNC operation. If performed *after* the CNC machining operation, these operations fall into the category of target actions.

Production control data collection

Many companies keep extensive records as to how much time is spent on each job done in the shop. For internal billing purposes, these companies require their manufacturing workers to report all work they do. The time reported is billed to the appropriate job. While most current production control data collection systems are quite easy for manufacturing people to use and are designed to allow fast input of information, the time a CNC person spends inputting data to the system is still nonproductive time for CNC machine tools and must be considered an important satellite element of the CNC environment.

Statistical process control data collection

Most companies today incorporate some form of statistical process control as a way of ensuring quality. With statistical process control, the CNC machine tool operator is required to report critical workpiece dimension information to the data collection system. Depending on the companies' specific SPC goals, reporting may be required for every workpiece or for a sampling of workpieces during the production run.

The time required for SPC data collection varies according to how much manual intervention is required of the person doing the reporting. With the best systems, automatic gaging is done on critical workpiece dimensions. The automatic gage is interfaced with the statistical process control system and will automatically report data to the system. At worst, the CNC operator may only be required to load workpieces into the automatic gage.

Unfortunately, not all SPC systems allow automatic entry of data. With some systems, measurements must be taken manually with conventional gaging equipment and entered manually into the data collection system. While various devices are used for this kind of data collection, usually some form of personal computer or remote terminal with a keyboard and display screen is involved.

If the CNC operator is the person responsible for reporting SPC data, the statistical process control system must be considered a very important satellite element of the CNC environment. Anything that detracts from the CNC operator's main function of operating the CNC machine tool can also detract from the utilization of the CNC machine itself.

TARGET MACHINERY, PERSONNEL, AND ACTIONS

Targets are machines, people, and actions that are affected by the use of CNC equipment.

The most common target elements are those related to secondary operations on workpieces after the CNC operations. In some cases, these are operations that a given CNC machine tool simply cannot perform. For example, a CNC machining center cannot perform grinding operations. If a surface being machined on a machining center must eventually be ground, this must be done subsequent to the machining center operation.

In other cases, it simply makes more sense to perform a given machining operation after the CNC operation, even though the CNC machine may be capable of performing the operation. Possibly the CNC operator has ample free time during the CNC machining cycle to perform secondary operations. This can minimize the overall time it takes to produce a given workpiece. For example, at many companies CNC operators perform tapping operations on workpieces on conventional machines while the CNC machine tool is working on the balance of the workpiece. This effectively eliminates tapping time from the overall machining cycle. The assumption here, of course, is that the CNC operator has time to perform the conventional operations *while* the CNC machine is in cycle.

Keep in mind that there are many other machines, personnel, and activities that deal with workpieces previously machined on CNC machine tools. Operations like cleaning, deburring, polishing, finishing, assembly, and packaging all fall into the category of target elements in your CNC environment.

Chapter Two

Applying Value-added Principles to Your CNC Environment

I t is important that your CNC machine tools (as well as any production equipment in your company) be utilized to their highest levels. Experts refer to the principles in this chapter as *enhancing value-added tasks*. Though we barely scratch the surface of value-added principles, our intent is to stress their importance to the utilization of your CNC machine tools. You should thoroughly understand the reasons *why* you must strive to utilize your CNC machines at their highest production levels.

These principles can help gage your current level of utilization as well as point to areas in need of improvement. Also, while these principles can be used to gage *any* company's level of efficiency (regardless of whether it uses CNC equipment), we focus on the value-added principles that have the most impact on the CNC environment.

YOUR COMPANY'S MOST BASIC FUNCTION

Ask yourself a simple question: *"What is it that our customers pay money for?"* Of course your company is in business to make money; the question has to do with naming specific things your customers pay money to get. Think about it. Put yourself in your customer's shoes. If you were buying a product from your own company, what would be important to you?

The larger the company, the less the typical manufacturing person is accustomed to thinking about a customer's needs or concerns. You may feel, for instance, that this question should be of concern only to salespeople, design engineers, and the company's upper management staff who work with customers on a regular basis. Yet your company's very survival is directly related to how well your company satisfies its customers' wants and needs. *Everyone* in your company should understand the needs of your customers, and should be willing to do whatever it takes to satisfy them.

Let's suppose your company manufactures valves for the oil industry. Your customer is an oil-well driller in need of oil flow control valves. Is your customer interested in paying money for your design engineer's time and efforts for developing the particular valves being purchased? You may say "yes, indirectly," but is your customer truly interested in buying the valve's design? No. Your customer is much more interested in buying a valve that holds up well in the field and, as long as it works, the customer probably couldn't care less about its design.

Along the same lines, is your customer interested in paying for the design of fixtures and cutting tools needed to machine the valves? No. Of course tooling must be designed and manufactured, and these costs will be reflected in the price of the valve, but your customer does not want to hear about tooling costs.

Is your customer concerned with paying money for the time it takes a process engineer to develop workable processes for the component parts of the valve? No. Certainly there will be a cost related to processing in order to make a quality valve, but your customer will be unconcerned with this process as long as the quality of the valve is sufficient.

Is the customer interested in paying money for your programmer's time for programming, the setup person's time for making the setup, or the time it takes to verify the CNC programs associated with each component? No. Again, costs will be incurred, but your accounts payable department cannot bill the customer for these charges.

Is the customer interested in paying for the time it takes to inspect workpieces and maintain statistical process control data entry? No. The customer may want to see your SPC data, but is unconcerned with any costs you incur to get the data.

Regardless of the product your company manufactures, similar questions will produce similar answers. Even if your company does not manufacture an end product, as would be the case with a contract shop producing specific workpieces for a wide variety of products, you can ask similar questions with essentially the same results. In all cases, your customers will not be concerned with *how* you produce acceptable products as long as what you produce is up to their standards. What they are paying for is *the production of the product*, not the manufacturing engineering required for making the product.

Remember that the same buying principles apply to consumer products. Whether the product is a tube of toothpaste, a jar of peanut butter, a television, or a valve for an oil well, customers will always be more concerned with the *product* than the methods by which it is manufactured.

If your customers are unconcerned with the methods by which your products are produced, when *is* your company making money? Is it while the design engineer develops new products? No. While a process engineer develops processes for component workpieces? No. While your programmers write programs? No. When your inspectors determine the quality of workpieces? No. When statistical process control data is collected? No.

The work your company does while *getting ready* to run production, while very important, does not bring in money. For the purpose of this text, we lump all of the tasks that do not make money for your company into a group called *necessary support tasks*. Though they may be essential to your company's ability to make money, the company does not profit directly from the work.

Here is a *partial* list of typical company activities that fall into the category of necessary support tasks:

- Accounting
- Accounts payable and receivable
- Design engineering
- Human resources
- Machine maintenance
- Manufacturing engineering
- Payroll
- Process engineering
- Production control
- Programming
- Purchasing
- Quality control and inspection
- Setup
- Shipping and receiving
- Tool crib
- Tool engineering
- Training

Admittedly, we still have not answered the question. The intent to this point has been to point out the many tasks that fall into the necessary support category, tasks that do not actually make money for your company. This does not suggest that these departments are not essential to the production of product: in fact, they are. Each person working for the company is important to the company's success. But unless production is being run, no money can be made.

The *only* time your company makes money is when products are being produced. This includes any operation that brings the product (or component workpieces) closer to completion during the production cycle—when products are being molded, cast, stamped, machined, ground, heat treated, painted, cleaned, degreased, sheared, formed, punched, welded, assembled, or by some other means fabricated with a manufacturing process.

We call tasks that actually make money for your company *value-added tasks*. In the case of the oil valve, value is added when castings are poured, when machining operations are taking place to produce component workpieces, and when component parts are assembled. Other value-adding operations that bring component workpieces closer to completion include such tasks as cleaning, deburring, and finishing (painting, plating, etc.).

VALUE-ADDED TASKS AND THE CNC ENVIRONMENT

As mentioned, we will limit our discussion of value-added tasks to include only the CNC environment. Many different types of machine tools and production machines populate the typical manufacturing company. While many of the points we make in this text can be easily applied to other types of production equipment, our focus will remain on those related to CNC equipment utilization.

There are only *two* ways to improve CNC equipment utilization. Both are tied directly to value-added principles. One is to *minimize* (or eliminate) necessary support tasks related to the CNC operation. The other is to *enhance* the value-added tasks. Everything we show in this text to improve CNC machine tool utilization falls into these two categories.

MINIMIZING THE NECESSARY SUPPORT TASKS

We discussed in Chapter One many necessary support tasks in the CNC environment. Though the implication is that these are tasks that must be performed in order for the CNC environment to function, it may be possible to minimize (or eliminate) some of them while still maintaining, or even improving, the company's CNC environment.

Here are some examples of necessary support tasks that exist in most CNC environments:

- Workpiece processing
- Programming
- Setup documentation
- Program loading
- Workholding setup
- Workpiece loading
- Cutting tool measurement (for tool length, etc.)
- Measurement of program zero location
- Program verification
- First piece inspection

Here we give a few specific examples of how some of these necessary support tasks can be minimized or eliminated. They represent the kinds of improvements that require constant vigilance.

Programming can be minimized in most cases if your company has similar workpieces to machine (families of parts), or if you machine similar operations on different workpieces (bolt hole patterns, grooving operations, pocket milling, etc.). A feature provided by most CNC controls, called *parametric programming*, allows one program (or subprogram) to run a number of similar operations. Long-term programming time can be dramatically reduced, and in some cases, programming time can be *eliminated*. In perfect applications for parametric programming, an operator can simply input variables to the control to specify the size and shape of the workpiece to be machined. Though many companies can benefit from the use of parametric programming, few programmers in the industry even know of its existence.

Setup documentation can be minimized for relatively simple setups if the programmer includes the setup instructions within the CNC program. All current-model CNC controls allow messages to be placed in the program. The programmer can use this feature to relay verbal instructions to the setup person on the display screen of the CNC control as soon as the program is displayed.

Program loading time can be minimized by using the most recent distributive numerical control systems. Many of these systems allow the setup person to completely transfer a CNC program within 30 seconds while at the CNC machine tool.

Workholding setup time can be minimized by ensuring that the setup person has all tools needed to make the setup in one central location as soon as the setup begins. This includes workholding tools (fixtures, blocks, vises, etc.) and hand tools (wrenches, screwdrivers, Allen wrenches, etc.). The setup person then does not have to search around the shop to find a needed tool.

Program verification time can be minimized if the programmer takes advantage of countless programming techniques to facilitate the running of the first workpiece. We cover many of these techniques in our discussion of setup time reduction later in this text.

Throughout this text, we give many suggestions regarding how you can minimize or eliminate specific necessary support tasks commonly performed in the CNC environment. However, it is not possible to cover every possible improvement. For this reason, you must be alert for other ways you can minimize or eliminate necessary support tasks in your own CNC environment.

ENHANCING THE VALUE-ADDED TASKS

As defined, value-added tasks in the CNC environment involve those things that happen while CNC machines are actually running workpieces, and include anything that influences the time it takes to machine workpieces. There are two ways to enhance value-added tasks.

- *Minimize anything that detracts from actually running workpieces.* Part loading, inspection (if done while the machine is not running), and manual operations such as applying tapping compound, breaking clamps loose for finishing, and turning the workpiece around in a turning center are examples of functions that detract from the task of actually running workpieces.

- *Reduce cycle time.* The process used to machine workpieces is commonly the largest single factor determining cycle time, and is directly related to the quantity of workpieces to be produced. Generally speaking, the higher the production quantities, the more thought and effort should be given to the process used to machine the workpiece. Another factor, and one often overlooked, is the method by which programs are formatted. A poorly formatted program can be a very inefficient program. Several techniques to reduce cycle time are described later, during our discussion of setup and cycle time reduction techniques.

IMPLEMENTING CHANGE TO ENHANCE VALUE-ADDED TASKS

Almost all actions you take to improve CNC machine tool utilization involve change of some kind. Unfortunately, humans tend to be somewhat reluctant to change. The saying, "If it ain't broke, don't fix it," emphasizes this reluctance. Just because something isn't broken doesn't mean it is functioning at its maximum potential. Instead of leaving well enough alone, successful companies instill in the minds of workers that processes free from change will eventually become stagnant. Stagnant processes will soon become obsolete. Obsolete processes result in obsolete products. Obsolete products result in a failing business. Failing businesses result in loss of jobs. The knowledge that jobs cannot be taken for granted should provide all the motivation needed to accept change and improve current methods. Only by inspiring workers to question the status quo can companies ensure freedom from obsolescence.

Though these principles can be applied to any facet of life, consider what they mean to your CNC environment. *Every* setup can be improved. *Every* machining cycle can be shortened. *Every* process can be made better. Indeed, every action that occurs within your CNC environment allows room for improvement. Only by actively pursuing change for the better can your people make your company the best that it can be.

Chapter Three

Defining Your Company's CNC Needs

I n Chapter One we noted that no two CNC environments are exactly the same. Given the number of manufacturing processes being performed on CNC equipment, the variety of workpiece configurations, the diversity of cutting tools, the variations in production quantities, and many other manufacturing-related criteria, it is unlikely that any two CNC environments will ever evolve in exactly the same manner. Even if two competitive companies are producing exactly the same product (possibly two contract shops doing work for the same manufacturer), the personalities of the parties involved will vary, causing dramatic differences in the use of CNC equipment.

Though this is the case, there are similarities in the way companies utilize their CNC equipment that lead to similarities in CNC environments since many companies face similar problems. By making some generalizations, we can begin discussing similarities.

In this chapter, we first present the three most basic types of companies using CNC equipment. Though CNC environments vary even within each type, these categories allow us to begin discussing similarities. We then discuss several important facets of the CNC environment from the perspective of each company type. Since we can present only tendencies and trends (not hard and fixed rules), you must be prepared for exceptions. Our intent in this chapter is to help you gain a perspective of different CNC environments. More likely, you will gain an insight into how your own CNC environment can be improved.

THE THREE BASIC COMPANY TYPES

Every company that uses CNC equipment can be classified into one of three basic categories: those that produce their own products, those that machine production workpieces for other companies, and those that produce tooling or prototype workpieces.

There will be some overlap with these three company types. Companies that produce their own products, for instance, may have their own toolrooms to produce tooling for their production machines. They may also have research and development departments to produce prototype workpieces. On the other hand, a contract shop that devotes the bulk of its work to producing workpieces for other companies may also have a product line of its own. Though we acknowledge this potential for overlap, we wish to begin isolating the criteria that contribute to wise decisions within the CNC environment. Company type is the most basic criterion that determines how decisions are made. The narrower your company's focus in this regard, the easier it will be to make wise decisions related to your CNC environment.

COMPANIES THAT PRODUCE A PRODUCT

This kind of company gets its revenue from the sale of a product. The CNC machine tools are used to machine component workpieces making up the product. The priorities related to the CNC environment in this kind of company are driven by the company's production control functions. Key decisions are based on production quantities, timely completion of component workpieces (lead time), ease of material handling, and other important production-related criteria.

Product-producing companies tend to view uptime (the percentage of time production equipment is actually producing component workpieces) as the key objective in the CNC environment. The higher the uptime, the more products can be produced. For this reason, these companies tend to make the necessary investment to acquire the best resources and keep uptime at maximum levels.

This willingness to invest whatever it takes to ensure maximum uptime is closely tied to the practice of factoring manufacturing costs into the selling price of products. While we are oversimplifying the practice, if the manufacturing methods needed to produce a product cost more than the product is worth (including profit), either another production method is needed or the product never makes it to the marketplace.

Since the sale of products pays for manufacturing costs, and since many manufacturing people do not have access to information related to a given product's sales success, it can be quite difficult for people in manufacturing to gage just how well their own CNC environment is performing. Many CNC people do not even know the shop rates (the cost in dollars per hour for a machine's use) for CNC machines used by the company. To make matters even worse, some companies consider this kind of cost information proprietary, not to be shared with manufacturing people in the CNC environment. This is one reason why product-producing companies tend to have the most complicated CNC environments. It can be very challenging to implement efficient manufacturing processes for the component workpieces of a product without direct access to product cost information.

Another reason product-producing companies tend to have complicated CNC environments is that most also have their own toolrooms for producing parts needed by production machine tools. Since more and more toolrooms are being equipped with CNC equipment, they present additional challenges to the company's overall CNC environment. As you will see throughout this text, toolroom applications and production applications for CNC equipment often conflict, which leads to confusion and controversy about how CNC machines are best applied.

COMPANIES THAT PRODUCE PRODUCTION WORKPIECES FOR OTHER COMPANIES

This kind of company receives its revenue from contracting to produce workpieces for other companies that produce a product. Profit is derived solely from producing *workpieces* for others. Unlike product-producing companies, this kind of company may not have the financial resources to purchase tooling, hire people, and in general, adequately support the CNC environment. In addition, most workpiece-producing contract shops can seldom assume they will see the same job more than once (competition for work among contract shops is fierce). This tends to dampen their willingness to invest capital for any single contract. Instead, they make do with what they currently have.

This makes their entire outlook about the CNC environment completely different from that of product-producing companies. It also helps to explain why there is so much controversy related to how CNC equipment should be used. What appears to be a perfect CNC-related solution to a manufacturing engineer working for a product-producing company may appear unfeasible to a person working for a contract shop that produces workpieces for others.

Though there are exceptions, workpiece-producing contract shops tend to be rather small. The majority employ fewer than 50 people and have fewer than 15 CNC machine tools. This is one reason why their CNC environments are usually simpler to manage than those of product-producing companies. Nevertheless, working for a workpiece-producing contract shop can still be quite challenging. A high degree of motivation, innovation, and improvisation is required on a daily basis. The contract shop must commonly machine the same workpieces as the customer currently produces but is too busy to handle. Not only must this shop produce quality workpieces at a price equal to or below what the product-producing company charges internally, it must do so with the limited resources at hand.

Another reason why workpiece-producing companies tend to have simpler CNC environments is that most specialize in the kind of manufacturing processes they perform. Most have only one or two different types of CNC equipment.

COMPANIES THAT PRODUCE TOOLING AND PROTOTYPES

Almost all forms of manufacturing production require tooling, and tooling can come in many forms. Cutting tools, gages, fixtures, molds, and extrusion dies are among the almost limitless forms of tooling used to produce workpieces on production machinery. While many product-producing companies have their own toolrooms to produce tooling for production equipment, many depend on outside vendors for at least part of the tooling requirements.

The tooling-producing company gets its revenue from the manufacture of tooling for product-producing companies or contract shops producing workpieces. While these companies tend to have more in common with workpiece-producing contract shops than with product-producing companies (they tend to be small and make do with what they have), they are also quite distinct. Since their CNC machine tools are not being used as production machines, the emphasis on uptime is not as strong. Though this kind of company still requires efficient machining, it is likely that no single machine will be running all of the time. More likely, certain (even very expensive) machines will sit idle for long periods of time while other operations are taking place.

The work in these shops can get very complicated, requiring highly skilled CNC people. For example, mold-producing shops must commonly manufacture very complicated 3-dimensional shapes requiring the use of a 3-dimensional computer-aided manufacturing system. Prototype-producing shops must also commonly produce 3-dimensional objects. Some of the most complex and innovative CNC techniques in existence today have been developed for tooling- and prototype-producing companies.

FACTORS THAT CONTRIBUTE TO A COMPANY PROFILE

Though there are only three basic company types, there are many factors that further determine the CNC needs of a given company. Some of these criteria expose reasons for major differences in the ways companies (even in the same category) develop their CNC environments.

PRODUCTION QUANTITIES

Almost all manufacturing decisions begin with *how many* workpieces must be produced, making the production quantity one of the most important criteria determining your company profile. Admittedly, there are companies that run varying lot sizes. However, most that do base the most important decisions on their largest quantities.

For the purpose of this text, we categorize companies into five levels with regard to the production quantities they run on CNC equipment.

1. Ultralow quantities: fewer than 10 workpieces.
2. Low quantities: 11–100 workpieces.
3. Medium quantities: 101–500 workpieces.
4. High quantities: 501–1000 workpieces.
5. Ultrahigh quantities: more than 1000 workpieces.

Product-producing companies vary widely with regard to the production quantities they typically run, making it difficult to generalize on that basis. Those that commonly run ultralow and low quantities tend to have a great deal in common with tooling producing companies and workpiece-producing contract shops. The higher production quantities grow, however, the more a product-producing company tends to take on its own look.

Contract companies that machine production workpieces for others tend to have ultralow, low, or medium production quantities. Since most product-producing companies have hard-and-fixed production methods for high and ultrahigh production quantities, they are usually unwilling to contract out this kind of work.

Companies producing tooling and prototypes tend to have the lowest production quantities. It is not uncommon for a die or mold shop, for example, to run only one workpiece on its CNC machine tools.

COMPLEXITY OF WORK

CNC machine tools are used for a wide variety of production tasks, ranging from very simple to highly complex. For example, a vertical machining center used by one company may do nothing more than drill and tap holes, while an identical machine in another company may be used to machine 3-dimensional shapes in plastic injection molds. The complexity of your work affects many areas of your CNC environment. As work becomes more complex, for example, the skill level required of CNC personnel must also grow, as must the complexity of your program preparation and transfer system.

Once again, product-producing companies vary dramatically with regard to complexity of work, making it difficult to generalize. Since they commonly have adequate resources, they tend to gear up for complex work with whatever is required to handle the work efficiently. Contract companies that produce workpieces for other companies also tend to vary in this regard. Some have customers who contract out only their most troublesome jobs, making the contractor's work quite complex. Others may bid only on the easiest jobs, keeping their work quite simple. Though tooling-producing companies do vary in this regard as well, they tend to have the most complex work (especially those that produce molds and other 3-dimensional shapes).

TOLERANCES HELD

Today's CNC machine tools boast amazing accuracy and repeatability specifications. Though they are capable of successfully machining to almost unbelievable tolerances, many CNC users do not require all that the machines can do in this regard. Additionally, what one company may consider a very close tolerance, another may consider wide open, leading to more confusion and controversy in the CNC environment. Tooling selection, gaging requirements, and even the initial choice of which machine tool you buy are influenced by the tolerances you expect to hold.

For our purpose here, we break tolerances into three levels. We'll call an *extremely tight tolerance* an overall tolerance of under 0.0005 inch (0.01 mm). A *tight tolerance* is between 0.0005 and 0.0015 inch (0.01 and 0.04 mm). *An open tolerance* is anything over 0.0015 inch (0.04 mm).

Companies in all three categories vary widely with regard to the tolerances they need to hold. While tooling-producing companies almost always need to hold tight tolerances, product- and workpiece-producing companies may also need to hold tight tolerances.

TYPICAL CYCLE TIMES

Another factor that varies dramatically from one company to another, *typical cycle time*, has a big impact on CNC machine utilization. Typical times for machining cycles determine how busy your CNC operators will be during production runs. The longer the cycle time, the more time the operator will have during the cycle for performing other tasks. With extremely short cycle times (e.g., less than 30 seconds), the operator may be so busy loading and unloading workpieces that there will be no time for anything else. As cycle times increase, the operator will be able to easily keep up with the loading and unloading functions and may even have extra time.

With very long cycle times, the operator may have a great deal of free time during the cycle. In this kind of environment, companies tend to utilize as much of their CNC operators' time as possible. A variety of tasks could be assigned to the operator during long production cycles. The operator could be deburring, cleaning, or polishing the workpieces while the machine is in cycle. Or the operator could be performing secondary operations on the workpieces, lowering the overall cycle time needed to produce the workpiece. If time during the machining cycle permits, companies may even use their CNC operators to run two or more CNC machines simultaneously, minimizing the number of people required in the CNC environment.

OTHER FACTORS THAT DEFINE COMPANY PROFILE

We have listed the four factors that we feel most influence the majority of decisions made in the CNC environment. There are many more factors that could further shape your company profile, and any single one could grow in importance and could even be the most important in the entire CNC environment.

For example, the *materials* a company typically machines is an important factor in helping to shape your company's profile, contributing to the choice of cutting tools, speeds and feeds, motor horsepower, and dust control system. The importance of this factor would escalate if a company machines highly flammable materials, like magnesium or titanium. In this case, the CNC environment would have to consider the safety of the people involved in every important decision being made. This factor will become the most important for companies that machine nuclear materials. Indeed, the entire CNC environment must be designed around the safe machining of highly toxic material and disposal of machining waste.

HOW COMPANY TYPE INFLUENCES KEY DECISIONS

Companies react differently to key CNC decisions. In this section, we take two common functions of the CNC environment and discuss how different companies will have different (and sometimes opposing) views of how the CNC environment should be run, with special attention to additional factors that contribute to the company profile. Though we only scratch the surface of the topic of each discussion, it should become clear how important it is to understand your own company's profile when making decisions for your CNC environment.

PROGRAMMING METHODS

The ability to efficiently develop CNC programs may be unimportant or it may be absolutely essential—or somewhere in between—depending on the company type and profile. On one hand, consider a product-producing company that runs ultra-high production quantities. This company may dedicate one or more of its CNC machine tools to machining only one workpiece. For this kind of company, once the CNC program for each machine is developed, verified, and optimized, there will be no need for further programming. While this kind of company may need someone knowledgeable about CNC programming on the staff to handle any engineering changes, program preparation will not have a very high priority in the CNC environment.

On the other hand, consider the contract shop that commonly runs production quantities of fewer than 50 workpieces. This shop may seldom see the same job a second time; therefore, almost every production run will require a new CNC program. For this kind of CNC company, efficient program preparation will be among the most important functions contributing to its success.

Complexity of work will influence a company's programming needs. A company that does little more than simple hole machining operations on machining centers will require very little programming. For this kind of simple work, the company may elect to employ a single manual programmer who uses nothing more than a simple CNC text editor to create CNC programs. Since most CNC machine tools incorporate highly powerful and helpful "canned" cycles to help with programming, a good manual programmer can easily and efficiently program for simple hole machining operations. Conversely, a company that performs more complex work, like 3-dimensional milling in injection molds, will have to purchase a sophisticated computer-aided manufacturing system to help the CNC programmer prepare programs.

Yet another factor affecting the way CNC programs are prepared is the available personnel in the CNC environment. Many product-producing companies staff their CNC environments with enough people to ensure maximum uptime. These companies tend to have one or more people preparing CNC programs while their CNC machines are in production (either manually or using some form of CAM system). On the other hand, contract shops may utilize one person to perform many of the tasks related to the CNC environment. It is not unusual for this kind of company to expect the CNC machine operator to prepare the CNC program, make the setup, verify the program, and run production. For this kind of company, anything that can be done to minimize the length of time needed to prepare CNC programs will effectively reduce setup time. This kind of company tends to prefer shop-floor programming on conversational controls.

DISTRIBUTIVE NUMERICAL CONTROL SYSTEMS

Distributive numerical control systems allow the CNC user to save and retrieve CNC programs for future use. Most require some form of separate device (personal computer, floppy drive unit, etc.) to actually store and retrieve CNC programs. The need for this kind of device is based directly on the company's profile.

Consider a company that dedicates its CNC machines to running only a few different workpieces. If all the programs a given CNC machine runs will fit into the memory of each CNC control, this company will have no need for an elaborate DNC system, since program transfers will almost never be required. Though they will need some way to *back up* their CNC programs in case of control failure, a very simple and inexpensive, though possibly somewhat cumbersome, portable program transfer device will suffice. Conversely, companies that run a great number of different programs, and especially those with a great deal of repeat business, have a strong need to minimize the amount of time required to transfer programs to and from the CNC control. For this kind of company, a sophisticated automatic DNC system may be essential to minimize downtime.

THE ROLE OF THE CNC COORDINATOR

All companies should have a CNC coordinator. This could be a single person or a group of people working together. Whether your company formally assigns someone to the position or people are simply assigned the related tasks, *someone* should be made responsible for coordinating functions in the CNC environment.

We cannot stress this enough. A company that has little or no coordination among people in its CNC environment cannot hope to utilize its CNC machine tools at their maximum potential. Just the contrary is true. A company that does not coordinate activities in the CNC environment pays the price by having a great deal of waste. The larger the CNC environment, the more potential for waste. Here we address several of the main objectives a CNC coordinator should undertake and offer specific tasks the coordinator should strive to accomplish.

CHOOSING THE CNC COORDINATOR

In most companies, the CNC coordinator can be selected from the workers in the CNC environment. This person should possess a firm knowledge of the entire

CNC environment and have a good working knowledge of what each person or department within the environment does. For this reason, selecting someone who already works for the company will eliminate the time required for a new hire to become familiar with your company's methods.

Equally as important as CNC knowledge are communication and negotiation skills. The CNC coordinator must be able to effectively mediate differences of opinion. Along the same lines, the CNC coordinator must even possess some sales skills to ensure that good ideas are implemented.

Once you determine who will act as CNC coordinator, keep in mind that authority must come with responsibility. A CNC coordinator who is given the responsibility to continually improve the CNC environment will also need the authority to carry through with the changes required to make improvements. This is especially necessary when it comes to settling disputes. The CNC coordinator must have the authority to make firm decisions.

Senior CNC programmers tend to make the best CNC coordinators. Since the typical CNC programmer must already understand the most basic needs of CNC people in order to prepare programs, it will be a relatively simple matter to expand this knowledge to include the workings of related departments.

IMPROVING COMMUNICATIONS IN THE CNC ENVIRONMENT

The greater the number of people in the CNC environment, the more difficult it is to ensure that everyone understands the needs of everyone else. The CNC coordinator can easily gain an understanding of each person's needs for the purpose of streamlining tasks. To illustrate, we include some specific examples of how improved communication leads to an improved CNC environment.

Eliminating duplications of effort

By simply getting people to communicate with one another, the CNC coordinator may discover instances when two or more people in the CNC environment are unnecessarily duplicating effort. Depending on the methods your design engineers use to dimension and tolerance workpieces, for example, time may be wasted while CNC programmers, CNC operators, and inspectors make (duplicate) calculations based on blueprint dimensions. Using datum surface dimensioning techniques and simple-to-read plus-or-minus tolerances are two examples of how the design engineer can make it much simpler for people in manufacturing to interpret drawings.

Soliciting new ideas and suggestions

Improved communication among people in the CNC environment opens the door to participation for everyone in the CNC environment, even those not commonly credited with the ability to make good suggestions. Some of the best ideas for improvement come from the people who are closest to the CNC machines, the

CNC operators. These are the people who see the CNC cycle run over and over again. With a minimum of experience, they will soon be able to spot ways to improve the CNC cycle. Yet many companies do little to solicit their ideas and suggestions.

PART
II

EDUCATING YOUR PEOPLE

It should come as no surprise that manufacturing in the United States has suffered in recent years. Competition from the Pacific Rim, Europe, and even third-world countries is fierce, and has forced gradual yet substantial cutbacks in U.S. companies. During any period of cutback, managers tend to look to what they consider nonessential areas of the company for potential spending cuts. Unfortunately, apprenticeship programs, which have played a very important role in developing the competence of our work force, have been among the areas hardest hit. Very few companies today offer even machinist or toolmaking apprenticeships.

Though the elimination of apprenticeship programs may have resulted in short-term cost reduction, it has created new problems related to the development of entry-level personnel. While technical schools have done their part by developing and maintaining excellent curriculums related to many areas of manufacturing, there is still a large gap between the needs of industry and the proficiency of entry-level manufacturing personnel. In the seminars I conduct for advanced CNC techniques, hardly a class goes by where some manager or shop owner does not voice concern about the difficulty of finding and keeping good, knowledgeable CNC people at a wage the company is able to pay.

The cause of our industry's entry-level work force problem is not limited to cutbacks brought on by the fluctuating state of our economy. With the best of intentions, many so-called expert job counselors tell our youth to avoid the manufacturing sector as a career option. With some shortsightedness, they perceive manufacturing in the United States to be in a long-term state of decline, and direct newcomers to the work force toward training for jobs in the service sector of our economy. In the long run, this could turn into a self-fulfilling prophecy. If we continue to see a decline in young, enthusiastic people entering the field of manufacturing, there will eventually be no one left to run our factories.

Another reason for the decline in the quality of manufacturing's entry-level work force is our country's general perception of what is involved in a manufacturing career. While our European counterparts enjoy a very high level of dignity, respect, and even admiration, manufacturing people in the United States, and especially any form of tradespeople, are perceived by many as among our lower class. Young people, who are easily influenced by peer pressure, bow to this notion and tend to avoid any job in which their hands may get dirty. Many who do enter the field of manufacturing do so as a last resort, after exhausting other career options. This fuels the problem yet further. Our manufacturing sector is not receiving the cream of the entry-level work force to begin with.

Regardless of the reasons for our entry-level work force problems, the fact remains that it must be dealt with. How well you tackle these problems in your own company may very well dictate whether your company survives the next decade. In many areas of the country, and especially in rural areas where the worker pool is limited, companies are forced to take the initiative and develop their own methods of training new people. In some of these companies, the in-plant training programs being implemented rival even the best of past apprenticeship programs.

The primary focus of Part II will be to address the problems companies face in training entry-level personnel. While there are many job responsibilities related to the CNC environment, we will concentrate on those areas that require the most training. By sheer numbers alone, CNC operators represent the greatest number of people in most CNC environments and will be heavily favored in our presentations. Rest assured, however, that we will address training for other important job responsibilities related to the CNC environment, including programming and setup.

Though it represents the major portion of manufacturing's training needs, training entry-level people is only one facet of the training problem. As a company's needs change, it is often necessary to develop employees' skills in other areas of the company. Additionally, one person may have to perform several functions for the company. In the CNC environment, for example, an operator may start out having the simple task of running one CNC turning center. The operator's responsibilities may be initially limited to simply loading and unloading workpieces and making simple offset changes to maintain workpiece size. As the company's needs change, this same operator may eventually wind up running several turning centers. At some future date, the operator may also be expected to run machining centers. After gaining experience with several types of equipment, the operator may be trained to make setups, and even develop CNC programs.

As a secondary objective of Part II, we intend to introduce you to the special concerns of cross training people in the CNC environment. We cannot overstress the importance of your workers' constant growth and improvement. A person whose job remains stagnant for a long period of time will eventually become unconcerned with the company's needs. Cross training one person for a variety of jobs not only improves the company's overall potential for growth, it gives that person an additional reason to come to work every day.

Part II will be especially helpful to readers who have no current training programs in place within their companies. We will show how to incorporate training, from the hiring of new people to actually developing your own in-plant CNC training program. This section will also be of value to readers who do currently have in-plant training programs in place. You will at least learn about other proven methods of in-plant training implementation. More likely, several of the suggestions we make will have value in your own current training environment.

Part II contains three chapters. In Chapter Four, we introduce you to the most important issues related to industrial training, including the importance of having well-trained personnel, the benefits of training, how to find and hire CNC people, and how to utilize training resources that help you make the most of your training program. In Chapter Five, we show you how to develop your own in-plant CNC training program, including how to limit the scope of training, how to create a good learning environment, and how to select your instructor. In Chapter Six, we show you a proven key concepts approach, including the actual course content your instructor can use to present training courses for CNC. This lengthy chapter will save your instructor from having to develop a training program from scratch. At the completion of Part II, you will have a clear understanding of how to approach the development of your own in-plant CNC training program.

Chapter Four

Introduction to Industrial Training Issues

E very aspect of your company's success is directly related to the proficiency level of your workers. The success of your SPC program depends on the proficiency of people inputting data to the system. The success of your company's new products is directly related to the proficiency of your design engineers. The success of your manufacturing processes is tied to the proficiency of your process engineers. Truly, there is no area of your company that is not dramatically affected by the proficiency of your staff. The CNC environment is no exception.

THE IMPORTANCE OF WELL-TRAINED PERSONNEL

In the preface to this text, we defined CNC equipment utilization. We said our primary focus throughout is optimum utilization of your CNC equipment. In *any* CNC environment, there is a direct correlation between optimum CNC equipment utilization and the proficiency of CNC staff. If your CNC programmers and process engineers are not proficient in tooling and cutting conditions, efficient machining will be impossible. If your CNC machine setup people are not proficient in workholding, inefficient (and possibly dangerous) setups will be made. If your CNC programmers are not proficient in efficient program formatting, cycle time will suffer. If your CNC operators are not proficient in reading gaging devices and adjusting tool offsets, scrap rate will be high. If *any* CNC utilization characteristic is not properly understood by the related people, *optimum utilization* of your CNC equipment *will be impossible.*

Given the severity of problems caused by underexperienced people, it is amazing how many companies pay little or no attention to the continuing development of their people. It is not uncommon, for example, for a company to spend hundreds of thousands of dollars to tool up for a new process with CNC equipment. The company will put in the required electrical service, pour the recommended foundation, develop atmospheric control, and spend the money required to buy the CNC machine tools. Yet the company may spend very little for the initial training to get the CNC staff up to speed, and even less for continuing education to ensure ongoing optimum utilization of the equipment.

Most CNC-using companies offer an orientation session, normally handled by the personnel department. This orientation tends to be rather general, simply acquainting new hires with company policies. When it comes to competence, many employers offer little in the way of training. They *expect* new hires to possess most or all of the talents and abilities required to successfully undertake their daily responsibilities. Stated another way, a new hire is commonly expected to simply walk into the company and start being productive on the first day. Given the limitations of our current work-force alternatives, the diversity of our manufacturing methods, and the variety of CNC machine and control types, this expectation is terribly unrealistic, and leads to serious underutilization of CNC machine tools.

Before you undertake any form of cost-reduction program, before you implement changes in your processes to reduce cycle time, before you look for *any* kind of improvement in your CNC environment with the goal of increased production, consider the current proficiency level of your CNC personnel. *The benefits that can be gained in CNC machine utilization by improving the proficiency of the current CNC staff will usually outweigh any other single improvement you can make in the CNC environment.* And only with a well-trained and highly motivated staff will your company be able to tackle the future challenges it will surely face.

HOW TRAINING INSPIRES THE RIGHT ATTITUDE

Though it may be somewhat difficult to measure the tangible benefits provided by workers with the correct mental attitude, having them is of the utmost importance in any positive work environment. A highly concerned worker is a productive worker. And *any* shortcoming in a worker's ability can be overcome if the worker has the right mental attitude.

Poor mental attitude among workers is a common symptom of insufficient training. If no guidance is available to help a worker with a job-related question or problem, the worker will likely become frustrated. Worse, the worker may make mistakes due to a poor understanding of the job's responsibilities. Mistakes lead to blame, and when blamed for mistakes, when no help is available, a worker's attitude will surely suffer.

The correct mental attitude cannot be taught. You must *inspire* it in your workers. The most powerful way to inspire the right mental attitude is through training.

All managers desire highly motivated, self-sufficient, concerned workers. But these qualities in a staff member cannot be taken for granted. While you may be very lucky in your recruiting efforts and find a few people who already possess these desirable qualities *before* hiring, most employers would agree that finding people with the right mental attitude is more the exception than the rule.

Instilling the right mental attitude in your workers can be difficult, especially if your company has paid little attention to training issues in the past. Since everyone *talks* about employee involvement, it is likely that your company's workers have heard hollow "rah-rah" speeches and may be skeptical when they hear of any new program aimed at getting employees more involved. Since actions speak louder than words, you will have to *demonstrate* that your new training program is real.

Stressing the worker's importance to the company

How many of your staff members truly understand their own worth to the company? In many companies, and especially in larger companies, people in the CNC environment (on the shop-floor and in manufacturing engineering) have no idea of their impact on the company's overall performance. Workers having no conception of their own worth commonly become apathetic to the company's needs.

The attitude of the uninformed employee can be likened to the attitude of nonvoters at election time. Most nonvoters avoid voting because they feel their vote will not affect the outcome of an election. In similar fashion, many shop people do not actively participate because they feel the company will perform exactly the same with or without them. And just as with nonvoters, it is impossible to inspire any kind of participation, motivation, or action from people when they have no conception of their own worth.

When nonvoters avoid an election, especially one that determines serious issues of concern to them, often they will regret the choice not to vote if the election is lost by a narrow margin. This regret should inspire them to participate in the *next* election. In similar fashion, when workers in the CNC environment are made aware of their worth to the company and told that their active participation has a measurable impact on the company's success (if not its very survival), concerned workers will regret their lack of active participation and strive to do better in the future. As you consider your own CNC environment, there are three questions you can ask yourself to help determine whether your company helps workers understand their own value.

Do I encourage involvement? You have heard the saying "actions speak louder than words." It is easy to tell someone how important he or she is to the company. But words cannot take the place of actions. Everyone in your CNC environment has a great deal to contribute to the success of your company. When you encourage active participation, you are *showing* your people that their suggestions and ideas have meaning and value to the company. And only when *everyone* is actively participating can you expect to achieve optimum CNC machine utilization.

Here is one example of an easy but commonly overlooked way to encourage involvement in the CNC environment. In most companies, too little credit is given to the people responsible for actually running production, the CNC machine operators. These people are commonly kept in the dark regarding how programs are developed and verified. Yet they are also the people on the front line, who monitor production cycles more closely than anyone else in the company. While monitoring cycles, they notice any inefficient movements a program makes. They notice that a given tool wears excessively. They notice an inefficient sequence of tooling. Because of their closeness to the machining cycle, they can often identify potential improvements to the cycle that a typical CNC programmer will overlook. If they are told to simply load and unload workpieces and are not asked on a regular basis what might be done to improve the machining cycle, a great resource of your company is being ignored.

Does everyone in the CNC environment know the worth of each CNC machine tool? It is difficult to gain an appreciation for the value of anything if its cost is unknown. As you make a purchase for yourself, you know your care and concern for your purchased item will be directly related to its cost. Disposable razors meant to be thrown away after a few uses will not inspire much in the way of care and maintenance. On the other hand, after purchasing a new automobile, you go to extreme lengths to care for and maintain it properly.

In like manner, any concerned worker in the CNC environment will have a much greater appreciation and concern for the CNC equipment if he or she knows its worth. With CNC machines costing from $50,000 to well over $500,000, these machines should command a great deal of care and respect from the people involved with their use.

Does everyone in the CNC environment know our CNC machine shop rates? All companies must charge for the use of their CNC machine tool use. In companies that perform contract machining of production workpieces, these charges are usually quite obvious and easy to calculate, since a customer is directly billed according to the amount of CNC machine use. On the other hand, in companies that manufacture a product, machine use charges will be less tangible since it is the sale of a *product* that generates revenue. In product-producing companies, machine use billing will be internal and used mostly for accounting purposes. *Whether the billing is internal or external, all companies track the cost of using every machine tool they own.* Most companies use an hourly rate, called the machine's *shop rate* as the basis for machine costs.

In smaller companies and especially job shops, the shop rate is usually easy to find and almost everyone knows it. Since direct communication between the company owner and workers occurs on a regular basis in small companies, each machine's shop rate is probably referred to on a regular basis.

In larger companies and especially companies that manufacture a product, the shop rate may not be readily available. Even manufacturing engineers may not know each machine's shop rate. Though it may be harder to find, *all* companies must account for their costs by one means or another, meaning *every* machine will have a shop rate.

Here are just two scenarios illustrating why *everyone* in the CNC environment should know the shop rate for each CNC machine your company uses:

1. Consider a vertical machining center having a shop rate of $100 per hour. Of course, this price can be realized only while the machine is producing workpieces. When the machine is down for any reason, its owner is losing $100 per hour of downtime. A concerned worker, knowing he or she controls the company's ability to acquire $100 per hour, should be motivated to do whatever it takes to keep this machine running.

2. It is human nature for everyone to place a high value on personal time. We tend to think of our own time as having more value than anything

else. However, if you compare the typical shop rate of any CNC machine to the wages of anyone in the CNC environment, it is likely that the CNC machine's shop rate will be at least three to four times the wage of any single worker. Knowing that a given CNC machine's time is worth much more than their own (in monetary terms), concerned workers are likely to put forth the effort to keep the machine running at optimum levels.

The importance of sincere concern

You must remember that the attitude of the typical worker in the CNC environment is simply a reflection of the company's overall attitude toward its personnel. If company management shows sincere concern for its workers, the feeling is generally reciprocated. On the other hand, if the company shows little regard for its workers, or worse, discourages participation and comments, the company can expect little in the way of employee involvement.

BENEFITS OF TRAINING

The benefits of implementing a training program are numerous. Here we list the more tangible ones you can expect to realize as you begin. Keep in mind that if your company will be implementing a new training program, some form of justification will likely be necessary. Though it may be somewhat difficult to measure the expected impact of some of these benefits in monetary terms before training begins, most training programs can pay for themselves in more efficient use of the CNC equipment, in minimized mistakes, and in suggestions for improvements that come through the training program.

Minimized mistakes

Mistakes can cost a company dearly. Wasted time, scrapped workpieces, lost customers, damage to tooling and equipment, and worst of all, injuries to workers all can result from mistakes. About the only good thing that comes from mistakes is that they help you gage the quality of the training your company provides. If mistakes are made repeatedly, it should be taken as a signal that your training needs to be improved.

While no amount of training can completely guarantee that mistakes will never be made, well-trained personnel are less apt to make mistakes than untrained personnel. Training can be a cost-effective way of avoiding losses.

Improved quality

Poor quality is often the result of ignorance. Ignorance of the best programming methods leads to poorly formatted programs and can make it difficult to produce good workpieces. Ignorance of workholding tooling leads to poor setups that make it impossible to machine workpieces to size. Ignorance of cutting tools and the right cutting conditions leads to unacceptable workpieces. Training is the most effective way to ensure that ignorance will not stand in the way of quality work.

Enhanced consistency

One of the most serious problems caused by insufficient training is inconsistency. Without guidance, people will develop their own unique techniques to handle specific problems.

When it comes to programming, for example, many format options are available. When left completely to their own devices, no two CNC programmers will ever end up formatting programs in exactly the same manner. While minor deviations may be relatively easy to understand and handle, very odd (yet still workable) CNC programs are commonly developed by people who have had little or no formal training. Inconsistency among programs commonly results in confusion, frustration, wasted time, arguments, and in general, underutilization of the CNC equipment.

In like manner, if any two CNC operators are left completely on their own to develop operation skills, the result will be two completely different ways to run the machine. In extreme cases, this inconsistency will make it impossible for one operator to move from one machine to another without having to spend a great deal of time getting comfortable with another person's way of doing things.

Training can eliminate inconsistencies that lead to underutilization. Though there may be a need for open discussions to allow differences of opinion to be voiced, once everyone agrees on the best method of handling each specific function, *everyone* will be doing things in essentially the same manner.

Improved efficiency

Well-trained people will be more efficient in their duties than untrained people. If clear-cut, consistent, and precisely documented procedures are set in place and presented well during training, no time will be wasted while CNC people try to figure out what they are supposed to do. While this may be as much a function of the company's documentation abilities as it is a function of training, new hires must be trained to use the company's documentation.

Additionally, a company may have several *unwritten rules* that everyone is simply expected to know. For example, the assignment of tool offset numbers is commonly unspecified. Most companies that use turning centers make the primary tool offset number for each tool the same as the tool station number. Many companies that utilize CNC machining centers make the tool length compensation offset the same as the tool station number. While this offset assignment is logical and easy to understand, it may be one of many such unwritten rules that new CNC operators must know *before* they can start running CNC machines.

Enhanced safety

Untrained CNC people are *dangerous* CNC people. Given the size, weight, and power available from today's CNC machine tools, mistakes of any kind can result in dangerous situations. Well-trained workers minimize the potential for danger.

The most dangerous element in any machine shop is ignorance. When a person is unaware of the potential for danger, very little respect is given to the equipment being operated. Consider, for example, a worker who has been working in a very safe area of the company performing light assembly work. This person is accustomed to a safe working environment. If this person is eventually assigned to operate a CNC turning center without the proper training, he or she may not even recognize the potential for dangerous situations. When faced with a serious problem, this operator may not exercise the caution necessary to safely solve the problem.

One classic CNC-related example of a dangerous situation caused by ignorance has to do with transferring operators from CNC wire electric-discharge-machining (EDM) equipment. Generally speaking, wire EDM machines are quite safe to run. If a mishap occurs, about the worst that happens is that the wire breaks. Dry runs are performed entirely for the purpose of checking whether the workpiece configuration looks correct (commonly by drawing the shape on a plotting table or on the CNC display screen's graphic function). But an operator of a CNC wire EDM machine, who is moved to a CNC machining or turning center without training, may not recognize the increased potential for danger. Remember that an operator proficient on a given machine tool still needs training when reassigned to an unfamiliar tool.

The three levels of safety. In any CNC environment, the very first priority is *operator safety*. Nothing about the way the program is written, the way workpieces are loaded, the way cutting conditions are developed, or anything else related to the operator's responsibilities should place the operator in a dangerous situation. The second priority is *machine safety*. Nothing about the machining cycle should place the machine in danger or cause undue wear and tear on the machine tool. The third priority is *workpiece safety*. It is, of course, necessary to run good workpieces in an efficient manner. The only way to ensure safety at all three levels is by having competent CNC people. And only training can ensure a competent work force.

Improved methods

Some of the best enhancements to CNC machine utilization come from the training process. During the original development of your training program, for example, when your people are determining what your company needs to teach during the training, they will be forced to take a very close look at current methods. This examination will likely turn up the potential for time- and cost-saving enhancements.

Additionally, once training is implemented, many suggestions for improvements will come from students in the training program. During my own in-plant training courses, I am constantly surprised by the quality of suggestions and comments students make, even entry-level students.

Letting training help pay for itself. It is important that the people in your company's training department track the suggestions made and implemented during training. Since training costs will be of continuing concern, trainers can continue to justify the existence of the training program by showing the impact of training on the success of the company. Figure 4-1 shows a simple form that can document and track suggestions made during training.

Suggestion Tracker							
Actual Date	Suggestion	By	Machine	Priority	Date to Implement	Expected Savings	Savings

Figure 4-1. Typical form used to track savings resulting from suggestions made during training.

TRAINING NEW CNC PEOPLE

In very basic terms, manufacturing companies develop in-plant training for one of only two reasons. First, training is required when *new* people must be brought into the CNC environment. Possibly new equipment is being purchased, requiring more people. Maybe an increase in business requires the company to go to two or three shifts. Or new people may be hired to replace people leaving the company. Regardless of the reason new people must be brought in, you have two choices of where the people will come from: within your own company or hired from outside. Since companies are finding it increasingly difficult to find and hire people with extensive CNC experience, training new people in the CNC environment almost always begins with CNC basics. Indeed, this is usually the primary reason why a company considers implementing a training program in the first place—to build the capability to train inexperienced new CNC people.

The other reason companies develop in-plant training is because they need to better utilize *existing* staff now working in the CNC environment. In your own company, possibly you wish to expand the abilities of your CNC operators, allowing them to operate two or more different types of CNC machine tools. Possibly you want your CNC operators to eventually become CNC setup people or programmers. Or you may wish to target a specific problem area for improvement, such as excessive scrap rate or inefficient workpiece loading. These kinds of problem areas typically require improving the competence level of the people involved.

By far, the most challenging CNC training program to develop is one designed to train entry-level people. Yet this tends to be the kind of training program needed most by manufacturing companies today. Since a good entry-level training program encompasses the major portion of information that must be transferred during *any* CNC training, and since the *selection* of people working in the CNC environment plays a major role in the success of training, we place a strong emphasis on how you find, hire, and train entry-level CNC people.

FINDING CNC PEOPLE

Remember that new people to be placed in key CNC positions do not have to come from outside the company. In fact, your best resource for filling CNC positions may be your own company. Companies that promote from within enjoy the option of selecting people with a proven track record. Though many of the points made in this section do assume you are selecting people whom you do not know, choosing from candidates you know removes the doubt surrounding the filling of open positions.

This personnel selection presentation assumes you have the freedom to choose the people that fill CNC positions. There may be restrictions within your company that do not allow this. Especially when selecting CNC operators, CNC setup people, and others who work on the shop floor, company management may not always have the luxury of directly choosing the people working in the CNC environment. In companies that have labor unions, seniority plays the major role in the placement of personnel. While company management may have the ability to question a person's placement through union proceedings, placement of personnel throughout the shop may be more a function of how long they have worked for the company than how well they qualify for a given job.

The available employment base

As with any type of employment, the typical proficiency level of available people will vary with your company's location. Key factors that affect the CNC proficiency you have locally available include the size of your area's manufacturing base, technical schools in your area, the size of your community, and the current business climate. Some areas are blessed with an abundance of competent CNC people while others (especially rural areas) have very little to choose from.

Your expectations of new hires should, of course, reflect the choices you have in your area. If you are in a large community with a strong manufacturing base, you may expect to hire experienced CNC people who require very little training to become productive. In a smaller, less industrialized area, you may be lucky to find anyone who has even heard of CNC, let alone anyone who knows how to use it. In this case, you must expect to offer extensive training just to get new people to a level at which they can begin to be productive.

How you approach the hiring process will depend on your hiring expectations. In a large community, you can expect to find several applicants qualified for your CNC positions. In this case, you can be quite picky when it comes to whom you hire. Your interviews can include very specific questions related to the machine tools you use and the processes you perform. In smaller communities, your expectations will likely not be so aggressive. You may be forced to hire people with limited (if any) CNC experience. You may be lucky to find people who have machining experience. Your primary intention during the interviewing process may be to select the people best suited to *learning* their job responsibilities.

As you begin the hiring process, keep in mind that the person best suited for the job may not always be the person with the most experience. Obstacles like personality conflicts and salary requirements may prohibit you from hiring the person most qualified for the job. Most managers would agree that motivation and enthusiasm are more important than current experience level. A highly motivated, enthusiastic person with less CNC experience will be willing to put forth whatever effort is required to fully master the job responsibilities. This kind of person will usually work out better in the long run than an expert who is not totally satisfied with the job to begin with.

CNC positions to fill

There are several job titles related to CNC. In manufacturing companies, the four most common are CNC operator, CNC setup person, CNC programmer, and CNC coordinator. In some companies, especially smaller ones, one person may be expected to perform functions related to two or more of these. However, to help you categorize your needs, we present each individually.

CNC operators are responsible for actually maintaining production. First and foremost for the CNC operator is keeping the machine running efficiently once a setup is complete. Primary responsibilities include loading and unloading workpieces, inspecting critical dimensions and entering statistical process control data, maintaining workpiece size with tool offset changes, and possibly some minor program editing.

Some in our industry feel that a CNC operator does not have to know much about CNC to simply maintain production. While this may be true in rare, highly automated applications, it has always been my contention that the more CNC operators know, the better operator they can be—and the more they can contribute to the CNC environment. Well-trained operators will find mistakes in programs. They will recognize and expose inefficient cutting conditions. They will come up with suggestions to improve workholding. In general, they will facilitate the highest level of CNC machine utilization. For these reasons, CNC operator training should stress the *whys* of CNC operation just as strongly as *how to* perform each operation task.

CNC setup people are responsible for changing over from one job to the next. They must possess a higher level of CNC knowledge than CNC operators. Additional responsibilities for the CNC setup person include assembling, measuring and loading cutting tools, locating and inputting the position of the program zero (origin) point, making workholding setups, loading the CNC program into the control, trial-running both new and proven CNC programs, modifying CNC programs to correct mistakes and optimize the cycle, and maintaining the setup during the production run should problems, such as tool breakage, occur.

CNC programmers must, of course, possess a very high level of knowledge about the CNC machine tools with which they are working. They must also possess the knowledge of basic machining practice as it relates to each machine tool. In most CNC environments, they must understand processing and have the ability to develop a sequence of operations for the program they are writing. Most programmers are also expected to plan and document the setup for each program they write. They must understand each machine's programmable functions and the programming commands related to each CNC machine. They must be able to format the programs in a safe, efficient manner. If the company owns a computer-aided manufacturing system, they must also be able to develop CNC programs with the help of a computer. During program verification and optimizing, most CNC programmers are expected to be available to monitor the machine's first few cycles and make corrections and improvements as required.

CNC coordinators are responsible for overseeing the entire CNC environment. Though not all companies formally specify a person to act as CNC coordinator, this very important person (sometimes a group of people) coordinates the efforts of the entire CNC environment with other areas of the company. In any company utilizing CNC, there must be a great deal of communication among almost all sections of the company. It is important that design engineering, tool engineering, production control, quality control, training, human resources, and process engineering all work together to facilitate the basic objectives of the company. The CNC coordinator will act as spokesperson for the CNC environment, ensuring that all CNC-related functions of the company flow as smoothly as possible.

Who needs training the most? In all but the smallest companies, CNC operators, by far, make up the largest percentage of CNC personnel. It is not unusual for a company to have many CNC operators yet only a few CNC setup people and programmers. In one medium-sized product-producing company that runs a large number of repeat jobs, for example, there are 10 CNC machine tools. This company employs 25 operators (to operate the machines during three shifts). They have relatively long-running jobs, so they require only four setup people (two for day shift and one each for other shifts). Since the same jobs are run over and over, this company needs only one CNC programmer. In this particular company, the CNC programmer also acts as the CNC coordinator.

This is indicative of the relationship of the number of CNC operators to others in the CNC environment. Since CNC operators make up the bulk of the CNC work force, we devote the most time and effort in this text discussing their hiring and training.

Where to look

If you are trying to hire new people from outside the company, your first step will be to locate suitable applicants. We suggest that you first discuss your employment needs with *current employees*. It is likely that someone in your company has an acquaintance, friend, or relative suitable for the job. Employee-recommended candidates usually work out quite well since they feel an obligation not only to satisfy their employer, but to live up to the statements made by the recommending employee.

Another avenue for finding applicants is the local newspaper *classified ads*. Depending on the job market in your area, your want ads may produce mixed results. While they almost always show *some* results, it can be very difficult to adequately screen people prior to holding an actual interview. For this reason, many CNC managers let their personnel department (human resources) screen applicants when newspaper want ads are placed. Only those qualified for the open positions will be granted interviews.

If you allow people in your human resources department to screen applicants responding to want ads, remember that these people become part of your company's CNC environment. In order to ensure that screening is fair and fruitful, the person doing the screening should have at least a working knowledge of what CNC machine tools do. That person will need to be able to communicate intelligently with the applicant (and with you). It can be very frustrating for applicants possessing the desired skills to prove their abilities to a screening person who has no understanding of CNC. One way of ensuring the screening staff's understanding of CNC is to enroll them in the in-plant training course new hires will attend.

If you are in a small community, your local newspaper's want ads cover only limited territory and will offer limited success. You can expand your coverage by placing your ad in the newspapers of larger cities in your area. This is especially true when trying to recruit people to fill more advanced CNC positions.

To expand your coverage yet further, keep in mind that there are several *trade journals* that offer classified ad sections for hiring. *Manufacturing Engineering*, published by the Society of Manufacturing Engineers (SME), for example, offers a special classified section called *Opportunities* in which available positions can be listed.

Another alternative is the *employment agencies and temporary services* in your area. A reputable employment agency will nicely handle much of your searching and screening process as long as its people understand your specific needs. Remember that people working for employment agencies or temporary services may have the same problems as a company's human resources department when it comes to understanding CNC. You'll want to confirm that your agency understands the special needs of manufacturing companies. With a good agency, you can be relatively confident that an applicant they represent has the necessary qualifications when called in for an interview. With employment agencies, there will usually be no charge until you actually hire someone. The fee is then typically a percentage of the new employee's salary for the first year.

If you are not sure your open position will be permanent, or if you would like to confirm a worker's abilities prior to permanent hiring, *temporary services* make an excellent choice. Most will charge a percentage of the worker's salary (per week or month) while the worker is in your employ. Since the worker is not in your permanent employ, there will be other reasons to use temporary services. For example, your company may be able to avoid many costs related to medical insurance, pension plans, and taxes with temporary workers.

Perhaps the best alternative for finding CNC applicants, though often completely overlooked, is the local *vocational school, technical school, junior college, or university*. Since schools are highly motivated to place their students after graduation, they will work very hard to suit the needs of local industry. In fact, many will work with you on a one-on-one basis and adjust their training programs to suit your company's specific needs.

Though it may not have occurred to you, *machine tool salespeople* in your area often know when CNC people are looking for work. Since they call on many companies in your area, they commonly know much about the business climate, including which companies are expanding and which are cutting back. Since they work directly with CNC people, they commonly know when a highly qualified person is unhappy in the current position. Because they make it their business to serve all the needs of their customers, they are usually quite willing to share this information if you ask.

The importance of motivation, aptitude, and proficiency

These are the three qualities all employers seek most in a new worker. *Motivation* will inspire a new worker to take the actions necessary to perform and excel. It is the most important of the three qualities. With enough motivation, an employee will overcome *any* shortcomings in aptitude and proficiency.

Aptitude is the applicant's natural tendency to easily understand concepts related to your CNC environment. It allows the newcomer to adjust and adapt to your company's particular ways of doing things. The higher the aptitude, the shorter the learning curve required for becoming proficient.

Proficiency is the state of the applicant's *current* abilities. It is the sum total of the worker's past employment experiences. We mention it last for a reason. While proficiency is important, many employers place too high an emphasis on proficiency. They would like to hire a person who can just step in and begin performing the assigned duties with absolutely no effort on the employer's part. An employer overly concerned with proficiency bases the decision to hire solely on how little effort will have to be expended getting the new worker to a level at which work can be performed adequately.

Hiring based solely on proficiency can be a drastic mistake. First of all, it is unrealistic to expect anyone to possess *all* of the talents your particular company will require (unless that person has previously worked for your company). Regardless of how large the employment base is in your area, it is unlikely that you will find a person who has worked directly with your own brands of CNC machine tools, controls, CAM system, and other devices related to your CNC environment. Even if you can find such a person, your own manufacturing processes and products will be completely new and will have to be learned, as will your methods of documentation, reporting, and inspection. There will *always* be some period of adjustment. *Motivation and aptitude* will carry the new worker through this learning period, hence their great importance.

Given the short period of time you will be spending with each applicant during the interview process, it can be difficult to gage motivation and aptitude levels. There is no clear-cut set of questions that will thoroughly expose these qualities in all people. This is why many employers impose a probation period during which the worker's motivation, aptitude, and proficiency can be much better measured.

CONTINUING EDUCATION

Though training programs usually are initially developed to allow a company to train entry-level new hires, once implemented they also work nicely to enhance the skills of your current staff. Here are four ways training can help your current staff.

IMPROVING *EVERYONE'S* WORKING KNOWLEDGE OF CNC

In the discussion of the elements of the CNC environment in Part I, you saw how almost everyone in your company is in some way affected by what CNC machine tools do. Even in areas of the company not directly associated with CNC, like design engineering, production control, and human resources, CNC equipment has an impact on how things get done. While people in these departments may never have to write a CNC program or operate a CNC machine tool, the more they know about CNC, the better they can perform their own duties.

We have already given some specific examples. If design engineers understand how CNC machine tools require coordinates to be specified, their dimensioning techniques can be adjusted accordingly, making it easier for the CNC programmer to write programs. If production control people understand how CNC machines can reduce lead time, they can better schedule jobs moving through the shop. If human resources people understand the basics of CNC, they can communicate intelligently with CNC applicants. There is almost no one in your company who will not benefit in some way by attending an introductory course on CNC.

CROSS TRAINING

It is not uncommon for a company to require its personnel to possess multiple skills. For example, a CNC operator who runs one type of machine is eventually expected to be able to operate other types of machines. Indeed, workers who continue to grow in their abilities become more and more valuable to their companies.

Companies that cross train for this purpose enjoy a great deal of flexibility. CNC operators can easily move from one machine type to another based on production requirements. CNC programmers can help out in any area of the company as problems come up. The scheduling of work time for everyone in the CNC environment becomes easier. One person can cover for another in case of sick time, vacation time, or personal time.

Additionally, the long-term needs of a company will probably change from time to time. To minimize inventories and move toward just-in-time (JIT) techniques, for instance, production quantities must drop. Since CNC machine tools are so good for handling small-to-medium production quantities, work that was previously done on highly dedicated equipment (like transfer lines and automatic screw machines) is now being done on CNC equipment. When it comes to work force, this translates into companies needing fewer and fewer dedicated-machine operators. They now need more and more CNC machine operators. Instead of terminating the employment of dedicated-machine personnel, many companies offer their employees the option of learning how to run CNC machine tools.

This need for cross training is not limited to the CNC environment, but applies to all areas of manufacturing. New technologies are constantly being developed that eliminate the need for older and more cumbersome methods. If companies are to stay competitive, they must acquire the new technologies. As they do, people must be cross trained.

TARGETING PROBLEM AREAS FOR IMPROVEMENT

Remember the saying, "A chain is only as strong as its weakest link." Identifying the weak links in your own CNC environment is the first step toward determining when this kind of training is necessary. Repeated mistakes are a common symptom of weak links. If, for example, the scrap rate from a given turning center operation is running high, and operators are having trouble setting offsets correctly to hold size on workpieces, it could be that your operators simply do not understand how to correctly adjust offsets. In this case, operator skill level is the weak link and can be corrected with more operator training (to read gages, to set offsets, to understand tolerances, etc.).

When evaluating your weak links, be careful not to jump to conclusions. In the previous example, you may find that the scrap rate problem is not at all related to the competence of your operators. A more serious problem, like a poor process, improper workholding, bad tooling, overly aggressive cutting conditions, or inadequate setup could be causing your operators to have the sizing problems. In this case, the programmer, tooling engineer, process engineer, or whoever is responsible for planning setups may be the weak link in need of more training.

Some mistakes are easy to define and spot. A turning center operator is expected to measure a diameter and, if it is not to size, adjust an offset accordingly. If a mistake is made while taking the measurement or inputting the offset, it will be obvious when the next workpiece is run. Fortunately, mistakes of this nature can be overcome with little more than additional practice.

Other mistakes are not so clear cut. For example, is it a mistake if the programmer does not take advantage of a special programming technique that can reduce cycle time? Is it a mistake if a programmer does not include trial machining techniques in the program to help reduce program verification time? Is it a mistake if a programmer does not include start-up commands at the beginning of each tool use to make it possible (and easy) for the operator to rerun tools? In all of these cases, production may run well and good workpieces may be produced. But because of a lack of understanding, production is not being run as efficiently as possible. I contend these are just as much mistakes as the more obvious kind. They are simply mistakes of ignorance. CNC workers who have never been exposed to the best and most efficient available techniques cannot take advantage of them. Unfortunately, no amount of practice will correct mistakes of this kind.

The first step to accomplishing anything is knowing it is possible. Before you can use the automatic dialing function of your telephone, you must first understand the feature and know that your particular telephone has it. Before you can begin using your car's cruise control feature, you must know what cruise control is and whether your car has it. Before you can fully master your scientific calculator, you must understand the special symbols on each key and know the implications of when each key should be used.

The field of CNC is filled with alternatives. Even the most basic features can be applied in different ways. Within almost every operation function, programming command, and control feature, there are multiple ways of doing things, one of which will best apply to your particular CNC applications. If your people have never been through formal training, you cannot expect them to know all of their alternatives, let alone that they can take advantage of the best techniques for your particular application. Only through training can your people learn the *best* ways of handling day-to-day problems. If left totally to their own devices, they may come up with a workable way to handle any given problem, but it will not be the only way (and probably not the best way) of solving the problem.

IMPLEMENTING NEW METHODS

Astute companies are on constant lookout for better ways of doing things. When new methods are found, a change in the way people do things is required. Training for the people involved will allow a smooth transition from old methods to new.

Training for this purpose is commonly required when new CNC equipment is purchased, especially when the new equipment requires a major change in current methods. Possibly a tool and die shop currently using completely manual means to produce tooling is purchasing a CNC wire electrical discharge machine. The original training for new equipment commonly comes directly from the machine tool builder. Keep in mind that the machine tool builder's training may be limited to only one or two of your people. Additionally, training of this nature will likely not be custom-tailored to suit your company's specific needs.

Training for new methods is also required when new CNC accessory devices are acquired. Examples of CNC accessories complicated enough to require training include tool length measuring gages, bar feeders, rotary devices, pallet changers, probing systems, postprocess gaging systems, and tool-life monitoring systems. Additionally, there are many computer software accessories related to CNC. CAM systems, tool-path plotting systems, CNC text editors, distributive numerical control systems, and statistical process control systems are all examples of software products that commonly require training.

TRAINING RESOURCES

You have several alternatives in training your people. While one of the goals of this text is to help you develop your own in-plant training program, there are many training resources available that can minimize the investment you need to make. And even companies that do not develop their own in-plant training programs can take advantage of these resources to further their staff's understanding of CNC.

PROVIDING INFORMATION THROUGH YOUR COMPANY LIBRARY

Your company owns many valuable publications related to the manufacturing you do, and many are directly related to your CNC equipment. At the very least, you own one set of training manuals that accompanied each CNC machine tool your company has purchased. You surely own handbooks for solving shop-related problems and developing cutting conditions (speeds and feeds). You probably own informative books related to technical subjects like tool design, statistical process control, manufacturing engineering, and any other topic of primary concern to your company. At least one person in your company probably subscribes to one or more of the trade journals related to manufacturing. And, if there are trade journals related to the product your company manufactures, surely someone subscribes to them as well. Catalogs from your key suppliers line the bookshelves of those doing the purchasing. You may have informative sales videos given to you by suppliers to help you understand their products. You may also own training videos to help your people learn how to better utilize your company's CNC machine tools. These are all valuable resources that have been acquired and paid for by your company.

Some of these publications may have value to a limited number of people. Supplier catalogs and sales videos, for example, may be needed only by your buyer. For these publications, it may make sense to retain only one copy of each and store them close to the person needing them on a regular basis. Others, like programming manuals, trade journals, and technical handbooks, may be important to many people in the company. It is only using good sense to make them easily accessible to anyone who needs them. Informative books on manufacturing may not be critical to your company's day-to-day activities, but may be of interest to almost everyone in the company. For those publications that have value to several people, a company library makes an excellent central storage site that is available to everyone in the company.

Given the importance of your company's publications and the fact that any one of these resources may be needed by *anyone* in the company at a given time, it is amazing how difficult it can be in many companies to locate a needed publication. In many companies, for example, the machine programming and operation manuals are kept in the programming department. If a CNC operator needs to refer to key information about the machine (perhaps related to an alarm the control has generated), the operator must contact the programming department to solve the problem. If the problem occurs on the second or third shift when no programmers are available, the machine may sit idle until a programmer is available to provide the needed manual.

Of equal importance is the need for your company staff to stay abreast of the state of the art in manufacturing. Trade journals are on the cutting edge of manufacturing technology, and contain all kinds of new ideas and new product information. For this reason, they should be required reading for *everyone* in manufacturing. However, it is likely that only a limited number of people in your company receive trade journals—and probably fewer yet actually read them. Since our industry changes at such a rapid pace, you never know when a new idea will be shown in a trade journal that could have a dramatic impact on the success of your company, *if someone reads it.*

Your company library can make this valuable information accessible to everyone in your company, not just the select few who subscribe. Though trade journals can provide extremely valuable ongoing reference information, it is amazing how many people simply discard them after only one reading. Having access to a trade journal's contents at some future date may mean the difference between successful incorporation of a new process and dreadful failure. Your company library offers an easy way of retaining each issue for future use.

CNC directories, commonly published by publishers of trade journals, offer listings of all suppliers in the field of CNC and can be invaluable when your company needs to purchase CNC-related products. They can easily help you find the suppliers of just about any CNC product or accessory. *Modern Machine Shop's CNC and Software Guide*, published yearly (April of each year) by Gardner Publications in Cincinnati, Ohio, for example, provides a comprehensive list of all suppliers having anything to do with CNC.

Developing a company library is simple. It involves little more than ensuring that at least one copy of all important manufacturing publications is centrally located, and tracking who currently has each publication. Only a small continuing effort is required to maintain the library.

LOCAL LIBRARY

If you have never browsed the manufacturing section of your local public library, you may be very surprised at what you find. Though you may not find the most up-to-date information on every topic (libraries cannot afford to buy every book printed), most maintain a respectable section on manufacturing. Be careful to check the copyright date for books on CNC, since this technology changes rapidly. CNC books more than 15 years old may confuse rather than help readers regarding current techniques.

Most libraries also maintain an excellent periodicals department. You can find microfilm copies of almost every magazine article ever printed, including those in the more popular trade journals. You will also find an extensive indexing system that will let you easily research key topics. You can use this valuable resource whenever you must do something new. Say, for example, you have a new machining center job that requires thread-milling. If you have never thread-milled, you will need to research the principles of thread-milling before you will be able to perform the operation successfully. By using the microfilm index wisely, you will easily be able to find every trade journal article written on the subject.

MACHINE TOOL SUPPLIERS

Your initial source of training and support when any new piece of CNC equipment is purchased is the supplier of the machine tool itself. In fact, the support you receive after the sale should be a major issue during the buying process. While machine tool suppliers vary as to how much training they offer free of charge, most offer excellent training programs to their CNC customers. But as we show in this section, there are some limitations that keep most CNC users from using machine suppliers as their sole source of training.

Who's doing the training?

It is the general perception in the industry that machine tool builders offer the most comprehensive training for CNC machine tools. This feeling is logical since it seems likely that the people who design and build the machine know the most about it. Most offer formal training not only on machine usage (programming and operation), but on service, maintenance, and upkeep as well.

Keep in mind, however, that not all machine tool builders maintain a full training and support staff in the United States. In fact, many builders from outside the United States have little more than a skeleton crew to handle administrative matters. These builders sell through importers, and rely heavily on the importer to provide all technical support and service, including customer training. As long as the importer's training and support staff has been well trained by the machine tool builder, the level of training quality related to machine usage from the importer is generally quite high. However, there may be little, if any, formal training available for service and maintenance.

Though some machine tool builders sell their machines directly to end users, most machine tool builders and almost all importers work through a network of machine tool distributors to sell their equipment. Distributors located close to the training facilities of machine tool builders and importers commonly rely on them for the training of their customers. Since their customers are already close to the training facility, any training offered by the distributor would be a duplication of effort. However, as the distance between the distributor and machine tool builder/importer grows, so does the pressure on the distributor to take on more of the training burden for its customers. For this reason, many machine tool distributors offer excellent training programs for customers. Some even exceed the quality of training available from the machine tool builder or importer.

Since machine tool distributors tend to be very aggressive when it comes to making sales, they also tend to be very flexible when it comes to support and training matters. Many will train in their training facility or in your own plant, making it your decision as to where and when training takes place. Additionally, most will accept several of your people for initial training, as well as ongoing training, free of charge. Again, since they are highly interested in making machine tool sales, most will be flexible about training matters if you negotiate the issue *before* you purchase the CNC equipment.

Typical training

Most suppliers allow their users to send at least two people through the training program free of charge. The length of the training period will vary from one supplier to the next. It is customary for one training course to last from three to five full days, usually during one week. Training is usually done at the machine tool supplier's facility, though special arrangements can often be made for on-site training. If training is done at the machine tool supplier's facility, the end user will be responsible for any travel and lodging expenses incurred during the training period. This means even though the training itself may be free, there will usually be travel and lodging expenses if training is not offered locally. If training is done on site, you will need to ensure that your people will be kept free from interruptions. Nothing is more frustrating for instructor and student than having to stop the course while a student "puts out fires" in the shop.

Many of these training issues are flexible and negotiable during the purchasing process. Remember that distributors offering training tend to be more flexible with their training than are machine tool builders and importers. Since machine tool distributors are the people actually doing the selling, they tend to be much more willing to offer special considerations in order to make the sale.

Limitations

The importance of the training and support available from your machine tool supplier cannot be overstressed. All CNC users should take advantage of what their suppliers have to offer, especially when purchasing a new type of CNC equipment. However, there are many limitations that keep most end users from relying on machine tool suppliers as their sole source of training.

Costs. Very few machine tool suppliers offer unlimited free training to their customers. As stated, most will allow only a limited number of people to attend their training courses free of charge. After that, the end user must pay for all training it receives. Typical charges for the machine tool supplier's standard courses (held at the supplier's facility) run from $500 to well over $1000 per student per course. If you expect training to take place in your own facility, it is not uncommon for machine tool suppliers to charge $500 to $1500 *per day* of training.

These charges are only part of your total training bill when you send people to machine tool supplier courses. All travel, lodging, and meal expenses must be paid by the end user. It is not uncommon to spend well in excess of $2500 for one person's attendance in a machine tool supplier's 5-day course if travel and lodging is required.

Time permitted. As stated, most machine tool supplier courses run for three to five days. While this may sound like an acceptable amount of time, not everyone can absorb the vast amount of information at the pace it is presented. In reality, the machine tool supplier is trying to cram as much information into five days as your local technical schools present in more than one full semester.

Pressure on the attendee to perform. Along the same lines, most attendees must be able to begin working with the CNC machine as soon as they return from training. This puts enormous pressure on the student to learn.

Attendee experience level. When your company sends people to training courses conducted by machine tool suppliers, your people will be attending the course with people from other companies.

I have found that, in almost every class, attendee experience level varies dramatically. In many classes, attendees range from absolute newcomers to people with several years of CNC experience. This makes it very difficult for the instructor presenting the course. If the instructor stays very basic in discussions so as not to lose the beginners, the experienced people will eventually complain that the course is too basic. If the instructor skims the basic material and prematurely discusses advanced topics to please the experienced people, the beginners will be lost for most of the course.

While good instructors will stick to a previously planned course outline, and while this outline is usually made available to all attendees *prior* to the course, the problem of varying experience levels remains. This problem also causes the quality of the course to vary from one offering to the next. If, for example, one offering of the course is made up of predominantly beginners, the instructor can keep things quite basic and make everyone happy. As the people return home, they say very good things about the course. When more people are sent to the next session, there may be people of all CNC experience levels. The second group of beginners may not get nearly as much from the course as those in the prior session.

Instructor level. While it may come as a bit of a surprise, most instructors working for machine tool suppliers have had no formal training to be instructors. Most are simply application engineers doubling as instructors. While application engineers who have a strong interest in teaching eventually become very good instructors, most machine tool suppliers use a *trial-by-fire* approach when an application engineer begins teaching. Their application engineer may attend only one or two courses taught by someone else before beginning to teach others.

Scope of training. Though the actual content of training programs will vary from one supplier to the next, the most basic goal of each is to get all attendees to a level at which they can *begin* using the equipment. Suppliers will teach what they feel is the safest, most proven method of utilizing their equipment. While this may sound good, keep in mind that there are many, many ways of utilizing CNC machine tools. Some are easy to understand and very safe to incorporate. Others sacrifice ease of use and safety to improve efficiency. Most machine tool suppliers will shy away from teaching any techniques that even border on compromising safety, even though they may admit that more efficient methods are possible than those they teach.

Also consider that no single method of CNC utilization will best suit the needs of all companies. Criteria such as production quantities, similarity among the workpieces machined, and the available cutting tools dramatically affect what may be best for a particular company. For example, there are several ways to determine and enter the program zero location. If the number of setups a given company makes is small because of high production quantities (say only one or two setups per month), just about any program-zero-determining technique will be acceptable. Any time wasted due to inefficient program zero assignment can be easily buried in the overall cost of the production run. If, on the other hand, a company runs small lot sizes and makes a great number of setups (8 to 10 per day), the method by which program zero is determined and entered becomes much more important. Any inefficiency will result in a great deal of wasted time.

This is but one specific example of numerous CNC usage techniques that must be determined for the user's own application of the machine tools. In most generic training courses conducted by machine tool suppliers, time does not permit the fine-tuning of any one CNC technique to each attendee's particular applications. When confronted with several methods of handling a given CNC function, most instructors will present the safest or easiest-to-understand method, though possibly not the most efficient.

Timeliness of courses. Though most CNC suppliers offer a schedule of CNC courses for the machine types they offer, the timing of these courses will not always correspond to your company's needs. Your supplier, for example, may not offer a course at the precise time a new person is hired. Also, most suppliers reserve the right to cancel a given course if a minimum number of attendees is not registered, meaning your new person may have to wait several weeks (or even months) before a training course is available.

Consistency. If you own several similar CNC machine tools manufactured by different machine tool builders, you will notice substantial differences in how the training departments of each of the suppliers teach your people. Each will present the methods it feels best suit its own machine tools. Often the methods will conflict, confusing students. Many instructors will not address another machine tool builder's methods, even though they may be better suited to your application.

CONTROL MANUFACTURERS

Most CNC control manufacturers also offer training courses. However, since a single control model can be applied to any number of different machine tools made by different machine tool builders, their programming and operation courses tend to be generic. Unless the CNC control manufacturer also manufactures the machine tool, it will not be able to make specific presentations about any function that may vary from one machine tool builder to another. Examples of programming topics CNC control manufacturers cannot discuss in great detail include the programming of a machining center's automatic tool changer, the programming of any M codes supplied by machine tool builders, and the programming of any accessory device (pallet changer, tailstock, probing system, etc.) supplied by the machine tool builder.

When it comes to operation, the control manufacturer's courses can address only the operation of the CNC *control*. While they may be able to relate the concepts of how CNC machine tools are operated, every machine tool builder determines what buttons and switches it will provide for operation of the machine.

For all of these reasons, most CNC users limit the use of control manufacturers to introductory programming and operation training only. While students can get an excellent working knowledge of CNC by attending these courses, they cannot typically get enough information to begin actually working with any one CNC machine tool.

However, CNC control manufacturers also offer courses unrelated to programming and operation. Most offer excellent service and maintenance courses that will teach your people how to maintain the CNC control and other electronic components of the machine tool. In fact, many machine tool builders will refer you to the control manufacturer for this kind of training.

Most control manufacturers also offer courses on certain special programming functions of their controls. If the control offers parametric programming capabilities, for example, the control manufacturer will probably offer courses on its use. If the control utilizes a programmable controller to handle the interface between the machine tool and control (as most current controls do), the control manufacturer will offer training on how this kind of programming is accomplished.

LOCAL TECHNICAL SCHOOLS

One excellent, yet often ignored, CNC training resource is your local technical school. These schools, including community colleges, vocational schools, technical institutes, colleges, and universities are found in all but the most rural of communities and should be included in *every* company's training program. You probably have one within easy driving distance of your company.

Most technical schools offer excellent entry-level courses in many facets of manufacturing, including basic machining practice, making them a valuable resource for training new hires. The standard manufacturing courses they offer usually are aimed at giving students a broad understanding of manufacturing functions, and may not be specific enough for your company's particular needs. For this reason, many technical schools are very flexible, willing to work with local industry to develop courses aimed more specifically at their students' needs. Some will even develop the training program you desire and present it in your own plant, using your own equipment.

SOCIETY OF MANUFACTURING ENGINEERS

The Society of Manufacturing Engineers is a nonprofit professional society dedicated to the advancement of scientific knowledge in the field of manufacturing engineering. SME provides public training seminars, clinics, and conferences, as well as in-plant custom training for all facets of manufacturing, including those related to CNC.

TRAINING SUPPLIERS

There are many companies that specialize in supplying CNC-related training. Most can supply books, videos, and other training materials and offer in-plant training courses as well. While these companies commonly offer a series of standard courses, most are willing to tailor their courses to your needs. Most can be found advertising in trade journals and are usually listed in CNC directories.

MANUFACTURING TECHNOLOGY CENTERS

Though not commonly known, there are many state and federally funded facilities aimed at helping manufacturing companies. Manufacturing technology centers offer all kinds of assistance, including help with research and design for new processes, programming for complicated work, and rapid prototyping processes. In this sense, they act almost as a cooperative, providing new technologies for companies that cannot afford it for themselves. Most offer CNC training courses.

ON-LINE SERVICES

With the growing popularity of computer-based information networks, more and more manufacturing-related bulletin boards are becoming available. The Society of Manufacturing Engineers' *SME ON-LINE* is one example of this kind of service that can supply reference information related to all facets of manufacturing, including CNC technology.

Chapter Five

How to Develop an In-plant Training Program

B ecause each training resource mentioned to this point requires the assistance of outsiders, and since none of the training can be perfectly customized to your company's own specific needs, you will not have total control of your training program if you rely solely on external resources. Though this may be the case, any combination of the resources previously mentioned can be used in conjunction with the techniques we present in this chapter. You can easily customize your training environment to take advantage of your own in-plant training capability in conjunction with outside help.

The largest single advantage of developing your own in-plant CNC training program is flexibility. Since you have total control of how training is done, you can tailor the training specifically to your needs. The time and location constraints attendant to contracting with outside trainers will not apply. You can train during the week or on weekends, during working hours or off, during the morning, afternoon, or evening. You control how long each training session runs. You control the length of time each course runs. And, of course, you completely control the course content.

Having total control of course content opens the door to doing things not commonly associated with any outside vendor courses. For example, you can combine training for company policies and practices, training for basic machining practice, and training for CNC all in one single course. Especially helpful when it comes to training new CNC operators, this gives you the flexibility to combine several very important topics in one training program. Conversely, most outsiders can specialize in only one of your training needs.

DETERMINING THE SCOPE OF NEEDED TRAINING

The first step to implementing any in-plant training program is assessing your company's training needs and determining how best to address those needs. To do this, you must target those CNC positions you wish to train for. In this section, we present your most important considerations.

THE FIVE LEVELS OF CNC PROFICIENCY

No single training course could adequately cover the vast amount of information related to the field of CNC. To begin to limit the scope of what must be presented during your own CNC training courses, we offer five *proficiency levels* related to CNC machine tool utilization. These are the levels at which people can be productive in their CNC related job responsibilities. In Chapter Six, we suggest a method of teaching all five proficiency levels with a single proven approach.

We assume that candidates for CNC training already possess some knowledge of basic machining practice related to the type of CNC machine being used. Depending on the level of proficiency you wish to address in a given training course, this may not always be the case. You may, for example, need to train entry level CNC operators who have no previous shop experience. In this case, you must begin by teaching students about the various machining operations your company performs, how to read and interpret blueprints, how to use and interpret gaging devices, and in general, discuss many other basic machining practice topics. Because basic machining practice topics are not in the scope of this text, we limit our presentations to developing CNC related training only.

The five proficiency levels are:

Level 1—ability to communicate about CNC intelligently.
Level 2—ability to operate CNC equipment.
Level 3—ability to make setups for CNC equipment.
Level 4—ability to program CNC equipment.
Level 5—ability to teach CNC.

Everyone in manufacturing should have the ability to communicate intelligently about CNC. Though many people in manufacturing will never have to program, set up, or operate CNC equipment (people like managers of CNC operators, design engineers, and process engineers), the more they know about CNC, the better they can communicate with others in the CNC environment.

To become a proficient CNC machine operator (commonly thought of as the entry-level CNC position), a person must possess knowledge of the machine's components and be able to manually activate important functions. Operators must be able to perform procedures related to running production (machine start-up, activating programs, changing tools and inserts, etc.). A knowledge of how tool offsets apply to the machine is also mandatory. These are but a few of the many functions a CNC operator must understand.

A CNC setup person must in addition understand how setups are made; how programs are verified; how workholding setups are made; how tools are assembled, loaded, and measured; how program zero is assigned; and how workpieces are inspected.

A CNC programmer must understand all of the machine's programmable features and functions. He or she must know how to plan setups, process the sequence of operations, perform required math calculations, and, of course, format programs. The programmer must also be able to help in the verification of the CNC programs. If the company uses computer software for the purpose of preparing and verifying CNC programs, the programmer must also be proficient with this software.

The CNC trainer must possess a thorough knowledge of all topics being presented during training. The trainer must additionally possess many "people" skills (including patience), communication skills, and the willingness to freely share knowledge.

Though we have listed only a few of the responsibilities related to each level of CNC proficiency, you should see just how large the overall scope of CNC training can be. As you begin to develop the training curriculum to be used in your own in-plant training courses, you *must* limit the scope of what you intend to accomplish during each course. Just as any good college breaks up a degree program into a number of related courses, so can your company break up the complex task of CNC training into small and easy-to-handle sections. The better you are at doing this, the easier it will be to develop each course and the better each course will be.

LET CNC UTILIZATION DRIVE THE TRAINING

Most companies do not need to develop a training program that encompasses the entire scope of CNC. Instead, they can target specific problem areas that need training the most. When it comes to new hire training, for example, CNC operators commonly make up the greatest number of people in the CNC environment. Since many companies have trouble finding and keeping CNC operators, one of the first in-plant CNC courses they implement is for entry-level CNC operators.

However, as stated in the preceding chapter, entry-level operator training is just one of the critical areas a training program should address. Continuing education can keep your current CNC people improving their CNC skills. When it comes to continuing education, the most basic criterion should always be *improving CNC machine utilization*.

CNC programmers, for example, have a very large impact on machine utilization. With every program they develop, numerous choices are made that affect how well the machine will be utilized. Poorly formatted programs result in underutilization. Companies should strive to continue the education of their CNC programmers to ensure that they are taking advantage of the most recent and efficient techniques possible.

Setup people, tool designers, process engineers, quality control people, and shop managers also have an impact on how well your CNC equipment is utilized. Be sure you consider everyone in your CNC environment as you make choices as to how your training program will be implemented.

SETTING GOALS

To reap the benefits of training discussed in the previous chapter, you *must* set goals. Your goal criteria will help you limit the scope of your training program and help you plan just what it is you wish to accomplish during a given course.

Your goals should be written and should make it clear to everyone involved in the training program what you expect to achieve during training. These goals will help the instructor understand the key points that all students must understand in order to complete the program. If shown to the students, your goals will act as guidelines to the course, providing an outline of the most important points. The student will know when to pay special attention as a goal-related topic is being presented.

We cannot stress enough the importance of goal setting. Without goals, information can be presented by an instructor and even understood by a student, but no clear-cut set of expectations can be achieved. With clearly defined goals, everyone in the learning environment will know the criteria for success.

Goal setting for CNC training requires planning by people familiar with the common problems associated with CNC equipment. Ideally, several experienced people should be involved with the goal-setting process for each course, since many of the ideas for specific goals come from past problems with the company's utilization of CNC equipment.

When you consider how many specific tasks a CNC person (in any position) must perform, it should not be difficult to develop a very specific list of goals. If you are having problems setting goals, however, simply watch your people and list the tasks they perform. Model your goals after those people who perform the most efficiently. Goal-setting criteria can include reducing scrap, improving consistency among workpieces, improving efficiency, or any other function that has a positive influence on the CNC environment.

EXAMPLES OF SPECIFIC CNC TRAINING GOALS

To further stress the importance of goal setting, we include some examples of goals for several types of CNC training. While no one set of goals will apply to all companies, this list should help you see how goals can be set and documented. Goal setting truly forces you to think through what your students must know at the completion of the training program.

Common goals for training entry-level turning center operators

This set of goals was developed for a medium-sized company with 12 CNC turning centers run by 19 operators. Though the company runs three shifts, some operators run as many as three machines. Production quantities are relatively high, ranging from 500 to more than 3000 workpieces. Operators are expected only to maintain production, their responsibilities being to load and unload workpieces, inspect workpieces, and adjust size with offsets. The company employs four setup people (two for first shift, one for second, and one for third) to change setups and verify new programs; no program editing is done by the turning center operator. This particular company is in a remote area, with a limited employment base. For this reason, new employees typically have no previous CNC experience and, in many cases, little or no basic machining practice experience.

Goals related to basic machining practice:

- Each student must understand the company's method of drawing and dimensioning workpieces (blueprint reading). We make sure that everyone understands our variable dimensioning techniques for shaft collars.
- Each student must be able to proficiently measure dimensions with the measuring tools our company uses (6-inch scale, dial caliper, micrometer, telescoping gages, thread gages, and depth micrometer).
- Each student must understand the tolerancing methods used by the company, including dimension tolerances (plus/minus) and surface relationship tolerances (geometric tolerancing). This understanding is essential if an operator is to make offset changes correctly.

Goals related to company procedures:

- Each student must understand our routing package, including print, routing sheet, and time recording entry forms.
- Each student must understand our statistical process control data retrieval system.
- Each student must understand our tool crib procedures for getting replacement tools and inserts.

Goals related to CNC machine operation:

- Each student must understand the dangers of machine operation and be able to apply safety procedures including the use of dry run, single block, rapid override, etc. Also, each student must understand the functions of the machine that require the most caution (chuck jaw open/close, tailstock quill movement, turret indexing, etc.).
- Each student must be able to start up and power down the machines.
- Everyone must be able to perform manual functions, including jog movements, handwheel movements, tailstock movements, spindle functions, and turret index.
- Everyone must be able to upload/download programs, call up the active program, scan within the program, and perform minor editing functions (for simple things like speeds and feeds).
- Everyone must be able to correctly set offsets, as well as determine how much out of tolerance any one dimension may be. Everyone must understand how we determine the offset number related to each tool, how to call up the work offset, wear offset, and geometry offset pages on the display screen. If a workpiece surface (length or diameter) is not within

the tolerance band, each person must understand how to make the correct offset entry to bring the dimension back to the middle of its tolerance band. Once production is running, each person must understand how to continue to adjust offsets for tool wear, and when a tool insert is replaced, how to adjust the offset back to its original value.

- Be sure everyone understands the function of each button and switch on the machine.

Notice how specific these goals are. Those who see this list will know exactly what are considered to be the most important topics for discussion. Also notice how goals can reflect past problems. In the first goal of the list, for example, a specific reference is made to a particularly confusing style of dimensioning the company uses. From past experience, the person setting these goals knows that this is a consistent problem area, one that must be clarified for each new operator.

While you may not totally agree with this particular list of goals based on *your* company's needs, at the very least you should see how setting goals helps you plan the in-plant training curriculum. As you can see, setting goals helps you plan what the students *must* know at the completion of the training program.

Common goals for training experienced machining center operators

The following set of training goals was developed for a small job shop that has three vertical machining centers. Since this company is small, and since production quantities are rather low (typically under 100 workpieces), each operator is expected to do more than simply run the machine and maintain production. Each must be able to make the workholding setup; load the program; assemble, measure, and load tools; and verify the program. While the programmer (the owner's son) will be available to help with specific problems, this company expects its operators to be highly self-sufficient. Because the company cannot afford to train people from scratch, and because it does a wide diversity of work for different companies, new hires must possess a high level of experience (at least one year) prior to being hired. Though training in this company is quite informal and mostly on the job, goal setting is still very important for determining just what will be expected after training is completed.

Goals related to basic machining practice:

- The company hires only experienced CNC operators, so very little is done during the training period with regard to basic machining practice. New hires are simply introduced to those machining operations we perform that are new to them. It is assumed that students will be quick to pick up this information.

Goals related to company procedures:

- New hires must be taught the company's operating procedures related to processing, tooling, and priority setting.

Goals related to CNC machine operation:

- Each student must be shown the differences between the machine functions with which they have had experience and the functions of our machines. This should go quite fast, since the student will be simply building on previous experience.
- We make sure new hires can perform common program manipulation procedures, including manual keyboard entry, upload/download programs, calling up the active program, scanning within the program, and performing editing functions.
- Each student must be able to make setups based on our workholding tooling and fixturing for the jobs we do on a regular basis. Since we are a job shop and it is hard to predict our future work, each student must be able to adapt to new setups as they come along.
- Each student must be able to assemble, measure, and load tools. Each must also be able to correctly enter tool length offsets.
- Students must be taught how to measure the program zero location for our setups. This includes corner finding with an edge finder as well as indicating the center of a hole with a dial indicator. They must also know how to enter the measured values into corresponding fixture offsets.
- Each student must be shown how to make enhancements to our programs for the purpose of improving efficiency. These modifications include feed and speed changes, eliminating inefficient motions, and minimizing air cutting time.

Notice how each set of goals matches the company's specific training needs. In the first set of goals, a high emphasis is placed on basics. In the second set, the company assumes a high level of knowledge prior to hiring.

IT'S NOT OVER UNTIL IT'S OVER

Industrial training differs from any other type of education in two important ways. First, at the completion of industrial training, the students must immediately be able to apply what has been learned. This means they must possess a firm understanding of *all* the materials presented. Compare this to a typical high school or college curriculum where years may go by before the student must directly apply learned information. While the gage of success in typical school courses is the grade the student earns, that for industrial training is how well the student can perform the tasks taught during training.

Second, the price of failure in an industrial training program can be very severe indeed. Given the high level of danger associated with industrial equipment, ineffective training can result in scrapped workpieces, damaged tooling, damaged machine tools, and injuries to workers. Compare this to most other forms of education, where failure (while emotionally painful) causes no physical damage.

For these reasons, be careful with how rigidly you schedule the time allowed to complete the training program (especially with new courses). Since the learning pace of students varies dramatically, and because of the urgency tied to their becoming proficient, you will want to schedule ample time to complete the training course. To overcome time constraints, many companies do not actually assign a fixed time limit to a course that involves learning tasks which expose people to potentially dangerous situations. This kind of course is not considered complete until *every* student shows proficiency in the material taught.

THE LEARNING ENVIRONMENT

The quality of your learning environment plays a big role in the success of your training program. Listed here are the six most important factors contributing to the learning process. Together they constitute the learning environment. The better the quality of each factor, the better the learning environment.

SIX FACTORS THAT CONTRIBUTE TO LEARNING

Motivation is the most important factor in any learning environment. Students must be highly motivated to learn. Motivation is the driving force that causes students to stick with it even when they are having trouble understanding what is being presented. Indeed, *any* problem with learning can be overcome if the student's motivation is high enough.

In similar fashion, the instructor must also be highly motivated. The instructor must have a burning desire to present course material in a way students can assimilate and comprehend. When students are having problems, the instructor must be motivated to spend the extra time it takes to ensure that the students eventually understand the material. Again, compared to other forms of education where an instructor simply grades the student at the end of the course, industrial instructors must have the drive to guarantee every student's understanding. This can be very challenging, as students' aptitude levels vary.

Aptitude will determine how quickly and easily learning will be acquired. The aptitude of the instructor for making attention-holding presentations, giving pertinent analogies, preparing illustrative visuals, designing realistic practice exercises, and, in general, keeping the students' interest levels high will help students catch on to new material quickly and easily. Instructors with high aptitude make it easy for students to learn. In similar fashion, the students' aptitudes for learning manufacturing-related functions—specifically CNC—also contribute to quick and easy learning. Students with high aptitude make the instructor's task easy.

Presentation is the heart of training. The better the instructor plans and prepares the presentation, the easier it will be for students to learn. Presentation can consist of many things, including the instructor's lectures, demonstrations, simulations, overhead and projector slides, videos, and any other medium that helps to convey an idea.

Repetition reinforces a student's understanding of learned information. Even students with extremely high aptitude will find it difficult to learn from presentations made only once. *All* training sessions should begin with a review of recent presentations. Depending on the frequency and duration of each training session, entire sessions should, at times, be devoted to reviewing what students should already know.

Reviews also help the instructor limit what is presented during each session. Knowing that certain information will be reviewed, the instructor can avoid getting too deeply into complicated topics the first time the information is presented. Only after students have a firm grasp of the basics will the instructor dive deeper and introduce more complicated variations.

Concentration is essential to the learning process, on the part of both the students and instructors. Regardless of how well presentations are made, complex topics such as the concepts of CNC require students to think. In similar fashion, despite the instructor's proficiency in preparing and making presentations, questions will inevitably come up that force the instructor to think on the fly. Therefore, a good learning environment should make it easy for everyone to concentrate. Training should be done in a quiet place free from interruptions.

Practice with reinforcement acts as the gage to judge the success of training. Well-designed practice exercises should be realistic, forcing the student to perform procedures exactly as they will be performed on the job. Reinforcement must come as the result of students' work. If a student demonstrates a firm understanding of the presented information, reinforcement consists of praising the success. On the other hand, if practice exposes a student's lack of understanding, reinforcement should come in the form of repeated presentations, review, and more practice, ensuring that the student eventually catches on.

SELECTING THE INSTRUCTOR

The person responsible for teaching is obviously one of the most important people in your training program. This does not mean, necessarily, that you must hire a special person to do the training. In many cases, you can find someone in your company who has the ability to instruct.

A common misconception in the selection of a CNC instructor is that the person with the most CNC experience will make the best instructor. While this may be true in some cases, a person's expertise in a given field is no guarantee that the person will make a good instructor. An experienced operator may, for example, thoroughly understand CNC and be totally competent in all job responsibilities— yet this person may have no desire (or ability) to transfer knowledge to others.

Compare this to any profession that involves learning complex concepts. A baseball player may bat .400, but may have no interest or ability to pass on batting skills to someone else. An author may have several books on the best-seller list, but have no desire to teach the basic skills of writing. An actor may star in several hit movies, but be unable to teach acting skills. The same problems apply to teaching CNC. First and foremost, your instructor must have the desire and ability to relate what he or she knows to others.

Teaching is a *portable skill*. By this we mean good instructors will have the ability to adapt to teaching different subjects. They will have the ability to effectively communicate *anything they know*. Given time to learn the topic and prepare teaching materials, a good instructor will be able to teach just about anything.

Given the choice between a CNC expert who has no teaching background or a teacher who has no CNC background, in most cases the teacher will adapt to CNC more easily than the CNC expert will adapt to teaching. If you are going to be hiring a special person to do the teaching, it may be difficult to find a person who has extensive experience with CNC as well as teaching skills. Always place your emphasis on *teaching skills*.

Instructor attributes

That a person has not taught before does not mean that person will not adapt to teaching and eventually become an excellent instructor. If you must choose your instructor from your current staff, keep in mind that there are several human-nature-related attributes you can easily spot in potential instructor candidates.

Patience. The most important instructor attribute is patience. It can be very frustrating when a student does not understand presented materials, especially when the instructor finds the information being presented easy to understand. The instructor may have to rethink the subject matter several times in order to present information in a way the student *can* understand. If the instructor shows even the slightest frustration, the student will sense it quickly. Frustration on the instructor's part leads to loss of confidence on the student's part.

Remember that learning is a *2-way street*. Some instructors may contend that if the student cannot understand information presented accurately, it's entirely the student's problem. However, good instructors know that if the student is having trouble understanding, it is at least half the instructor's problem. Accurate presentation is no guarantee that the student will understand. To simply repeat the information in the same manner will do nothing to improve the situation. Instead of becoming frustrated, the instructor should view students having problems as a challenge to come up with a new twist in the presentation. Possibly an analogy is called for that makes it easier for the student to understand.

Patience is an easy attribute to gage in a person's personality. If people show patience when handling any difficult situation, it is likely they will also show patience in the learning environment.

People skills. The instructor must have the ability to communicate subject matter to *all* students. This means they must enjoy working with people and be willing to ignore personality conflicts during training. You can easily judge this attribute in a person by observing how the person interacts with others in your company.

Desire and ability to relate knowledge. Instructors must freely share what they know with others. This means there can be no secrets in the learning environment. In some companies, job security is measured by how much more one person knows than another. People in this kind of environment become unsure and unwilling to share anything they know. Instructors must feel that they can present information to the limits of their ability with no fear of consequences.

This desire to relate knowledge must go further than simply not having to worry about job security. The instructor should gain a great deal of satisfaction and take pride in students who complete training and go on to be successful in the CNC environment.

THE CLASSROOM

A good classroom will facilitate the instructor's ability to teach as well as the students' ability to learn. While almost any quiet, out-of-the-way room will suffice in a pinch, the classroom should allow the instructor to present material without serious constraints. While oral presentations require nothing special of the classroom, remember that visuals are sometimes more important for making points. At the very least, the classroom should be equipped with some form of chalkboard or easel so that the instructor can make sketches.

Well-prepared instructors minimize the amount of time taken during training to make illustrations on the fly by preparing visuals *before* training begins. Overhead projector slides, for example, are easy to make and inexpensive, and the cost of the overhead projector is relatively low. Photographs and other colorful illustrations can be easily made on 35-mm slides with little expense.

More and more, professional presenters are using computer-based presentation software to prepare visuals for training. With today's technology, it is relatively simple and inexpensive to connect a personal computer to a television monitor or other video equipment through a *scan converter*. The scan converter translates the VGA graphics computer signals into video that can be played on any television monitor. While these devices are not inexpensive, they allow a high degree of flexibility and creativeness during training, which can dramatically improve student attention spans.

Videos, including training videos, demonstration videos, and videos taken within the company to show current processes can be very helpful in training. If videos are to be used, the classroom must of course be equipped with a VCR and television.

The classroom components discussed to this point all help the instructor to better make presentations. However, there may also be a need for devices to help students learn. CNC control simulators and personal computers for practicing with the software your company uses are among other devices you may need to include in your classroom.

It almost goes without saying that a well-designed classroom should enhance the students' ability to concentrate on subject matter. The room should be spacious, tables and chairs should be comfortable, visuals and illustrations should be large enough for everyone to see, and lighting intensity should be sufficient to discourage napping.

While we have just described the ideal classroom, many companies simply do not have the resources to create a perfect classroom. Again, just about any room can be used as a classroom, but the points just made should illustrate the importance of setting up an environment conducive to learning. For companies that do not have adequate facilities, keep in mind that most local hotels can provide meeting rooms furnished with all the audio-visual equipment your people need. You will find the cost of most hotel meeting rooms to be quite reasonable. By the way, getting people out of your company for training also eliminates the possibility of interruptions during training sessions.

COURSE MATERIALS

Ideal training courses provide students with materials that complement instructor presentations. Course materials should allow students to review presented information *between sessions* and help get ready for the next session. Better yet, course materials should help students learn *more* about course topics of interest than the course outline intends.

At the very least, each student should have a pencil and paper for taking notes so that information can be reviewed between sessions. However, course manuals developed by the instructor can minimize, if not eliminate, the students' need to take notes and provide a much more comprehensive way to review information. CNC textbooks, while usually not specific enough to meet the needs of most industrial instructors, do offer a way for the student to gain background information about the subject being presented (probably more than is needed to complete the course being presented), and should be used if the instructor has not developed a training manual designed specifically for the course. Truly, *anything* that promotes the students' ability to learn between sessions, including copies of presentation materials such as charts, graphs, and photographs, can be included with the course materials to make learning easier.

WHEN TO TRAIN

One of the main benefits of developing your own in-plant training program is that you control *when* training takes place. You also control the length and frequency of training sessions. This lets you custom-tailor your training to take place when your people need it.

One of the disadvantages we discussed earlier of machine tool builder and control manufacturer training programs is that the entire course commonly takes place during three to five days. This makes training *very* intensive for attendees. Your own in-plant training program should not be so aggressive when it comes to completing a given course.

Ideally, most courses should be broken into short sessions of no more than two to three hours with adequate time between sessions to allow students to absorb and digest information presented during each session. However, the type of training you do may dictate the frequency and duration of training. It is commonly necessary to provide new-hire training, for example, on a daily basis with long periods of time in each session to get new people up to speed as quickly as possible. In these cases, special care must be given during these intensive sessions to ensure the students' understanding of presented material. The longer the session, the more important it is that the students be given breaks. No more than one hour of training should go by without a short break period.

Remember that *every student* has a saturation point. If more information is presented than the student can absorb in a given period of time, the student will simply not understand information presented after a certain point. This leads to frustration and wasted time.

Training should be scheduled during a time when students and instructor are likely to be fresh and ready to work. While this time will vary from one student to the next, it has been my experience that most people are freshest first thing in the morning (as long as everyone works first shift). However, some companies need to maximize production, and schedule training during nonworking hours. For these companies, it may be necessary to train during evening hours and/or on weekends. Again, with in-plant training, *you* determine what time frame best suits your needs.

GAGING SUCCESS

As with anything new, it is quite likely that your first attempt at training will not be perfect. While you may be pleased with the overall training effort, and students may be performing satisfactorily in their jobs, there will probably be room for improvement. A good training program will be flexible, allowing improvements to be made when necessary.

As stated earlier, it is easy to gage the success of an industrial program. By simply monitoring the students' work after the course, you can easily tell how well students understand the information presented during the course. By seeing where their strengths and weaknesses lie, your training people can make modifications in the training program to improve future courses.

At the completion of each training course, be sure to have all students complete an evaluation of the course, letting them voice their opinions as to what *they* feel should be done to improve the course. Some of the best suggestions for improvement will come from these student evaluations.

Chapter Six

Course Content:
The Key Concepts Approach
To Teaching CNC

T he scope of computer numerical control includes a vast amount of information. So much, in fact, it is quite likely that no single person will *ever*
completely master it all. For this reason, CNC instructors must have the
ability to extract and organize small portions of this information to be presented
during a given training course, and limit the number of things a student must
understand to become proficient in the material being taught. For beginning instructors, this in itself is no easy task.

UNDERSTANDING THE KEY CONCEPTS APPROACH

Any experienced instructor will agree that preparation and organization are necessary before teaching can begin. Training without preparation and organization
can be likened to a ship lost in the ocean without navigational equipment. Just as
the captain of a lost ship cannot control the ship's course, an ill-prepared instructor cannot control the direction the training program will take.

We freely admit that there are many ways to teach CNC. Our intent here is to
offer *one* proven method to organize and transfer CNC information in a logical
and easy-to-understand manner. If you are a beginning instructor, this should minimize the amount of outlining and organization you must do on your own. If you
are an experienced instructor in search of new ways to present information, this
will suggest a host of teaching alternatives.

There are many benefits to using the key concepts approach. First and foremost, this approach allows instructors to stress the concepts related to using CNC
equipment, giving them a way to present *why* things are done with just as much
emphasis as *how* they are done. Once students understand why a given function of
CNC is required and works as it does, it will be relatively easy for them to comprehend the specific tasks related to making the function work.

This ability to stress the basic concepts of CNC leads to two other benefits of
the key concepts approach. First, since many of the basic principles of CNC remain essentially the same from one machine type to another, these same concepts

can be applied to *any* form of CNC equipment. The instructor can actually teach students about any kind of CNC machine tool with one set of key concepts. Of course there are major differences in the actual usage of different CNC machine types, and presentations must be tailored accordingly, but with the key concepts approach, an instructor can dramatically minimize the number of different presentations needed for training.

This consistent approach to teaching makes it easy for both instructor and student. Because an instructor can easily teach about any kind of CNC machine, the student, understanding how the key concepts are applied to one type of machine, can more easily learn about another. Though the presentations we give here are limited to teaching the usage of CNC machining centers and CNC turning centers, you should be able to grasp how easy it will be to alter the given outlines and presentations to teach CNC wire EDM machines, CNC turret punch presses, CNC press brakes, CNC flame and laser cutting machines, and all other types of CNC machine tools.

Second, since the key concepts approach makes it easy for the instructor to present everything from basic principles through highly complex ideas, it is highly conducive to teaching *all five levels* of CNC proficiency introduced in Chapter Five.

- Level 1: ability to communicate about CNC intelligently.
- Level 2: ability to operate CNC equipment.
- Level 3: ability to make setups for CNC equipment.
- Level 4: ability to program CNC equipment.
- Level 5: ability to teach CNC.

With the key concepts approach, an instructor can teach everyone, from people just trying to communicate intelligently about CNC to experienced CNC instructors. In our discussions here, however, we limit our focus to only the first four levels of proficiency, since it is unlikely that any beginning CNC instructor will be teaching other CNC instructors.

As you will see, these key concepts make it very easy to introduce, discuss, review, and elaborate on *all* important CNC features. The higher the proficiency level you are trying to teach, the more detailed your presentations must be. For example, when teaching people just trying to learn to communicate intelligently, you simply present the reasoning behind each key concept. When instructing CNC operators and setup people, you must take it a step further, showing the operational implications of each key concept. When teaching programmers, you must go further yet, presenting the programming words and techniques related to each key concept.

The ability to train different proficiency levels with one presentation method even lets you minimize the number of different courses you must hold. Under ideal circumstances, of course, it is always best if all students are at essentially the same level. For example, if everyone to be trained is at entry level, presentations can be highly focused. However, it may not always be possible (or feasible) to hold a special course for each proficiency level. For example, if you wish to train only one design engineer interested in learning more about CNC (proficiency level 1), it may not be possible to justify the time required to design and/or hold a special course for one person.

In this case, it may be necessary to include the design engineer in a CNC course for CNC operators. While the information presented during a CNC operator's course will be much more detailed than required for the design engineer's purpose, the key concepts approach will allow the instructor to easily relate those things of importance to each proficiency level. Since each key concept is presented in a sequence that parallels the proficiency levels, the design engineer can simply ignore presentations made after those relevant to proficiency level 1.

A good training program always puts a *light at the end of the tunnel* for students, giving them a way of gaging their own progress. The key concepts approach lets the instructor easily outline the entire training program for the beginner to see before training begins. Students will know, each step of the way, just how much more they must learn in order to complete the course.

Well-prepared training programs also make learning easy. In ideal situations, students should be surprised at how easy it is to learn presented material. The key concepts approach makes it *seem* easy to learn about CNC. This perceived simplicity, if not abused, will help build the student's confidence. With this approach, for example, there are only 10 key concepts. If the student understands that it is necessary to master only 10 basic concepts to become proficient in CNC, learning CNC will seem relatively simple.

THE 10 KEY CONCEPTS

The material in this section is highly detailed. We offer *many* opinions on how we feel CNC is best taught, opinions that have been formulated during 15 years of teaching CNC at all levels of CNC proficiency. The operative thought, however, is that they are still only *opinions*. You will find that the learning environment changes with every new group of students you teach. What works well in one class may not be so successful in the next. For this reason, and since no two instructors will totally agree on every point related to how CNC should be taught, you must be prepared to tailor these suggestions to match your own training needs. At the very least, the outlines and suggestions presented here should give you an excellent place to start. Successful instructors are always alert to new and better ways to relate information; therefore, the potential to improve on our suggestions is always present.

The first of the six key concepts are related to programming, the last four to operation. To get the most from the key concepts approach, *everyone* should be exposed to *all 10* key concepts. Many presentations are made during the programming key concepts, for example, that a CNC operator must understand. In fact, the more operators know about the CNC programming for the machines they operate, the better CNC operators they can be. In similar fashion, there are presentations during the operation key concepts training that all CNC programmers should understand, especially those presentations on how programs are run, verified, and optimized.

Additionally, there are certain CNC *features* that people from all proficiency levels must understand. The programming-related key concepts make it easy to present many of these features, even if the student is not going to be a CNC programmer. For example, for CNC operators to run a CNC machining center, they must understand the programming-related feature called *tool length compensation*. They must understand why tool length compensation is required, how to measure tool length compensation values, and how to enter tool length compensation offsets. Though they will not actually give the commands to invoke tool length compensation in a CNC program, the more they know about this feature, the easier it will be to learn the operation tasks related to its use.

THE IMPORTANCE OF REVIEW

As mentioned, few students have the ability to completely master the complex functions of CNC by hearing presentations only once. If you are the instructor, you *must* review previously presented information. As noted in the previous chapter, industrial training of any kind *demands* success. Failure can result in scrapped workpieces, damaged machines, and injured personnel. If done correctly, your reviews will help turn up areas in need of more presentation and discussion.

It is good practice to review for *at least* the first 10 percent of each training session. Good instructors know that it makes little sense to present new information if students do not understand previously presented information. Given the importance of industrial training success, do not be afraid to review for longer periods of time if students are confused.

Instead of simply repeating information during your reviews, involve students. Since *they* should understand the information being reviewed, ask them many questions. Three important benefits come from involving students in this manner.

1. You can confirm their understanding. For questions they cannot answer, you know they need more help.
2. By hearing students putting the key concepts into their own words, other students may better understand. Sometimes their peers' explanations will sink in better than your own.
3. New ideas will emerge. You will be surprised at the number of ideas you get from your students while reviewing in this manner.

Since you will likely have to hold the course for others in the same proficiency level over and over again, you can incorporate your students' best explanations into your own future presentations. The saying "The best way to learn something is to teach it" truly applies. At the completion of the discussions of how to present each key concept, we offer several important questions you can ask during review. You can use our suggested questions and incorporate some of your own.

Reviews also allow you to expand your presentations of important yet diffi-cult-to-understand material. It may be impossible to relate every implication of how a given complex CNC feature is applied the very first time you present it. You may be lucky to have the students understand the basic function of the feature. However, during review, when your students grasp the basics, you can expand. The turning center's *constant surface speed* is one such feature. During your first presentation, you may be lucky and get the students to understand why constant surface speed is required and how it is programmed. You will probably *not* be able to make them understand how and why constant surface speed can waste cycle time if it is not efficiently programmed. Attempting to do this the first time may confuse your students more than help them. You may wish to reserve your presen-tation of the cycle time implications of constant surface speed until your first or second review, once your students thoroughly understand the feature.

(Expanding in this manner during reviews makes even *very* complex ideas seem easy to the student. If done properly, the student will find your expansion of a topic seem like a natural progression. I have heard students say "We already knew that!" many times when expanding presentations in this manner.)

COURSE OUTLINES

We offer comprehensive course outlines for teaching the two most popular forms of CNC equipment, the CNC machining center and CNC turning center. Notice that the key concepts remain exactly the same from one type to the other, as do many of the actual presentations.

Machining center course outline

Programming related key concepts

I. Key Concept 1: Know your machine
    (from a programmer's standpoint)
    A. Review of basic machining practice
    B. Machine configurations
        1. Vertical
            a. Components
            b. Directions of motion
        2. Horizontal
            a. Components
            b. Directions of motion
        3. Programmable functions of machining centers
    C. General flow of CNC process
        1. Deciding which CNC machine is to be used
        2. Machining process is developed
        3. Tooling is ordered and checked
        4. Program is developed
        5. Setup documentation is made

   6. Program is loaded into CNC control
   7. Setup is made
   8. Program is verified and optimized
   9. Production is run
   10. Corrected version of CNC program is saved
  D. Visualizing a CNC program's execution
   1. Program makeup
    a. Command structure
    b. Word structure
    c. Sequential method of program execution
   2. Visualizing movements
  E. Understanding the rectangular coordinate system
   1. Analogy to making and reading graphs
    a. Base lines
    b. Origin of graph
    c. Increments of each base line
   2. Axes of a machining center
    a. Increments of each axis
    b. XY, XZ, and YZ planes
    c. Understanding plus and minus coordinates
   3. Where to place the program zero point
   4. Absolute versus incremental motion
    a. Why absolute is better for beginners
  F. Measuring the program zero point
   1. Why program zero must be specified
   2. The machine's reference position
   3. Measuring alternatives
    a. For X and Y axe
    b. For the Z axis
   4. How to avoid program zero measurements
  G. How program zero is assigned
   1. Assigning program zero in the program
   2. Assigning program zero with offsets
  H. Introduction to programming words
   1. More on word format
    a. Decimal point programming versus fixed format
   2. Word types
    a. Brief discussion of word meanings (O, N, G, X, Y, Z, A, B, C, R, I, J, K, etc.)
II. Key Concept 2: You must prepare
  A. How preparation enhances safety

    B.   Steps in preparing to manually write a CNC program
- 1. Prepare the machining process
- 2. Do any math required and mark up the print
- 3. Check the required tooling
- 4. Plan the setup

III.   Key Concept 3: You must understand the motion types

    A.   Discussion of interpolation
- 1. How the control interprets motion commands

    B.   Things all motion types have in common
- 1. All modal
- 2. End point is specified in each command
- 3. Only moving axes need be specified

    C.   Rapid motion (G00)
- 1. Motion rate
- 2. How movement takes place
- 3. When to use
- 4. Words related

    D.   Straight-line cutting motion (G01)
- 1. How motion rate is specified (feed rate)
- 2. How motion takes place
- 3. When to use
- 4. Words related

    E.   Circular motion (G02, G03)
- 1. How motion takes place
- 2. When to use
- 3. How to specify the arc radius
  - a.   With the R word
  - b.   With I, J, and K
- 4. Arc limitations
- 5. Making a full circle in one command
- 6. Words related

    F.   Helical interpolation (G02, G03)
- 1. How the motion takes place
- 2. When to use (thread milling)
- 3. How to specify the arc
- 4. How to specify the Z-axis departure

    G.   Example programs showing motion types being used

IV.   Key Concept 4: You must understand the compensation types

    A.   Introduction to compensation
- 1. Reason why compensation is needed

        a.   Tooling unknowns
             Length of cutting tools
             Radii of certain cutting tools
             Location of workholding devices
        b.   Allowing for tool wear
        c.   Allowing the operator to trial machine
    2. How known offsets can be programmed

B.   Tool length compensation
    1. Why tool length compensation is required
    2. Related programming words
    3. When to instate
    4. Two ways to use tool length compensation
        a.   Tool length is offset
        b.   Distance from tool tip to program zero is offset
        c.   Why using the tool length as offset is better
    5. Measuring the tool length compensation value
        a.   On the CNC machining center
        b.   Off line
    6. Example program stressing use

C.   Cutter radius compensation
    1. Why cutter radius compensation is required
        a.   Easier program coordinate calculations
        b.   Range of cutter sizes
        c.   Workpiece sizing
        d.   Roughing and finishing with same coordinates
    2. Words related to cutter radius compensation
    3. The two ways to use cutter radius compensation
        a.   Offset is cutter radius (or diameter)
        b.   Offset is the difference between planned cutter size and
            actual cutter size
    4. Steps to programming cutter radius compensation
        a.   Instating cutter radius compensation
            Determining whether to use G41 or G42
            Determining the offset number to use
            Making the prior positioning move
        b.   Moving the tool while using cutter radius compensation
        c.   Canceling cutter radius compensation

D.   Fixture offsets
    1. Why fixture offsets are required
    2. Words related to fixture offsets
    3. Using only one fixture offset
    4. Using multiple fixture offsets
    5. The two ways of entering fixture offset values

        a.   Unknown distances between program zero points

        b.   Known distances between program zero points

    6. Fixture offsets with horizontal machining centers

    7. How known fixture offset values can be programmed

V.   Key Concept 5: You must understand how programs are formatted

    A. Introduction to program formatting

       1. Reasons for strict formatting of CNC programs

          a.   Familiarization

          b.   Consistency

          c.   Rerunning tools in the program

    B. Four types of program format

       1. Program start-up format

       2. Tool ending format

       3. Tool start-up format

       4. Program ending format

    C. How to use the formats

    D. Differences from one machine to another

       1. Vertical versus horizontal machining centers

       2. M-code differences

VI.   Key Concept 6: You have special features to simplify programming

    A. Hole machining canned cycles

       1. Understanding the cycle types

       2. Things all canned cycles have in common

       3. Types of canned cycles

       4. Words used in canned cycles

       5. How to clear clamps between holes

       6. Canned cycles with different Z surfaces

    B. Subprogramming techniques

       1. How subprograms work

       2. When to use subprogramming techniques

       3. Words used with canned cycles

       4. Example applications

          a.   Multiple identical machining operations

          b.   Multiple operations on holes

          c.   Rough and finish contour milling

          d.   Multiple identical workpieces

          e.   Control programs

    C. Other special features for programming

       1. Dwell command

       2. Scaling command

       3. Single direction positioning command

       4. Coordinate rotation command

       5. Optional block skip command (slash code)

      6. Mirror image commands
      7. Helical motion and thread milling
      8. Understanding all other G and M codes
  D. Rotary device programming
      1. Reasons for rotary devices
      2. The two types of rotary devices
        a. Indexers
            1°, 5°, and 90° indexers
            Methods of programming
        b. Rotary tables
            How to command rotary departures
            Comparison to other axes
            Understanding incremental versus absolute
            Rapid and straight-line cutting motion
            How to calculate feed rate in degrees per minute
        c. Approach to rotary device programming
            Program zero selection
            Additional preparation steps for programming

Operation-related key concepts

VII. Key Concept 7: Know your machine (from an operator's standpoint)
  A. Levels of operator responsibility
      1. Maintain production
      2. Make minor program edits as needed
      3. Completely set up and run production
  B. Directions of motion (from an operator's standpoint)
  C. The two basic operation panels
      1. Control panel
      2. Machine tool panel
  D. Understanding buttons and switches on the control panel
      1. Alphanumeric keyboard
      2. Other important keys
      3. Display screen functions
        a. Position functions
        b. Program functions
        c. Offset functions
        d. Setting functions
        e. Program check functions
        f. Alarm and message functions
  E. Machine panel functions
      1. Inconsistencies among machine tool builders
      2. Important buttons and switches
        a. Mode switch
        b. Cycle start

       c.   Feed hold
       d.   Conditional switches (dry run, feed hold, etc.)
       e.   Feed rate and spindle speed override
       f.   Rapid override
       g.   Jog functions
       h.   Handwheel functions
       i.   Other buttons and switches on the machine panel

VIII. Key Concept 8: There are three modes of operation
    A.   How all buttons and switches are related to one of three modes
    B.   The manual mode
       1. For completely manual operations
          a.   Jog, handwheel, spindle control, etc.
    C.   The manual data input (MDI) mode
       1. For manual operation not given by the machine tool builder
          a.   Tool changing, spindle starting, etc.
    D.   The program operation mode
       1. To run CNC programs
          a.   From memory or through the communications port

IX.   Key Concept 9: You must understand the key operation procedures
    A.   How procedures make operation easy
       1. Step by step
       2. Often-used procedures will eventually become memorized
    B.   Manual procedures
       1. Turn the machine on and off
       2. Manually start the spindle
       3. Manually jog the axes
       4. Use the handwheel
       5. Load tools into the spindle
       6. Load tools into the tool changer magazine
       7. Turn coolant on and off
       8. Reset the position displays
       9. Enter and change tool offsets
      10. Activate mirror image
      11. Select between inch and metric modes
    C.   Manual data input procedures
       1. Change tools
       2. Activate spindle
       3. Send the machine to its reference position
    D.   Program loading and saving procedures
       1. Load programs by tape
       2. Load programs through the keyboard
       3. Load programs from the communications port
       4. Send programs from the control for saving

    E.   Program display and edit procedures
        1. Display a directory of CNC programs
        2. Delete a program from memory
        3. Search to other programs
        4. Search words within a program
        5. Alter words within a program
        6. Delete words within a program
        7. Insert words within a program
    F.   Setup procedures
        1. Make vise setups
        2. Make fixture setups
        3. Measure tool lengths
        4. Measure the program zero point

X.   Key Concept 10: You must know how to safely verify, optimize, and run CNC programs
    A.   Safety priorities
        1. Operator safety
        2. Machine safety
        3. Workpiece safety
    B.   Common mistakes made by manual programmers
    C.   Procedures for verifying CNC programs
        1. Machine lock dry run
        2. Free-flowing dry run
        3. Free-flowing normal run
        4. Normal air cutting cycle
    D.   Running the first workpiece
    E.   Running verified programs
    F.   Rerunning tools within a program
    G.   Program optimizing

This comprehensive outline requires further explanation (coming up after the turning center course outline). Even then, you may not agree with every detail that is included. Remember that we are simply trying to *help* with the development of your own course outline and have included those items found to be most important through 15 years of teaching CNC courses.

Turning center course outline

Notice how similar this outline is to the one for a machining center course. The key concepts remain the same, as do many of the actual presentations. Only those differences that are specific to turning centers are changed.

Programming-related key concepts

I.   Key Concept 1: Know your machine (from a programmer's standpoint)
    A.   Review of basic machining practice

B. Machine configurations (components, axes)
    1. Universal style
    2. Chucker style
    3. Bar style
    4. Vertical style
    5. Twin-spindle style
    6. Engine-lathe style
    7. Gang style
    8. Swiss style
C. How the X axis specifies diameter
D. Programmable functions of turning centers
E. The two ways to control spindle speed
    1. Constant-surface-speed mode
    2. Rpm mode
    3. Advantages of constant-surface-speed mode
        a. Simpler programming
        b. Consistent finish on workpiece
        c. Longer tool life
    4. Two times when rpm mode must be used
        a. Center cutting tools (drills, taps, etc.)
        b. Thread chasing
    5. How to program spindle modes
F. The two ways to control feed rate
    1. Feed per minute
    2. Feed per revolution
    3. How to program feed rate modes
G. General flow of CNC process
    1. Deciding which CNC machine is to be used
    2. Machining process is developed
    3. Tooling is ordered and checked
    4. Program is developed
    5. Setup documentation is made
    6. Program is loaded into CNC control
    7. Setup is made
    8. Program is verified and optimized
    9. Production is run
    10. Corrected version of CNC program is saved
H. Visualizing a CNC program's execution
    1. Program makeup
        a. Command structure
        b. Word structure
        c. Sequential method of program execution
    2. Visualizing movements

    I.   Understanding the rectangular coordinate system
        1. Analogy to making and reading graphs
           a.   Base lines
           b.   Origin of graph
           c.   Increments of each base line
        2. Axes of a turning center
           a.   Increments of each axis
           b.   Understanding plus and minus coordinates
        3. Where to place the program zero point
        4. Absolute versus incremental motion
           a.   Why absolute is better for beginners
    J.   Measuring the program zero point
        1. Why program zero must be specified
        2. The machine's reference position
        3. Measuring alternatives
        4. How to minimize program zero measurements
    K. How program zero is assigned
        1. Assigning program zero in the program
        2. Assigning program zero with offsets
    L.   Introduction to programming words
        1. More on word format
           a.   Decimal point programming versus fixed format
        2. Word types
           a.   Brief discussion of word meanings (O, N, G, X, Z, A, C, R, etc.)
    M. Understanding special attributes of CNC words
        1. Modal words versus one-shot words
        2. Initialized words
        3. Words that allow a decimal point
 II.   Key Concept 2: You must prepare
    A. How preparation enhances safety
    B. Steps in preparing to manually write a CNC program
        1. Prepare the machining process
        2. Do any math required and mark up the print
        3. Check the required tooling
        4. Plan the setup
III. Key Concept 3: You must understand the motion commands
    A. Discussion of interpolation
        1. How the control interprets motion commands
    B. Things all motion types have in common
        1. All modal
        2. End point is specified in each command
        3. Only moving axes need be specified

C. Rapid motion (G00)
  1. Motion rate
  2. How movement takes place
  3. When to use
  4. Words related
D. Straight-line cutting motion (G01)
  1. How motion rate is specified (feed rate)
  2. How motion takes place
  3. When to use
  4. Words related
E. Circular motion (G02, G03)
  1. How motion takes place
  2. When to use
  3. How to specify the arc radius
    a. With the R word
    b. With I and K
  4. Words related
F. Polar coordinate interpolation (for machines with C axes)
G. Example programs showing motion types being used
IV. Key Concept 4: You must understand the compensation types
A. Introduction to compensation
  1. Reason why compensation is needed
    a. Tooling unknowns
            Incorrect positioning of cutting tools
            Nose radii of cutting tools
            Position of tool relative to program zero
    b. Allowing for tool wear
    c. Allowing the operator to trial machine
  2. How known offsets can be programmed
B. Dimensional tool offsets
  1. Why tool offsets are required
    a. To allow for tool wear
    b. To allow trial machining
  2. Related programming words
  3. When to instate and cancel
C. Tool-nose radius compensation
  1. Why tool-nose radius compensation is required
    a. Easier program coordinate calculations
    b. Range of insert radius sizes
  2. Words related to tool-nose radius compensation
  3. Steps to programming cutter radius compensation
    a. Instating tool-nose radius compensation
            Determining whether to use G41 or G42

        b.   Tool motions with tool-nose radius compensation
        c.   Canceling tool-nose radius compensation
    D.   Geometry offsets
        1.   Advantages of geometry offsets over assigning program zero
             in the program
             a.   Safe operation
             b.   Keeps program zero assignment separate from program
             c.   Enhances efficiency
             d.   Facilitates rerunning of tools
        2.   How to measure geometry offsets
             a.   Using the measure function
  V.   Key Concept 5: You must understand how programs are formatted
    A.   Introduction to program formatting
        1.   Reasons for strict formatting of CNC programs
             a.   Familiarization
             b.   Consistency
             c.   Rerunning tools in the program
    B.   Four types of program format
        1.   Program start-up format
        2.   Tool ending format
        3.   Tool start-up format
        4.   Program ending format
    C.   How to use the formats
    D.   Differences from one machine to another
        1.   Based on machine type
        2.   M-code differences
  VI. Key Concept 6: You have special features to simplify programming
    A.   Simple canned cycles
        1.   One-pass turning and boring
        2.   One-pass threading
        3.   One-pass facing
    B.   Multiple repetitive cycles
        1.   Rough turning and boring
        2.   Rough facing
        3.   Finish turning, facing, and boring
        4.   Grooving
        5.   Drilling
        6.   Threading
    C.   Hole machining canned cycles
        1.   Understanding the cycle types
        2.   Things all canned cycles have in common
        3.   Types of canned cycles
        4.   Words used in canned cycles

      5. How to clear clamps between holes

      6. Canned cycles with different Z surfaces

  D. Subprogramming techniques

      1. How subprograms work

      2. When to use subprogramming techniques

      3. Words used with canned cycles

      4. Example applications

          a. Multiple identical machining operations

          b. Multiple operations on holes

          c. Rough and finish contour milling

          d. Multiple identical workpieces

          e. Control programs

  E. Other special features for programming

      1. Dwell command

      2. Scaling command

      3. Single-direction positioning command

      4. Coordinate rotation command

      5. Optional block skip command (slash code)

      6. Mirror image commands

      7. Live tooling commands

      8. Understanding all other G and M codes

  F. Rotary device programming

      1. Reasons for rotary devices

      2. The two types of rotary devices

          a. Indexers

              $1°$, $5°$, and $90°$ indexers

              Methods of programming

          b. Rotary tables

              How to command rotary departures

              Comparison to other axes

              Understanding incremental versus absolute

              Rapid and straight-line cutting motion

              How to calculate feed rate in degrees per minute

          c. Approach to rotary device programming

              Program zero selection

              Additional preparation steps for programming

              Understanding polar coordinate interpolation

Operation-related key concepts

VII. Key Concept 7: Know your machine (from an operator's standpoint)

  A. Levels of operator responsibility

      1. Maintain production

      2. Make minor program edits as needed

      3. Completely set up and run production

    B.   Directions of motion (from an operator's standpoint)

    C.   The two basic operation panels

        1. Control panel

        2. Machine tool panel

    D.   Understanding buttons and switches on the control panel

        1. Alphanumeric keyboard

        2. Other important keys

        3. Display screen functions

            a.   Position functions

            b.   Program functions

            c.   Offset functions

            d.   Setting functions

            e.   Program check functions

            f.   Alarm and message functions

    E.   Machine panel functions

        1. Inconsistencies among machine tool builders

        2. Important buttons and switches

            a.   Mode switch

            b.   Cycle start

            c.   Feed hold

            d.   Conditional switches (dry run, feed hold, etc.)

            e.   Feed rate and spindle speed override

            f.   Rapid override

            g.   Jog functions

            h.   Handwheel functions

            i.   Other buttons and switches on the machine panel

VIII. Key Concept 8: There are three modes of operation

    A.   How all buttons and switches are related to one of three modes

    B.   The manual mode

        1. For completely manual operations

            a.   Jog, handwheel, spindle control, etc.

    C.   The manual data input (MDI) mode

        1. For manual operation not given by the machine tool builder

            a.   Tool indexing, spindle starting, etc.

    D.   The program operation mode

        1. To run CNC programs

            a.   From memory or through the communications port

IX.   Key Concept 9: You must understand the key operation procedures

    A.   How procedures make operation easy

        1. Step by step

        2. Often-used procedures will eventually become memorized

    B.   Manual procedures

        1. Turn the machine on and off

  2. Manually start the spindle

  3. Manually jog the axes

  4. Use the handwheel

  5. Load tools into the turret

  6. Index the turret

  7. Turn coolant on and off

  8. Reset the position displays

  9. Enter and change tool offsets

  10. Select between inch and metric modes

  11. Activate the tailstock body and quill

 C. Manual data input procedures

  1. Index the turret

  2. Activate spindle

  3. Send the machine to its reference position

  4. Activate the tailstock

 D. Program loading and saving procedures

  1. Load programs by tape

  2. Load programs through the keyboard

  3. Load programs from the communications port

  4. Send programs from the control for saving

 E. Program display and edit procedures

  1. Display a directory of CNC programs

  2. Delete a program from memory

  3. Search to other programs

  4. Search words within a program

  5. Alter words within a program

  6. Delete words within a program

  7. Insert words within a program

 F. Setup procedures

  1. Load and bore soft jaws

  2. Load hard jaws

  3. Measure geometry offset values

  4. Adjust tailstock position

X. Key Concept 10: You must understand how to safely verify, optimize, and run CNC programs

 A. Safety priorities

  1. Operator safety

  2. Machine safety

  3. Workpiece safety

 B. Common mistakes made by manual programmers

 C. Procedures for verifying CNC programs

  1. Machine lock dry run

  2. Free-flowing dry run

3. Free-flowing normal run
4. Normal air-cutting cycle
D. Running the first workpiece
E. Running verified programs
F. Rerunning tools within a program
G. Program optimizing

DISCUSSION OF EACH KEY CONCEPT

In these discussions, we assume that, as the instructor, *you* have a firm understanding of CNC principles and will focus on how to *present* what you know to your students. In the discussion of each key concept, we will first offer suggestions for how to communicate the most important ideas related to the concept. We will include helpful analogies you can use to help make key points. If different presentations are required for machining centers versus turning centers, we will discuss them separately. We will then present a list of questions you can use during your review of each key concept. Finally, we will offer suggestions as to how you can expand your discussion to include complex ideas for each key concept during your reviews.

KEY CONCEPT 1: KNOW YOUR MACHINE (FROM A PROGRAMMER'S VIEWPOINT)

The first key concept is the longest in terms of presentation time. Since the key concepts use a building block approach, the first is also one of the most important. It is very important to ensure that all students get off on the right foot.

Begin in a very basic manner. Consider what you know of your students' current CNC proficiency level; step back a little further than that to begin. It is always best to *work from what students already know toward what they do not*.

During this key concept, you must stress the importance of understanding the machine tool from a programmer's viewpoint. While this presentation will be most important to CNC programmers, all CNC people must understand many of the points made during this presentation.

The importance of basic machining practice

While basic machining practice is not usually within the scope of a CNC training course, the more a person understands it, the easier it will be to begin a CNC career. Depending on the current proficiency level of your students, you may wish to confirm their knowledge of basic machining practice as it applies to the kind of machine tool you are teaching.

Even people who simply need to be able to communicate intelligently in the CNC environment must understand basic machining practice before they can do so. CNC operators who are expected to adjust machining conditions must, at the very least, understand how to determine speeds, feeds, and depths of cut for the various machining operations being performed. Operators will also be expected to use gaging equipment and adjust tool offsets on the basis of measurements. In addition, CNC programmers must know how to develop the machining process to allow safe, efficient machining within their programs. They must also understand the capabilities of the various cutting tools used in their programs in order to program tool motions correctly.

Since basic machining practice is a prerequisite, and not within the scope of the courses I teach, I limit my presentations to specifying what the students *should* know, confirming their understanding. In cases when all students are weak in basic machining practice, you may be forced to begin with a lengthy presentation on this very important topic. This, of course, means the CNC training course may have to be lengthened to include your basic machining practice presentations.

Machine configurations

In this presentation, you will need to use graphics to explain the most important components of the CNC machine tools being taught. Line drawings or photographs work nicely for this purpose. Better yet, walk your students out to one of your machines to demonstrate its configuration.

Limit your discussions to include only those machines your students must use. If you are teaching a course for the use of vertical machining centers, for example, keep training as focused as possible and do not discuss horizontal machining centers. Unfortunately, it is likely that you will have to teach about several machine variations in the same course. You may, for example, need to discuss both slant-bed-style and gang-style turning centers in one course. Since you will be teaching about only the CNC machine tools your company owns, you can still stay relatively focused in your discussion of machine components.

For machining centers, describe the purpose and location of the machine's bed, column, spindle, table, ways, automatic tool changer, control panel, and other components you feel are important. In similar fashion, this turning center presentation should include a discussion of headstock, spindle, turret, tailstock, control panel, and any other accessory components like the bar feeder, part catcher, and tailstock.

While discussing machine configurations, also explain the directions of motion (axes) of the machines you are presenting. For machining center courses, and especially when *table movement* constitutes the axis motion (as it does for the X and Y axes of many vertical machining centers), limit your presentation at this time to include only the axis directions (X, Y, Z, etc.), and do not discuss the polarity of each axis (plus or minus directions). If the table of a vertical machining center forms the X and/or Y axis, for example, it can be quite difficult for beginners to visualize plus and minus directions, since the cutting tool is not moving along with the axis. The same problem commonly exists on horizontal machining centers if table motion forms one or more axes. For this reason, I recommend delaying the discussion of axis polarity until the presentation of program zero, especially when teaching programmers. If the cutting tool moves along with the axis, as it does on almost all turning centers, you can easily relate plus and minus directions.

As part of your axis directions presentation for turning centers, remember to stress that the *X* axis specifies a *diameter* value on most turning centers. In the inch mode and when using absolute programming techniques, an *X* value of X3.0 specifies that the tool moves to a *diameter* of three inches.

Your machine configurations presentation must also include a discussion of the programmable functions of each machine. Explain that for any true machining center and turning center, almost everything is programmable. At the very least, tool changes, spindle speed and direction, feed rate, and coolant control are programmable. For your particular machines, you will likely need to explain even more features that are programmable, especially if your machines are equipped with special accessory devices (pallet changers, bar feeders, tailstocks, etc.).

For turning centers, you will also need to include material about how spindle speed and feed rate are controlled. Since spindle speeds can be programmed in direct rpm or with constant surface speed, you will have to explain the differences. Because constant surface speed allows three major advantages (simpler programming, consistent finish, and longer tool life), I recommend teaching programmers to use constant surface speed for all machining operations except those when rpm *must* be used (center cutting operations, threading, knurling, etc.). Introduce the G codes related to spindle control and give examples of both methods.

You must relate the two ways to handle feed rate for turning centers. Since almost all machining operations can be programmed in a feed-per-revolution mode, I recommend teaching turning center programmers to use this method of feed-rate control for all machining operations. Stress that the only time feed per minute should be used with turning centers is when an axis must move at a given feed rate while the spindle is stopped (as would be the case when a bar puller is used to advance the bar). Introduce the G codes and give examples of both methods of feed rate control.

In addressing the different proficiency levels, you can present as much information about machine configurations as necessary. If you are teaching people who are just trying to communicate intelligently, little more than a cursory view of machine components, directions of motions, and programmable functions will be necessary. CNC operators and setup people, however, must additionally memorize axis names and directions. CNC programmers must have an even greater understanding, encompassing knowledge of all programmable functions.

The machine's reference position. If your CNC machines utilize a *reference position* (also called *zero return, machine zero*, and *grid zero*), introduce this position as part of your discussion of machine configurations. Since most machines require the operator to send the machine to its reference position every time the machine is powered up, and because the reference position is commonly involved in the program zero assignment, it is wise to get students to understand this position early on in the class. Explain that the reference position is a very precise location somewhere along the machine's travels, and describe just where it is for your particular machine tools. Point out that this location is very important for several operation functions, and let students know this position can be reached both manually and through programmed commands.

General flow of the CNC process

I find it easier to relate information throughout the course if students understand the big picture of how jobs are processed through the CNC environment. Just as it is easier for the tire-changing members of a race car's pit crew to perform their duties if they understand all that goes on during a pit stop, so is it easier for those in the CNC environment to be productive in their responsibilities if they understand their place in the overall scheme of the CNC environment.

Though this foundation-building presentation may be somewhat elementary for students in the higher proficiency levels, it is very important for beginning CNC programmers, operators, and setup people, and can be quite helpful for people learning how to communicate about CNC intelligently. The topics listed in our outlines are rather generic, given for a typical CNC environment. For your own in-plant training, you can be much more specific when it comes to relating the various steps to processing jobs going through your own CNC environment.

Visualizing a CNC program's execution

Though this presentation is especially important for beginning CNC programmers, it does not hurt any student to know the frame of mind required to write CNC programs. Since a CNC program is nothing more than a sequential set of step-by-step instructions, and since most manual programs must be prepared with nothing more than a blueprint, a pencil, and a piece of paper, programmers must be able to see the program being executed *in their minds* as it is written.

An analogy helps drive this point home. Have your students consider what it takes to write down a set of step-by-step travel instructions. Before the instructions can be written, the developer must be able to *visualize* the route from the starting point to the destination. Only then can the instructions be written. While writing each step of the instructions, the developer must check for mistakes. If mistakes are made, the person following the instructions will, of course, get lost. In similar fashion, CNC programmers must be able to visualize the motions made by a CNC program. They must be able to see holes being drilled, surfaces being milled, diameters being turned, and threads being chased *in their minds*. As each command is given, they must check and double check for mistakes. If motion mistakes are made in a CNC program, the machine tool will still follow the program's commands as written, possibly resulting in a dangerous situation.

To stress visualization yet further, ask each student to create a list, in step-by-step order, of exactly what it takes to simply drill a hole in a workpiece on a drill press. It is amazing how difficult some *experienced* machinists find this simple task when they are taken out of the machine shop. With the completed list of steps necessary to drill a hole on a drill press, translate their English sentences into

CNC commands that do the same thing on a CNC machining center. While the CNC words and commands will not make much sense yet, this will truly stress the need for a CNC programmer to be able to visualize the movements a CNC program will cause.

Further, I recommend stressing the *sequential* method by which a CNC machine executes a CNC program. Let your students know that a CNC machine will follow the commands in a CNC program just as a person will follow a set of travel instructions. First it reads, interprets, and executes the first command. Only then will it go on to the next command. This sequential execution order will be followed until the last command in the CNC program is executed.

This is also a good time to introduce the basic structure of the CNC program itself. With the sample program for drilling a hole still being shown, discuss how the CNC commands (blocks) tell the machine what to do at a given time. Remind them that they can be compared to the English sentences that make up the step-by-step travel instructions. Show that within each command there are CNC words. Just like English words that make up sentences, CNC words have a special meaning to the CNC control. Explain that each CNC word has a letter address (N, G, X, Y, Z, etc.) that specifies the word type and a numerical value that specifies the word amount.

Understanding the rectangular coordinate system

While this presentation is most important to beginning CNC programmers, CNC operators and setup people should also understand how CNC coordinates are calculated for use within the CNC program. Truly, this is at the heart of how CNC motions are commanded. There may be times when operators and setup people may have to make motion changes within the program. For example, the program may cause a hole to be machined out of location. With the ability to calculate coordinates, the CNC operator or setup person will be able to make the change without having to contact the CNC programmer.

I like to begin this presentation by asking students how many rotations of the *X*- axis drive motor they think it will take to make the table of a machining center move precisely one inch (25.4 mm) along the *X* axis. While they are pondering this question, ask them at what rpm rate the axis drive motor will have to rotate to cause a feed rate of five inches (127 mm) per minute. Unless you have some highly knowledgeable machine designers in your class, no one will know the answer to either of these questions. The point is to get students thinking about *how* the machine makes its motions. Stress that it is the *rectangular coordinate system* that allows the programmer to make axis motions without having to think about the internal workings of the machine.

Another application for the rectangular coordinate system to which *everyone* has been exposed is graphs. To begin discussing the rectangular coordinate system, show a simple 2-axis graph. My example graph illustrates a company's productivity for last year. The vertical base line represents productivity and is broken into 10-percent increments. The horizontal base line represents time and is broken into 1-month increments. The point at which the lines intersect (the graph's origin) represents January and 0-percent productivity. Show how points are plotted for each month, based on the productivity level for each month.

This is all basic and easy to understand for most students, since everyone has had to interpret a graph. Now point out that the CNC machine's linear axes can be viewed as a graph's base lines. Instead of representing theoretical concepts like time and productivity, the base line will represent a linear axis. Instead of increments of productivity percentage and time in months, the increments now represent units of inch or metric value. In the inch mode, each increment is 0.0001 inch for most machines. In metric, each increment is 0.001 mm. For machining centers, show a 3-dimensional rectangular coordinate system that represents the axes on your particular machining centers, with each base line representing one of the three linear axes (X, Y, and Z). For 2-axis turning centers, show a 2-axis graph with each base line representing one of the two axes of the turning center (X and Z).

Point out that where the base lines intersect is the origin of the rectangular coordinate system used for CNC coordinate calculations. This point of intersection can be thought of as the origin of any graph (where the base lines meet). Stress that this point is commonly referred to as the *program zero point*.

With the help of a visual to stress the CNC machine's coordinate system, have the students plot a few points. Have them designate each plotted point with the X, Y, and Z or X and Z coordinates. Stress that the points being plotted in X, Y, and Z are positions to which we can command the cutting tool to move. Make sure they understand that it is because of the rectangular coordinate system that a CNC programmer need never be concerned with how many revolutions of the axis drive motor are necessary to cause a given motion. Instead, points are simply plotted in the rectangular coordinate system. The machine moves the tool through these plotted points.

This is a good time to introduce the polarity related to each axis. Going back to the graph analogy for a company's productivity for last year, point out that all points happen to be plotted *after* January and *above* 0-percent productivity. In the example showing our plotted points for the machining center, all points happen to fall to the right of program zero in X and above program zero in Y. You must stress that this will not always be the case for CNC equipment. To illustrate this point, show a ring that must have a bolt hole circle machined on its face (on a machining center). Make the center of the ring the program zero point. With this example, you can easily illustrate that some of the holes to be machined fall to the left of the program zero point in X and some fall to the right. In Y, some holes fall above program zero and some fall below. Stress that coordinates that fall to the left of program zero in X and below program zero in Y are in negative territory, and must be designated as *minus* values.

You will need to elaborate on how plus and minus directions are specified if the programmer is working from the program zero point. Stress that this is the *absolute* mode of programming. When working in the absolute mode, the programmer need not consider which *way* an axis must move in order to get the tool to its programmed position. All that really matters is how far the desired end point is *from the program zero point*. Stress that this differs from another mode of programming called *incremental mode*. Also stress the benefits of working in the absolute mode (coordinates are easier to calculate, no compounded mistakes, and programmed coordinates make sense) and that beginning programmers should concentrate on working in only the absolute mode.

Be sure to warn beginning programmers to avoid the natural tendency of *thinking incrementally*. Tell beginners calculating coordinates to ask themselves *to what position* they wish the tool to move. In the absolute mode, this position is always relative to the program zero point. Beginners tend to ask themselves the wrong question. They tend to ask themselves *how far the tool should move*. This is thinking incrementally. Have them practice calculating coordinates in the absolute mode often, as beginners often have to struggle with this important concept.

You must make it clear that the CNC programmer determines the location of the program zero point, and that a wise choice for program zero will dramatically simplify the calculations required to determine the coordinates needed in the program. In many cases, if program zero is chosen wisely, coordinates going into the program can be taken *right from the print*. Show how the selection of program zero is based on print dimensioning. Stress that the surfaces from which dimensions are taken make excellent program zero point surfaces. If your students are experienced machinists, and if they are having problems determining where to place the program zero point, tell them that the program zero surfaces are normally the same as the workpiece surfaces a machinist would use to locate the workpiece in a setup.

The student must also understand that the program zero assignment is not magic. The fact that the programmer wishes a certain location on the workpiece to be the program zero point does not mean the CNC control will automatically know where this location is positioned within the machine's travels. Though you may not wish to dig too deeply into how program zero is assigned this early in the class (this is presented in greater detail during Key Concept 4), you might want to give a brief example of how program zero can be measured once the setup is made. Be sure to match your presentation to the method your company actually uses to measure and assign the program zero point.

Introduction to all programming words

While this presentation can be made at any one of several points in the course, this is as good a time as any to introduce the students to the various words involved with programming. Going by letter address (N, G, X, Y, etc.), introduce the student to the general meanings of each programming word. Stress that they do not have to memorize every word type at this time. You are simply introducing the words. You can also stress that since the typical CNC program contains fewer than 50 different word types, learning CNC programming can be likened to learning a foreign language that contains less than 50 words. Be sure to give the students a separate handout that lists all programming words, including G codes and M codes, for reference purposes.

Admittedly, there are a great number of words related to CNC, and beginners tend to be intimidated by the number of different CNC words. You must put their minds at ease to some extent when this presentation is made. Be sure they understand that during Key Concept 5, while discussing program formatting, you will show an easy way to help remember the meanings of the most-used programming words. Very few words will have to be memorized before beginning programmers can begin writing CNC programs.

This also makes a good time to present the special nature of certain CNC words. For example, discuss those words that allow *decimal point programming* rather than integer format. Show the older fixed format for words if the decimal point is not used. This should let you stress how important it is that beginning programmers remember to include a decimal point in all words that allow it, since the CNC control will not correctly interpret the CNC word if a needed decimal point is missing. Also discuss the difference between *modal* and *nonmodal* (1-shot) words and mention that some words are *initialized* at power-up.

Review questions for Key Concept 1

As stated earlier, reviews are *mandatory* to ensure a student's understanding and retention of presented information. Instead of simply repeating presentations, I recommend involving students by asking them to answer many questions as you review. Their responses either confirm understanding or demonstrate areas in need of further presentation.

Questions for all CNC students:

1. Why must all CNC people understand basic machining practice?
2. Describe the basic configuration and name the most important components of the CNC machines you will be working with.
3. Describe the directions of motion for the machine you will be working with.
4. Name the most important programmable functions of the machine you will be working with.
5. What is the reference position? Why is it important? Where is it located on the machines you will be working with?
6. Name the basic steps for processing jobs coming through our CNC environment.
7. Why is it important for a programmer to be able to visualize a CNC program's execution?
8. Name the basic elements that make up a CNC program.
9. Name the order in which a CNC control will execute a CNC program.
10. What is program zero? Why must it be used?
11. In what mode of programming is the programmer working when working from program zero?

12. How is the absolute mode specified for the machines you will be working with?
13. What is the incremental mode? Why is the absolute mode so much better than the incremental mode?
14. Name the steps related to measuring and assigning the program zero point for your particular machine.
15. Name some words that allow decimal point programming. Name some words that require integer format.
16. What does *modal* mean? *Nonmodal*?
17. What does initialize mean?

Questions for reviewing turning center presentations:

1. What is special about the *X*-axis designation for turning centers?
2. Name the two ways to control spindle speed. Name the related G codes. Describe constant surface speed. Which is better for most machining operations? Why? Name two instances when the rpm mode must be used.
3. Name the two ways to control feed rate. Name the related G codes. Which is better for most machining operations? Why? Name the only time that feed per minute must be used.

Topics for further discussion during review

A good review should push students past their current level of understanding. While students are demonstrating a firm understanding of previously presented information, you should seize the opportunity to help them go past what they already know. Here are some suggestions for topics in Key Concept 1 that may require further discussion.

More on constant surface speed. As experienced turning center people should know, constant surface speed can be a time waster if not programmed efficiently. Since the control will constantly respond to changes in diameter for the purpose of maintaining the speed you specify, a positioning movement to a large diameter for the purpose of tool changing will cause the spindle to slow. It is likely that the rapid motion to the tool change position will take longer than the spindle slow down. The reverse is true when positioning to a small diameter.

This may be difficult to discuss during your initial presentation of Key Concept 1, especially if your students just barely comprehend the nature of constant surface speed. However, your students will eventually show that they understand constant surface speed through the reviews you conduct. When they do, explain this problem and involve them in discussing how the problem can be solved.

More on the incremental mode. Though beginning programmers will have little use for the incremental mode, once they firmly understand how to use the absolute mode, you may at least wish to explain the few times when it can be helpful to use the incremental mode. Explain how, when subprogramming techniques are used, and when multiple identical machining operations must be performed, it can be very helpful to write the subprogram in the incremental mode to keep from having to repeat very similar movement commands over and over again.

More on rotary axes. During your initial presentation of the rectangular coordinate system, it may be difficult to relate how coordinates are specified along *linear* axes. Since many users use rotary axes as little more than simple indexers, you may wish to reserve your description of how rotary axis motion is commanded until students thoroughly understand motions with linear axes. When they do, and if your machines incorporate rotary axes, discuss the differences between linear and rotary axes.

More on program zero assignment. In your first presentation of Key Concept 1, it may be difficult for students to understand the many ways program zero can be assigned. While I recommend at least introducing students to one way of assigning program zero during the first presentation of Key Concept 1, you must be prepared to elaborate on this topic during reviews.

KEY CONCEPT 2: YOU MUST PREPARE

There are many things students must do in preparation for any job function they perform. While this key concept was originally developed to help CNC programmers prepare to write manual programs, and we limit the scope of our discussions to preparation for CNC program preparation, you can expand this discussion to include any task required in the CNC environment. For example, a well-prepared CNC setup person who has gathered all components necessary to make a setup will not waste time frantically searching the shop for tools while the CNC machine is down between production runs. In similar fashion, a well-prepared CNC operator who has organized all hand tools required for cutting tool maintenance will be able to change cutting tool inserts in an efficient manner.

Stress that even though this key concept contains no information about actual CNC functions or commands, it is one of the most important key concepts for three reasons:

1. Any complex task can be simplified if it is broken into smaller and easier-to-handle segments. The preparation a CNC person does allows the student to segment the tasks related to CNC.
2. During each segment, preparation allows a beginning programmer to concentrate on only the task at hand. For example, CNC programmers must wear several hats during the programming process. They must put on the process engineer's hat when developing the sequence of operations. They must put on the tooling engineer's hat when determining which workholding and cutting tools will be used. They must put on the mathematician's hat when calculating coordinates needed within the CNC program. In essence, preparation keeps the beginning programmer from trying to wear too many hats at the same time.

3. Since well-prepared CNC people are less apt to make mistakes, safety is enhanced by applying the principles of this concept. Ill-prepared CNC people are *dangerous* CNC people. You must stress that there is no excuse for accidents in the CNC environment, especially for as poor a reason as lack of preparation.

Prepare the machining process

Regardless of the method by which the CNC program is prepared (manually, with the help of a computer-aided manufacturing [CAM] system, or through shop-floor-programmed CNC controls), the programmer must have a clear understanding of the sequence by which the machining operations must be prepared. Students must understand that the process is of paramount importance to the success of the job. Stress that even a poorly formatted CNC program can eventually be made to function if the machining process is good. On the other hand, even a perfectly formatted CNC program (one that does exactly what the CNC programmer intends) will fail with a poor process.

I recommend that beginners write down each operation to be performed by the CNC program. This allows the programmer to develop the process separately from the CNC program. Stress that beginning CNC programmers (especially manual programmers) will be so confused with the various CNC commands while writing their first few CNC programs that they may forget to perform some critical operation during the program (forgetting to drill holes before tapping, forgetting to rough-bore before finish boring, etc.). With the step-by-step sequence of operations prepared, the beginning programmer can concentrate solely on programming while creating the CNC program.

Emphasize that the most difficult part of the CNC programming task is usually in preparing the process. Let beginners know that writing the CNC program is a *simple* matter of translating the process from English to the language a CNC machine can understand. While beginners may disagree at first, assure them that once they have written a few programs, programming commands will soon be memorized and used almost without conscious effort. Telling the CNC machine what they wish it to do will eventually become easy. However, the continuing challenge for CNC programmers is in developing the best, most efficient processes for the variety of workpieces their company machines.

While developing the sequence of operations, also stress that anything the beginner can do to help during programming should be done. For example, this makes an excellent time to determine the cutting tool station numbers and calculate the feed and speed for each cutting tool. This information will keep the beginning programmer from losing the train of thought during the programming process. The completed sequence of operations is like having an English version of the CNC program, and will make for excellent documentation for future use. Anyone needing to work on the CNC program in the future will be able to see what process the program uses to machine the workpiece.

Do the required math

Since computer-aided manufacturing system programmers and programmers who use shop-floor programming controls have little (if any) math to do, this presentation applies only to teaching manual programmers. Beginning manual programmers must understand that almost every positioning command in a CNC program requires calculations. These calculations require concentration. While experienced manual programmers may be able to perform calculations during the course of programming, beginners can maintain their train of thought while programming if the math is done up front.

There are two good ways to teach the documentation of the calculations. If prints are not crowded and cluttered, have your students write calculated program coordinates right on the print next to each coordinate's position. This keeps the coordinate calculation documentation from getting lost. However, if the drawing is too cluttered to allow math documentation right on the print, have your students place a numbered dot on the drawing at each coordinate position. On a separate piece of paper (I call it a *coordinate sheet*), have them write the coordinate values for each point in each axis. With either method, this allows the programmer to separate math calculations from CNC programming. While programming, the programmer can simply copy previously calculated coordinate values.

Check the tooling

It must be clear to beginning programmers that some of the most common causes of wasted time during setup are related to tooling, and most can be avoided with careful planning. Before they write a program, beginning programmers should check the availability of the cutting tools and workholding tools they intend to use. Tools not currently in inventory must be ordered or alternative tools have to be used. Again, nothing is more frustrating and wasteful than having written a perfect CNC program only to find that tools required by the program are not available.

I also recommend that the tool's length be checked by the machining center programmer to ensure that it is sufficient to reach the surfaces being machined without interference. Too many programmers completely ignore tool length constraints until *after* the setup is made. If a tool reach problem is found at this time, precious production time will be wasted while the setup person changes the tool length. Though you may wish to hold this more complex presentation for reviews, you must eventually show beginning programmers how they can determine the minimum length each tool must have in order to reach the surfaces being machined.

Plan the setup

In many cases, and especially with machining center programming, it will be impossible for the CNC programmer to complete the CNC program until the setup is planned. In order to avoid cutting tool interference during the program's execution, for example, the programmer must know the location of clamps and other obstructions on the workholding device. Additionally, completely planning the setup (including making the setup drawing) will help the beginning programmer visualize the setup during programming.

Make it clear that planning the setup will also expose mistakes in the process. It may be impossible, for example, to machine the workpiece using the planned process with the workholding setup provided. There may be no physical way to machine a particular surface, given the position of clamps. Finding mistakes in processing prior to programming saves wasted programming time.

Review questions for Key Concept 2

1. Name three reasons why it is important to prepare for creating CNC programs.
2. Name the four recommended steps for preparing to write a CNC program.
3. What other information can be included with your sequence of operations that will aid in programming?
4. Name two ways you can document your math.
5. Name two reasons why you should check tooling prior to creating a CNC program.
6. Why should you prepare the setup before you create a CNC program?

Topics for further discussion during review

For machining center presentations, it may be difficult to explain how programmers can check for certain tool length limitations during the first presentation of Key Concept 2. During review, however, you can more easily show how a minimum tool length can be determined, for each tool on a vertical machining center, from the minimum distance the machine's spindle nose can approach the table top. Say, for example, this minimum distance is nine inches (229 mm) for a large machining center. If a workpiece is held flush with the top of a short vise, say three inches (76 mm) high, every cutting tool must be at least six inches (152 mm) long just to reach the *top* of the workpiece. Any shorter, and a Z-axis overtravel will surely occur.

In similar fashion, every horizontal machining center will have a minimum distance the spindle nose can approach the table centerline. Depending on how the workholding device holds the workpiece, the programmer can confirm how long each tool must be to reach the surfaces being machined. In both cases, this simple check can eliminate wasted setup time while tools are lengthened.

KEY CONCEPT 3: YOU MUST UNDERSTAND THE MOTION TYPES

You can begin this presentation in essentially the same manner regardless of the type of CNC machine you are teaching. However, as your presentation continues, you must include actual examples of the motion commands. Any examples must, of course, be specific to the type of machine you are teaching.

To beginners in the field, it can be quite intimidating to watch a CNC machine in operation for the first time. One of the reasons for this is that the motions a CNC machine makes appear to be quite complex. To relieve this intimidation, stress that motion types have been developed by control manufacturers to *simplify* programming, and that there are only a limited number of ways to cause motion. Emphasize that most machining centers, for example, have just four ways to cause motion, and of these only three are commonly used. Most turning centers have only three ways to cause motion.

What is interpolation?

Begin your presentation by discussing *interpolation*. Explain why early NC machines that did not have interpolation were difficult to program. Tell them that any multiaxis cutting movement required a series of very tiny single-axis departures. I demonstrate this with a simple graphic showing the stair-stepping method that early NC controls required for even a simple, straight-line, 2-axis motion program.

Stress that because it is commonly necessary to make straight-line cutting movements, CNC control manufacturers developed *linear interpolation* to simplify the programming of straight-line multiaxis movements. With linear interpolation, only one command is necessary per straight-line movement, regardless of the motion's length. From a small amount of information in the linear interpolation command (the end point of the motion and the feed rate), the control automatically figures out how to make the straight-line motion.

In similar fashion, emphasize that control manufacturers continue to develop different forms of interpolation for commonly needed motion types. Circular interpolation, for example, allows a circular path in two axes to be easily commanded. Helical interpolation, used on machining centers for thread milling, makes it easy to command a kind of spiraling movement. Two of the axes (usually X and Y) move in a circular manner. The third axis (usually Z) moves in a linear manner. The resulting movement allows the thread milling cutter to cut helical threads in thread milling. Be sure to also introduce any special interpolation types your particular CNC machine tools allow.

Things all motion types share in common

Beginners can more easily learn the motion types if they understand their commonalities. This is information that needs to be learned only once and can be applied to any of the motion types. Stress that there are four important things all motion types share in common:

1. They are all modal. Subsequent commands of the same type do not require the repeating of the invoking G word.
2. All motion types require the programming of the *end point* for the motion. By one means or another, the tool is already at the beginning point for the motion prior to the actual motion command. I find that beginning programmers have problems with this concept. Be sure to stress the point-to-point nature required for programming motion commands.
3. All motion types require only the moving axes to be included in the motion command. Axes that are not moving should be left out of the motion command to shorten the program's length and avoid the possibility of

typing mistakes. Beginners tend to include too much information in motion commands, repeating positions in all axes for every command.

4. All motion types are affected by the current positioning mode. If the programmer is working in the absolute mode, end points included in motion commands must be given relative to program zero. If the programmer elects to work in the incremental mode, motions must be specified from the tool's current position.

Rapid motion

All types of CNC machines have some form of *positioning* motion command, commonly called *rapid motion*. Learners must first know why rapid motion is required. Stress that almost all noncutting movements should be done at a rapid rate to minimize program execution time. These motions include cutting tool movements to and from the workpiece as well as any noncutting movements internal to each tool. Also make clear that since the rapid rate for most machining centers is extremely fast (commonly at least 400 inches [10,000 mm] per minute [ipm]), there are safety functions that allow CNC users to override the machine's rapid rate during operation (commonly needed during a program's verification).

You will also need to explain how rapid motion occurs. If multiaxis rapid commands are given, for example, most CNC machines will allow *all* programmed axes to move at the same rapid rate until the end point for each axis is reached. This means that during multiaxis rapid commands, one axis will reach its end point before the others, and the machine will tend to dogleg into position. Beginning CNC programmers must know this in order to avoid obstructions during rapid positioning movements.

At this point, introduce the words involved with rapid (commonly G00 along with the moving axes) and give examples of rapid motion commands as they are required for your particular CNC machine tools.

Straight-line cutting motion

Explain that all CNC machining centers and turning centers also have the ability to perform straight-line cutting movements. Stress that this command can be used any time the programmer wishes to machine with a tool along a straight line at a specified motion rate. Examples of operations requiring straight-line cutting movements include machining straight turns, bores, and tapers on turning centers and milling straight and tapered surfaces on machining centers.

Students must also understand that the motion rate can be precisely controlled during straight-line cutting commands. Explain that the feed-rate word (specified with the letter address F) is used to specify the rate of motion. For machining centers, make clear that most machines allow the motion rate to be specified only in *feed per minute* (either inches per minute or millimeters per minute). This means that determining feed rates for machining centers usually involves calculations (spindle rpm x the desired feed rate in inches [millimeters] per revolution). Turning centers also allow feed rate to be specified directly in feed per revolution (inches or millimeters per revolution), making it easier to directly specify feed rates found in cutting condition reference data. Your students should know that if both methods of feed-rate control are possible, two G codes will be used to specify the desired feed-rate mode. Examples help stress the point. Also point out that feed rate is modal. The current feed rate will remain in effect until changed.

Finally, discuss the words related to straight-line cutting motion (commonly G01, the F word, and axis words) and give an example showing both rapid and straight-line cutting motions for your particular machine tools. (You can easily show a few holes being drilled on a machining center or a few diameters being turned on a turning center.).

Circular motion

Since all machining centers and turning centers are also provided with circular motion, you can begin this session in the same manner regardless of which type of machine you are teaching. Begin by explaining why circular motions are required (for milling circular contours on machining centers and machining radii on turning centers) and that, like straight-line cutting commands, the motion rate is controlled by an F word. Make sure they understand that the feed rate for a straight-line cutting command will remain in effect even for subsequent circular commands and vice versa.

Explain that circular motion commands require a little more information than linear motion commands. The programmer must specify the direction in which the tool is moving around the arc (clockwise or counterclockwise) and the size of the arc, by one means or another. To correctly specify circular motion, the programmer must first be able to determine whether the arc is being formed in a clockwise or counterclockwise direction. Describe how this is done for your particular controls. Most controls use G02 to specify clockwise direction and G03 to specify counterclockwise direction.

When it comes to arc radius size, explain that different controls require different methods of specifying arc size. Newer controls allow the arc size to be easily specified with a simple radius word (specified by the letter address R). If your controls allow the R word to specify arc size, you can keep instruction quite simple. Unfortunately, some controls require more cumbersome *directional vectors* to specify the arc center position. Most use the letter addresses I, J, and K for this purpose. Control manufacturers even differ on how to use I, J, and K. Some require the programmer to specify the absolute position of the arc center, while others require an incremental distance from the start point of the arc to the arc center. This can make teaching much more difficult to plan and to understand. Regardless of how your controls require arc size to be specified, you can at least tailor your presentations to include only those techniques required for the controls your company uses.

Most controls have limitations related to how much of an arc can be done in one command. The rules for maximum allowable arc portion vary from one control manufacturer to another. Since a typical beginning CNC programmer will rarely break these rules during normal programming, I recommend reserving your discussions of these arc limitation rules for review periods, once you have confirmed that students understand how to make basic circular motion commands.

At this point, show the words related to circular commands (G02, G03, R, I, J, K, and axis words) and show an example program. In fact, this in an excellent time to show a full example program, stressing the use of all three motion types discussed to this point.

Other motion types

You must be prepared to make presentations about any other motion types your particular CNC machines allow. However, most other motion types are used for special purposes and are not required in every CNC program. Most CNC programs can be completely developed with only rapid, straight-line cutting, and circular commands. For this reason, you may wish to simply introduce any other motion types your controls allow and reserve your detailed instruction of special motion types until much later in the course. For example, when teaching machining center programming, I recommend introducing the motion type called helical motion that is used for thread milling. But you should wait to present detailed information on how to command thread milling until Key Concept 6, which addresses special programming features.

In similar fashion for turning centers, if your turning centers have a rotary axis internal to the spindle drive system (commonly called a C axis), you can introduce the motion type called *polar coordinate interpolation* that allows contour milling operations on the end of a workpiece to be easily commanded (in the X/C plane). However, you can wait until Key Concept 6 to make more detailed presentations on how to use this form of interpolation.

Review questions for Key Concept 3

1. Explain interpolation. Why is it so important?
2. Name the motion types on the CNC machines you will be working with, and give the related invoking G codes.
3. Name four things all motion types share in common.
4. How fast is the rapid rate on the machine you will be working with?
5. What types of movements will you make with rapid motion commands?
6. How does your machine move under the influence of rapid motion?
7. What additional word is required when giving straight-line cutting commands?
8. Give examples of when straight-line cutting commands are used.
9. How do you determine which circular command to give, G02 or G03?
10. How do you specify the arc size on your machine?
11. Describe any other motion types used on your machine.

Topics for further discussions during review

Several important points about the motion types may be difficult to make during your first presentation of this key concept.

More on simplifying motion commands. Beginners will soon realize that if all they have available to cause motion is rapid, straight-line, and circular motion, programs will be very cumbersome. You may wish to elaborate on how certain special features of programming (to be discussed during Key Concept 6) minimize tedious commands. Hole machining with *canned cycles* on machining centers, for example, combines rapid and straight-line cutting motion to make programming holes much easier. Point out that once a hole machining cycle is commanded, the programmer need only list the centerline coordinates of each subsequent hole to be machined (one command per hole, regardless of the cycle type).

In similar fashion, the commands needed to rough turn, face, and bore on turning centers are dramatically simplified by the use of something called *multiple repetitive cycles*. In all cases, stress that these special features do not actually add motion types; they simply streamline the programming of those available.

More on arc limitations. It may be difficult to cover all the rules governing arc limitations the first time you present circular interpolation. Once you confirm that students truly understand how to give circular commands, you will want to ensure that they know the arc size rules. Moreover, for machining centers, if your controls allow you to order an entire circle to be made in a single command (as most do), you will need to relate this information as well.

More on arc direction. For machining centers, most arcs are made in the *XY* plane, and determining whether the arc is clockwise or counterclockwise is easy. The programmer simply looks at the arc from the same direction as the spindle of the machine tool. However, there may be times when the programmer will have to command an arc in the *XZ* or *YZ* plane. In this case, the programmer must know the plane selection commands required to switch planes (commonly G17 for *XY*, G18 for *XZ*, and G19 for *YZ*), as well as how to determine whether the arc is clockwise or counterclockwise. Explain that arc direction is determined by viewing the motion *from the plus side of the uninvolved axis*. In making an *XZ* circular movement, for example, the motion must be viewed from the plus side of the *Y* axis. Presenting this material during review also requires that you introduce the plane selection commands (commonly G17, G18, and G19).

KEY CONCEPT 4: YOU MUST UNDERSTAND THE COMPENSATION TYPES

Though the application of compensation varies dramatically from machining centers to turning centers (and among all other forms of CNC equipment), the reasons *why* compensation is required remain remarkably consistent among all forms of CNC machine tools. For this reason, you can make the same opening presentation for this key concept regardless of the type of CNC machine you are teaching.

I recommend that you begin by discussing common forms of compensation that students are currently familiar with. As analogies, you can discuss how an airplane pilot must compensate for wind velocity and direction while setting headings, how a race car driver must compensate for track conditions when entering a turn, and how a marksman must compensate for the distance to the target when adjusting the sight of a rifle. Since it makes it easy to continue the analogy later, when discussing offsets I prefer the marksman analogy.

Relate that, in all cases, compensation is required to allow for unpredictable and possibly changing variables. In the case of the airplane pilot, wind velocity and direction are unpredictable variables that change with every flight. For the race car driver, the changing track and weather conditions are unpredictable variables that change with every race. For the marksman, the distance to the target is a changing variable that may change with every shot. Compensation types for CNC equipment are no exception. With CNC equipment, and especially turning centers and machining centers, the unpredictable and possibly changing variables are all *tooling related*.

Introduction to offsets

As part of your introduction to compensation, you will also need to relate just *how* adjustments are made on CNC machines based on the unpredictable and changing tooling-related variables. In our analogy, a marksman adjusts a sight to compensate for the distance to the target. The amount the marksman moves the sight controls how much compensation is used. For CNC compensation types, make clear to your students that *offsets* are used to control the *amount* of compensation used for unpredictable, and possibly changing, tooling-related variables. However, instead of making a physical mechanical adjustment, as is the case when adjusting a rifle sight, CNC offsets are simply numerical values entered by an operator or setup person to tell the CNC control how much compensation is required for the tooling being used.

Explain that most controls have many offsets, and that the values within each offset are typically invoked by an offset number. Compare this to electronic calculators having multiple memories. If the calculator has several memories, each memory is commonly specified by a number. The user can, for example, place values in memories 1, 2, and 3. When needing the value stored in memory 3, the calculator user can simply press one or two buttons to invoke the value stored in memory 3 and the value will return. In similar fashion, explain that when programmers need to reference a value stored in a particular offset, they will do so by specifying the offset's number. Make sure you explain that offset numbers must be chosen in a logical manner. For most cutting tool applications for offsets, the offset number is made the same as the tool station number.

You can take the marksman analogy one step further in helping beginners understand CNC machine compensation. When the marksman adjusts the rifle sight, he will not know for sure whether the amount of adjustment is perfect. There may be other considerations, like wind, that affect the quality of the adjustment. Until the marksman fires the first shot, he will not know for sure that the sight is adjusted perfectly. Once the first shot *is* fired, he will know how much more minor adjustment, if any, is necessary to fine-tune the sight setting.

In similar fashion, a CNC setup person or operator will do his or her best to set offsets. Based on the type of compensation being dealt with, the CNC person will attempt to perfectly input the tool-related offset value. However, there may be other considerations that affect the quality of the settings. Tool pressure, rigidity of cutting tools, rigidity of workholding devices, and the amount of material being removed during the machining operation are just a few of the many factors that can affect the quality of initial offset setting. Just like the marksman, the CNC person will not know for sure whether each machining operation will perfectly machine the workpiece to size until *after* machining occurs. And just as the marksman needs to fine-tune the sight adjustment after firing the first shot, so must the CNC person fine-tune each offset value after machining and measuring the first workpiece. As you present each form of compensation, you can continue to bring up the marksman analogy whenever you need to elaborate on the use of offsets for a given compensation type.

Tool length compensation

When teaching machining center programming, I suggest that you begin your presentation of the actual compensation types with tool length compensation. Since tool length compensation is used for *every tool in every machining center program,* it is mandatory that all students thoroughly understand its use.

Point out that the unpredictable and changing variable with tool length compensation is the length of each tool; the CNC programmer will not know the precise length of each tool while developing the program. Instead, a command will be given for each tool telling the control which tool offset contains the tool-length compensation value. Explain that the instating command is commonly done during each tool's first Z-axis approach to the workpiece (commonly in a G43 command with an H word that specifies the number of the offset containing the tool-length compensation value). During setup, the setup person will measure each tool's length and enter the tool-length compensation offset values accordingly.

You may also want to stress that even if the setup person perfectly measures and enters the tool-length compensation offset for a given tool, there is no guarantee that the tool will machine the workpiece perfectly the first time. Tool pressure during the machining operation may cause the surface being machined to deflect slightly from the tool. The setup person may have to fine-tune each offset value after machining with each tool, especially where close tolerances are to be held.

To this point, all that you have presented will apply nicely to all student proficiency levels. From this point on, however, you will need to tailor your presentation to the students in your course. If teaching programmers, you may want to discuss that there are actually two ways to utilize tool length compensation. With one method, the tool's length is used as the offset value. With the other, the distance from the tip of each tool to the Z program zero surface is used as the offset value. Since the first method allows tool lengths to be measured off line, and since tools can be used from one production run to another without their offset values being changed, I strongly recommend teaching beginners to use the tool length as the offset value. At this point, you must also explain the words involved with tool length compensation and show examples emphasizing how your particular CNC machining centers require its programming.

For CNC operators and setup people, you will need to demonstrate how tool-length compensation values are measured for the particular machining centers you own. Also demonstrate how to access tool offsets and have students practice measuring and inputting tool length compensation values.

Cutter radius compensation

Since cutter radius compensation is used only for milling cutters, and only when milling on the periphery of the tool, you may elect to limit your presentation of cutter radius compensation at this time to simply making sure students understand *why* it is needed. Once the beginning programmer has written some practice programs and is beginning to feel comfortable with developing CNC programs, you can make a more detailed presentation about how cutter radius compensation is programmed (possibly during a review of this key concept).

To help build on what the student already knows about tool length compensation, you can simply emphasize that just as tool length compensation allows the programmer to forget about tool lengths during programming, so does cutter radius compensation let the programmer forget about a milling cutter's size (radius or diameter) while programming. The unpredictable changing variable is now the cutter's radius. In essence, cutter radius compensation allows a *range* of cutter sizes to be used. If the CNC setup person or operator does not have the planned cutter size, a different size cutter can easily be used without having to change the CNC program. You will also want to stress the other benefits of using cutter radius compensation (easier calculation of program coordinates, easy sizing for problems caused by tool pressure, and the ability to use the same programmed path for both rough milling and finish milling).

Stress that the offset used with cutter radius compensation now contains the milling cutter's radius (or diameter on some controls). During setup, the setup person tries to perfectly measure and enter the offset value. However, as with other forms of compensation, because of tool pressure or a nonconcentric tool holder, it is possible that the milling cutter will not machine the workpiece perfectly to size. Fine tuning of the offset value may be necessary after machining.

Keep in mind that, like tool length compensation, the cutter-radius compensation offset can be handled in two ways. Most manual programmers prefer to program work-surface coordinates (they are simpler to calculate) and have the operator enter the tool's radius in the offset. Teach this method if your students will be programming manually. Most CAM system programmers prefer to have the CAM system output a CNC program for the milling cutter's centerline path, based on a planned cutter size. In this case, the offset value will be the radial difference from the planned cutter size to the cutter size being used. Tailor your presentation to your company's way of handling cutter-radius compensation offsets.

The balance of your cutter radius compensation presentation will be based on the proficiency level of your students. If training CNC programmers, you will need to relate the details of how cutter radius compensation is programmed. To organize your presentation, stress that there are three distinct steps to programming cutter radius compensation:

1. It must be instated. You can present the words involved with instating cutter radius compensation (G41, G42, D word, etc.). As you know, G41 is cutter left and G42 is cutter right. While you should describe the method by which use of G41 or G42 is determined, most machinists find it easier to remember that G41 is used for climb milling and G42 is used for conventional milling (when a right-hand cutter is used). If there are any rules relating to how the cutter must be positioned prior to instating cutter radius compensation on your controls, you must call attention to them as well. Keep in mind that this prior positioning movement commonly determines how large the cutter can be without overcutting. Explain how the programmer must document the range of cutter sizes an operator or setup person can use to avoid delays during setup.

2. The programmer must drive the tool through its tool path. This is done by simply commanding the motions necessary to machine the contour. However, explain that programmers must be on the lookout for times when the programmed tool path violates some rule of cutter radius compensation. For example, it is possible to program a path during which a movement in one area of the contour will violate another surface of the contour, especially if the cutter used is larger than planned.

3. Cutter radius compensation must be canceled (usually with a G40 command). Explain any special rules programmers must follow for your controls during cancellation.

For CNC operators and setup people, demonstrate the use of cutter radius compensation on the machine tool. Show how offset values are measured and entered, as well as how sizing can be done after machining.

Fixture offsets

Unlike tool length compensation and cutter radius compensation, which are used to compensate for cutting tools, fixture offsets are used to compensate for the position of the workholding setup. Since many programmers will not know the position of the setup while programming (though it may be possible to calculate this position if fixturing is used that locates from table slots), fixture offsets allow the programmer to forget about the location of the workholding setup during programming. Once the setup is made and the setup person determines the location of the program zero point, the values locating each program zero point can be entered into the corresponding fixture offset.

This discussion should parallel what you said during Key Concept 1 when you introduced your students to the importance of program zero, but said little about how program zero is actually assigned. Now you can relate that using fixture offsets is the most popular method for assigning program zero on machining centers (replacing an older method requiring program zero to be assigned in the program).

Depending on the age of your controls, you may have to teach assigning program zero in the program (commonly with G92). However, if any of your machining centers allow fixture offsets, I urge that you teach the use of fixture offsets for those machines that allow them. If you have machines requiring the use of both methods, be ready to explain the advantages of using fixture offsets over using a command in the program to assign program zero (safer operation, easier rerunning of tools, more efficient programs, etc.).

Since machine tool builders recommend different ways of handling the program zero assignment, you will have to develop your presentation specifically for your company's ways of handling the program zero assignment. On some machines (especially horizontal machining centers), the program zero assignment can be almost transparent to the setup person. On others, the setup person actually measures the position of the program zero point for every setup made. Regardless of how easy or difficult it is on your machining centers to determine the program zero assignment values, you will need to confirm the student's understanding of *how* the related values are determined for your particular machine tools. I tend to present a very simple way, one by which the setup person uses an edge finder or dial indicator to manually locate the program zero surfaces relative to the machine's reference position. While this is a crude method of finding the program zero assignment values, it makes it very easy for students to visualize how program zero values can be determined.

For programmers, you will need to show the words related to fixture offsets (commonly G54 through G59), as well as the rules related to using more than one program zero point within a program. For operators and setup people, you must demonstrate your company's method of determining the program zero assignment values and demonstrate how they are entered into your CNC controls.

Dimensional tool offsets

All turning center operators must understand that this compensation type, which is also called simply *offsets*, is needed to compensate for cutting tools for three reasons:

1. It is difficult for a setup person to place every cutting tool into a turning center perfectly. Even if tools are perfectly placed, tool pressure may cause the tool or workpiece to deflect, causing sizing problems. Having dimensional tool offsets is like being able to physically move the cutting tool in its holder by tiny amounts to adjust incorrect positioning.

2. As any cutting tool continues to machine workpieces (especially single-point turning tools and boring bars), tool wear will dramatically affect the size of the workpiece. While the tool may not be completely dull, adjustments must be made to keep it machining properly.

3. Explain that if a turning tool is mispositioned by as little as 0.001 inch (0.025 mm), the diameter it machines will be off by 0.002 inch (0.051 mm), meaning only a small imperfection in tool placement will commonly cause the machining of a scrap workpiece. Let students know that tool offsets allow the setup person or operator to trial-cut with each tool to confirm the workpiece will not be scrapped. After trial machining with an offset value that forces excess stock to be left for finishing, the workpiece can be measured to determine how much the offset must be changed for perfect machining.

Students must also know that there are two values related to each dimensional offset, one for the X axis and one for the Z axis. This gives the operator control of sizing in both directions of motion.

Programmers must additionally know how to instate and cancel dimensional tool offset, commonly done with a four-digit T word close to the beginning of each tool in the program. The first two digits are the tool station number and the last two digits are the offset number. T0202, for example, indexes the turret to station number 2 while instating offset number 2. A subsequent T0200 cancels offset number 2.

To CNC operators, you must explain much more about dimensional tool offsets. They will be using dimensional offsets on a regular basis, as the result of almost every measurement they take. This ability to correctly adjust offsets is tied directly to the CNC operator's ability to use gaging equipment and understand tolerances. Before an offset can be correctly adjusted, the operator must be able to determine the *amount* of offset needed. You can easily develop group practice sessions right in the classroom. Use some workpieces taken from production and have the students practice taking measurements. From their measurements, and knowing what the mean value of the measured dimension should be, they can practice determining the direction and amount to adjust the related offset.

Tool-nose radius compensation

With this turning center compensation type, first explain that the unpredictable changing variable requiring compensation is the nose radius of single-point tools such as turning tools, boring bars, and grooving tools. Explain also that tool nose compensation is required only for tools that have a nose radius. Students must understand that if programmers use work-surface coordinates for programming, the nose radius of single-point tools will cause imperfections in any tapered or radiused surface being machined. You can easily develop visuals to show the imperfections caused by the tool-nose radius.

Stress that the imperfection caused by the tool-nose radius is quite small. At worst, it will be about half the tool-nose radius size (in a 45° chamfer or in a 90° arc) and will almost always leave excess stock on surfaces being machined. For this reason, tool-nose radius compensation is normally *not* used in roughing operations. In most cases, the finishing tools will easily machine the small amount of excess material left by the tool-nose radius of the roughing tool.

Like cutter radius compensation used on machining centers, tool-nose radius compensation lets the programmer forget about the radius of the cutting tool while programming. The setup person will input the tool-nose radius value (possibly with a code number specifying the tool type) into the corresponding tool offset during setup.

While tool-nose radius compensation is a great manual programmer's tool (it truly minimizes programming calculations), make it very clear to CAM system programmers that most CAM systems can output a compensated tool path based on a specified tool-nose radius just as easily as they can output the work surface path. Since the tool-nose radius will not change from one cutting tool insert to another (if 1/32-inch-[0.795 mm-] radius insert dulls, it will be replaced with another 1/32-inch-radius insert), CAM system programmers can eliminate the need for CNC-based tool-nose radius compensation altogether, minimizing the number of offsets a setup person must be concerned with during setup.

At this point you can discuss the words related to using tool-nose radius compensation as well as the steps needed for programming. Because tool-nose radius compensation is so very similar to cutter radius compensation, the balance of your instruction for programming will be very similar for both types of radius compensation.

Geometry offsets

Unlike the dimensional tool offsets and tool-nose radius compensation used to compensate for cutting tools, this turning center compensation type is used to compensate for the position of the program zero point in the workholding setup. Since most programmers will not know the values needed to assign the program zero point, geometry offsets allow the programmer to forget about the workholding setup while programming. Once the setup is made and the setup person determines the values needed to assign program zero for each tool, the values locating program zero can be placed into the corresponding geometry offset.

As with your presentation of fixture offsets for machining centers, this session should parallel what is said during Key Concept 1. In that key concept, students are introduced to the importance of program zero, but little is said about how program zero is actually assigned. Geometry offset is the most popular method used to assign program zero on machining centers (replacing an older method requiring program zero to be assigned in the program).

Machine tool builders recommend different ways of handling the program zero assignment, and this will require you to compose your presentation specifically for your ways of handling the program zero assignment. On some turning centers, the program zero assignment is almost transparent, requiring little of the setup person. Some use tool touch-off probes to simplify the program zero assignment. On others, the setup person actually measures the position of the program zero point for every setup made. Regardless of how easy or difficult it is on your machining centers to determine the program zero assignment values, you will need to confirm the student's understanding of *how* the values are determined for each tool type. I recommend presenting a very simple way, one by which the setup person uses the position display screen and skim cuts with each tool to manually locate the program zero surfaces. While this is a crude method of finding the program zero values, it makes it very easy for students to visualize how program zero values are determined.

One concept the novice tends to have problems with is that every tool in a turning center requires a program zero assignment. (With machining centers, there is only one program zero assignment for all tools.) This is because the cutting edges of each cutting tool held in the turret of the turning center can be in a different position tool to tool. On most turning centers, the tip of a turning tool will be in a different position than the tip of a boring bar or drill. Add to this the reality that not all CNC users use qualified tooling, meaning there may even be minor variations in cutting tool tip position among similar or supposedly identical tool types. Most control manufacturers handle this variation in cutting tool tip position by requiring that program zero be assigned for every tool being used. (If students have machining center experience, compare this to machining center program zero assignment, where the center of the spindle is used as the reference position from which to assign program zero. Since the spindle centerline position does not change from tool to tool, and since tool length compensation handles any variance in tool length, one program zero position can be used for all tools.)

For programmers, you will need to discuss how geometry offsets are invoked, commonly with the first two digits of a 4-digit T word. For operators, you will need to demonstrate the measurement of program zero, assigning values for each tool type and describing how they are entered into the CNC control.

Do your turning centers have tool touch-off probes? More and more turning centers are being supplied with tool touch-off probes to simplify the measurement and entering of program zero setting numbers. If any of your turning centers have this feature, you will need to explain how it can dramatically simplify the setup person's task of locating the program zero point for each tool. Be sure to demonstrate its use on the turning centers that have it.

Questions for review of Key Concept 4

1. Why is compensation needed on CNC equipment?
2. What are tool offsets? How are offset numbers determined?
3. Name the three compensation types for the machines you will be working with.

Questions for machining center review.

1. Why is tool length compensation required?
2. How do you determine the offset number used with each tool?
3. For what kinds of tools is tool length compensation used?
4. When is tool length compensation instated in the program? With what G code?

5. How is the offset number specified in the instating command?
6. Describe the method by which tool lengths are measured.
7. For what kinds of cutting tools is cutter radius compensation used?
8. Name four reasons why cutter radius compensation is required.
9. What G codes are used with cutter radius compensation?
10. How is the offset number used with cutter radius compensation determined? How is it specified?
11. Name the three steps programming cutter radius compensation.
12. How do you determine whether to use G41 or G42 to instate?
13. How must the tool be positioned prior to instating cutter radius compensation?
14. Why are fixture offsets required?
15. How do you determine the values that go into each fixture offset?
16. What are the programming words related to fixture offsets?

Questions for turning center review.

1. Name three reasons why dimensional tool offsets are required.
2. How are dimensional tool offsets invoked?
3. How is the offset number determined for each tool?
4. You have just finish-turned a diameter and find it to be 0.0024 inch (0.061 mm) oversize. What would you do to the dimensional tool offset to make the tool cut to size?
5. Before running your first workpiece, you notice a tight tolerance on a diameter machined by a finish turning tool. What could you do to the offset value for this tool to ensure that the diameter will be machined with excess stock?
6. Why is tool-nose radius compensation required?
7. For what kinds of cutting tools should tool-nose radius compensation be used?
8. What are the steps to programming it?
9. What information must be entered into the offset when tool-nose radius compensation is used?
10. How can the need for tool-nose radius compensation be eliminated?
11. How is program zero assigned on turning centers?
12. Why does each turning center cutting tool require its own program zero assignment?
13. How are geometry offset values determined?

Topics for further discussion during review

This is one of the more complex key concepts, and students will have to think through their lessons carefully. For this reason, you may wish to limit your initial presentation to include the most important facets. We mentioned during the discussion of cutter radius compensation, for example, that it may be wise to hold your extensive presentation of programming cutter radius compensation until after the student has written a few practice programs. Additionally, there are many important rules and even some safety-related considerations that may be difficult to relate during your initial presentation of cutter radius compensation. For these reasons, I recommend elaborating during reviews to confirm the students' understanding of these important issues.

More on mistakes with tool length compensation. Since mistakes made with the use of tool length compensation can result in disaster, ask students a few questions to make them think about what will happen if mistakes are made.

1. What will happen if the programmer forgets to instate tool length compensation for a given tool?
2. What will happen if the programmer specifies the wrong offset number when instating tool length compensation?
3. What will happen if an operator or setup person forgets to enter a tool length compensation offset value?
4. What will happen if an operator or setup person incorrectly enters a tool length value?

More on sizing with tool length compensation. It is important that students understand how sizing can be done with tool length compensation. A setup person or operator must, for example, be able to adjust the tool length compensation offset to adjust for minor imperfections in the surface being machined by a tool. Set up a few scenarios including problems for the students to solve. Here is an example:

- After running a program to machine a workpiece, you discover that a pocket milled by the 0.500-inch-(12.7-mm-) diameter end mill in station 3 is not quite deep enough. According to the print, it should be machined to a depth of 0.375 inch (9.53 mm), but instead is only 0.374 inch (9.51 mm) deep. What would you do to the tool length compensation offset for this tool to make the end mill machine the pocket to the correct depth?

  While this ability to size *after the fact* is important, there are times when very critical tolerances must be held. Setup people and operators must also understand how to *trial machine* in a way that will ensure that excess stock is left on a critical surface *before* machining for the first time. Here is another scenario you can present to get students thinking about how trial machining is done.

- Before machining your first workpiece, you discover a very tight tolerance on a 0.375-inch-depth pocket to be machined by a 0.500-inch-diameter end mill in station 3. You are worried that if you just let the end mill run, any imperfection in your tool length compensation value for this tool might cause the pocket to be machined too deep, ruining the workpiece. What could you do to the tool length compensation offset value for this tool to ensure that the pocket is definitely machined with excess stock? After machining with excess stock, say the pocket comes out to a depth of 0.364 inch (9.25 mm). What would you do to the tool length compensation value to make the tool machine perfectly to size?

More on sizing with cutter radius compensation. In similar fashion, students must know how sizing is done with cutter radius compensation. You can use similar scenarios to test their knowledge.

- You have just milled a counterbore with an end mill under the influence of cutter radius compensation. The size of the counterbore should be precisely 3.000 inches (76.2 mm) in diameter. However, the milled diameter comes out to 2.998 inches (76.5 mm). What would you do to the cutter radius compensation offset value in order to make the end mill machine the counterbore to size?

- Before running your first workpiece, you discover a tight tolerance on a contour to be milled with cutter radius compensation. What could you do to the offset value for cutter radius compensation to ensure that excess stock will be left on the milled surface?

More on dimensional tool offsets for long running jobs. When running high production quantities, CNC operators must understand that tool insert life will remain amazingly consistent from one tool insert to the next. An experienced operator will eventually learn the trend of each tool insert for the purpose of adjusting offsets to compensate for tool wear. For example, a new turning tool insert may machine 50 workpieces before a -0.0001-inch (-0.0025-mm) offset is required to compensate for tool wear. After that, a 0.0001-inch (0.025-mm) offset may be required after every 30 workpieces. After a total of 0.001-inch offset adjustment, the operator needs to replace the insert and reset the offset to its original value. Stress that once the offsetting trend for each tool is learned, setting offsets for a long production run becomes a very simple task for the operator. (In fact, keep in mind that offset changes can be *programmed* on most controls. You may wish to teach operators to alert the programmer once offset trends are learned. When offset changes are programmed, the operator can be freed to perform other tasks in the CNC environment.)

More on dimensional tool offsets from one production run to the next. Cutting tools used on turning centers are commonly used from one production run to the next. Setup people must understand that, if the offset is adjusted properly for the last workpiece of the most recent production run, it will remain correct for the first workpiece of the next production run. (The only exception to this will be if a dramatic change in material machinability is encountered.) Setup people should *not* arbitrarily clear all offsets (set them to zero) when making setups. Teach your setup people to nurture the habit of clearing offsets *as cutting tools are removed from the turret.* This way, the setup person can rest assured that offsets will remain correct for any tool currently in the turret, eliminating the need for trial machining for these tools when used in future production runs.

KEY CONCEPT 5: YOU MUST UNDERSTAND HOW CNC PROGRAMS ARE FORMATTED

It can be quite intimidating for a learner to look at a complex program for the first time. All of the strange commands seem to run together, forming a foreign-looking language. As the CNC instructor, you will need to set the students' minds at ease, letting them know that there is a great deal of structure to CNC programming. And once the structure is understood, reading and writing CNC programs will not be difficult.

To this point, you have presented many of the building blocks necessary to write CNC programs. While it is likely that you have shown several example programs prior to this key concept, your students have probably not in reality completely written one on their own. Key Concept 5 gives you a way to easily put it all together, drawing on the things your students already know.

Even though this key concept will be of the most importance to CNC programmers, CNC operators and setup people must also know how to read CNC programs. All setup people and most operators will be called on to make minor changes to the CNC program from time to time. At the very least, setup people and operators should be able to make basic edits, like speed and feed changes, to keep from having to involve the CNC programmer every time a minor change must be made.

Admittedly, there are many successful ways to structure the formatting for CNC programs. I tend to stress a very safe, easy-to-understand, and easy-to-use method of formatting. However, my recommended method is not the most efficient. You must structure your program-formatting presentation according to your own CNC programmers' ways of structuring programs.

Reasons for program formatting

Before digging into actual program formatting, it is important to stress the three reasons *why* programs must be strictly formatted. First, explain that strict formatting gives the learner an easy way to become familiar with CNC programming. Second, point out that if all programmers format programs in the same manner, there will be a high degree of consistency among all programs for each CNC machine tool. Anyone can easily read any program the company uses, regardless of who wrote it. Third, and most important, make sure your students understand that CNC programs must be strictly formatted so that all tools can be run independently of the rest of the program. Emphasize that there will be many times when the CNC setup person or operator must rerun just one of the tools. With properly formatted programs, doing so is as easy as scanning to the start-up command for the tool being rerun.

Also include in your discussion any machines in your shop that require different program formats (M-code differences, accessory device differences, etc.). Doing so before discussing the formats will minimize confusion as discrepancies come up.

The four types of program format

When it comes to actually introducing the formats, I recommend presenting four different types of program format. *Program start-up format* includes all information to get the machine up and running. *Tool ending format* includes all information needed to end use of each tool. *Tool start-up format*, like program start-up format, includes all information needed to get the machine ready to run the tool independently of the rest of the program. *Program ending format* includes everything necessary to close the program.

Beginners tend to misunderstand the need to include all information at the beginning of each tool for the purpose of rerunning tools. I recommend setting up a few scenarios of programming mistakes to drive the point home.

For example, say the programmer has two tools that run consecutively. Both tools require a spindle speed of 300 rpm. Since the S word is modal, the second tool will run at the proper rpm if the program is run in the normal manner. A programmer, knowing this, may be tempted to leave out the S300 word in the tool start-up information for the second tool that runs at 300 rpm. Ask your students if they can think of a time when this will cause a problem. (The answer is, of course, when the second tool running at 300 rpm must be rerun. If there is no S300 word in the tool start-up format for this tool, it will run at the same rpm as the last tool run. If the program had been completely run before the operator discovered a need to rerun this tool, it would run at the same speed as the last tool in the program.) I recommend setting up several scenarios like this to reinforce the students understanding. Include repeated axis positions, feeds and speeds, and other current modal information in your examples.

Also note that using the recommended formats makes it very easy to write CNC programs. Instead of having to memorize every word used in CNC programming, the beginner can easily use the formats as a crutch. When a person sees a CNC word, it is likely its meaning will be clear. Compare this to the road signs a driver sees on the highway. Very few drivers can recite every road sign from memory. However, any driver can easily tell you the meaning of each road sign once he or she sees it.

Make it clear that a large percentage of all CNC programs is nothing more than format. Only the actual cutting movements needed for each tool must be developed without the help of the format crutch.

At this point, you must go over the four types of program format for each type of machine you are presenting. Explain what each command and word in your format is doing. Develop your formats from current CNC programs your programmers have already written (as long as they adhere to the formatting rules just discussed). This will ensure that beginning programmers continue formatting programs for your company in the same manner as current programmers.

An example of program format for vertical machining centers. Following are the formats I use when teaching vertical machining center programming for a popular machining center control. It is quite safe and easy to understand, doing only a limited number of functions per command. However, it is not the most efficient way to format machining center programs. As stated, you must base your formats on your own programmers' ways of structuring programs.

Program start-up format:

O0001 (Program number)

N005 G91 G28 Z0 M19 (Ensure that the machine is at its tool-changing position)

N010 T01 M06 (Place the first tool into the spindle)

N015 G90 G54 S500 M03 T02 (Select absolute mode, coordinate system, start spindle clockwise at 500, and get the next tool ready)

N020 G00 X3.0 Y2.0 (Rapid to first X and Y position)

N025 G43 H01 Z.1 (Instate tool length compensation, rapid to first Z position)

N030 G01 ... F5.0 (Feed rate is part of program start-up)

Tool ending format:

N150 G91 G28 Z0 M19 (Rapid to tool change position, orient spindle)

N155 M01 (Optional stop)

Tool start-up format:

N205 T02 M06 (Place next tool into the spindle)

N210 G90 G54 S700 M03 T03 (Select absolute mode, coordinate system, start spindle CW at 700 rpm, and get the next tool ready)

N215 G00 X4.0 Y2.7 (Rapid to first X and Y position)

N220 G43 H02 Z.1 (Instate tool length compensation, rapid to first Z position)

N225 G01 ... F7.0 (Feed rate is part of tool start-up)

Program ending format:

N405 G91 G28 Z0 M19 (Return to tool change position, orient spindle)

N410 M30 (End of program)

To use these formats, a beginning programmer will start every program using the program start-up format as a crutch. Of course, actual numbers may need to be changed for speed, feed, axis position, tool stations, etc., but the basic structure will be correct for every new program.

After writing the program start-up format, beginning programmers will be on their own to develop the cutting movements needed for the first tool. When finished driving the first tool through its cutting movements, as long as there are more tools, they will follow the tool ending format. They will then follow the tool start-up format for the second tool. Next they come up with the cutting movements. From this point they will simply toggle between tool ending, tool start-up, and cutting information until finished cutting with the last tool, at which time they follow the program ending format. As you should now see, using program formatting in this manner lets the beginning programmer easily build a CNC program, one tool at a time.

Questions for review of Key Concept 5

1. Name three reasons why it is important to format programs in a strict manner.
2. Discuss why certain words and commands must be repeated at the beginning of each tool.
3. Name the four types of program formatting.
4. (To the instructor): List key words used in program formatting for your machines and have students explain their meaning.

Topics for further discussion during review

If you are like me, you try to keep program formatting easy to understand by breaking up each major task required in formatting, one per command. For example, your machining center program start-up format may have the tool moving in X and Y in one command and then moving in Z in the next. While this format is very safe and easy to understand, it is not very efficient. Relative to your particular format, see if students can make recommendations for improvement. Whenever efficiency is improved, there is usually some kind of tradeoff, possibly related to safety. Be sure they can spot the tradeoffs as well.

While reviewing the reasons why programs must be strictly formatted, you may wish to get operators thinking about how to rerun tools. In an example program (ideally one they have written on their own), see if they can pick out the start-up command for each tool.

KEY CONCEPT 6: YOU HAVE SEVERAL SPECIAL FEATURES TO SIMPLIFY PROGRAMMING

While Key Concept 5 gets beginning programmers to the point where they can *begin* writing their own programs, they have probably already figured out that programming (especially manual programming) with only the commands shown to this point would be very tedious indeed. With only G00, G01, G02, and G03 available to cause motion, for example, every hole drilled on a machining center would require at least three commands. In similar fashion, every rough turning pass required for turning center programming would require four commands.

Stress that Key Concept 6 will minimize the difficulty related to manual programming as well as shorten the typical program's length. While this key concept is the most important to CNC programmers, CNC operators and setup people should at least be exposed to the types of special programming features available, especially those the company's programmers use on a regular basis. Some of these features, like hole machining canned cycles for machining centers and multiple repetitive cycles for turning centers, can be learned with relative ease. A setup person or operator who understands them can minimize the number of times a programmer must be called on to make program changes.

Keep in mind that we only include information in this text for the most common special features of programming. With advances in CNC control technology occurring on a regular basis, it is likely that your particular CNC machines have more special features that you will want to include in your own presentation of Key Concept 6. Incidentally, if your company uses a CAM system to prepare CNC programs, your presentation of Key Concept 6 fits nicely with the introduction of CAM.

Hole machining canned cycles

All programmers must know that most machining centers and many turning centers come with a group of cycles that are very helpful for programming hole machining operations. These cycles, commonly called *canned cycles*, make it easy to command that holes be machined. The programmer simply instates the canned cycle on the first hole to be machined, and then simply lists the *X* and *Y* coordinates of other holes being machined. This means only one command is required per hole, regardless of what kind of machining cycle is commanded.

Begin by introducing the types of cycles that are available (drilling, peck drilling, tapping, boring, reaming, etc.) and explain that each is commanded by a special G code (G81, G82, G83, etc.). Let students know that all hole machining canned cycles are programmed in essentially the same manner. Truly, a student who understands how one canned cycle is programmed can easily adapt to the other kinds. Stress the things that all hole machining canned cycles have in common, including the fact that they are modal (canceled by G80 on most controls), that many of the words used in one canned cycle mean essentially the same thing in others, and that motions will always occur in the same basic manner from one canned cycle type to the next.

Next introduce the word types used with canned cycles. Be sure the students understand the special meanings for words used in canned cycles. Make it clear that some words may have a new meaning when it comes to canned cycle usage. For example, many controls use the letter address R for the rapid plane with canned cycles as well as to designate a radius size with circular movements. If your controls allow special obstruction clearing commands to avoid clamps and other obstructions between holes (G98 and G99 on some controls), you will need to discuss how they are used. Certain canned cycles may require additional presentations. When presenting the tapping cycle, for example, you will need to discuss how to calculate speed and feed for tapping. If your controls allow rigid tapping (also called *synchronous tapping*), you must discuss its use as well. Finally, give several examples of how canned cycles are used.

Subprogramming

While this presentation may not apply directly to CNC operators and setup people, make sure that programmers know that all CNC controls allow a series of CNC commands to be repeated using subprogramming techniques (also called *subroutines*). Used mostly by manual programmers to minimize the number of commands that must be included in the program, subprogramming can be used whenever a group of commands must be repeated. Discuss the most common applications for subprogramming, especially those your own company regularly uses (multiple machining operations on holes, rough and finish milling using the same set of coordinates, multiple identical machining operations, and control programs). Finally, talk about the commands that are used for subprogramming on your particular controls and show some examples.

Parametric programming

Many excellent applications for parametric programming are completely overlooked simply because CNC users cannot spot them. For this reason, I recommend at least exposing students to parametric programming, even during basic CNC courses. At the very least, get your students to the point where they can spot good parametric programming applications. Admittedly, parametric programming can become quite complex. For this reason, and since it can take a complete course to cover all the functions of parametric programming, I recommend limiting the scope of your presentation to a simple description of what can be done with parametric programming, not how to use it.

First, describe what parametric programming is. Compare it to the language BASIC (or any other computer programming language) used on personal computers and state that parametric programming includes many computer programming features. This computer programming capability within CNC programs opens the door to many amazing things. Point out that most versions of parametric programming also include many CNC-related features that cannot be found in normal manual CNC programming commands.

Next describe the four application categories for parametric programming:

1. *Families of parts*—the most useful. Since almost all workpieces machined by a given company bear at least some similarities, almost every CNC user has at least some applications for parametric programming. With parametric programming, all workpieces included in a family can be machined with one program. In this application, parametric programming can be compared to having the ability to write a general-purpose subprogram. A series of variables passed to the parametric program specifies the special attributes of a given workpiece in the family currently being machined.

2. *Complex geometric shapes*. While most CAM systems can outperform parametric programming for workpieces in this category (especially when compared to actually developing the parametric program), there are still some good applications for parametric programming. One screw machine cam manufacturer, for example, uses a parametric program to machine the complex rises and falls around the circular cam. With one parametric program permanently stored in the control, an operator can easily define the values related to each rise and fall in one parametric

program calling command. Dimensions are taken right from the print. While there are CAM systems that can do this, none will beat the speed and ease of doing so right at the CNC machine tool.

3. *Driving optional devices.* Certain devices, like probes, post process gages, and some spindle probe digitizing systems, require a higher level of programming than can be found in CNC manual programming. Parametric programming gives the accessory device manufacturer a way to interface the device to the CNC control.

4. *Utilities.* A world of functions can be added to CNC controls with parametric programming techniques. Part counters, cycle-time meters, tool-life managers, and manual probing using edge finders are a few among many added functions a CNC user can enjoy with parametric programming techniques. In many cases, applications in this category give the end user ways to accomplish tasks previously thought impossible.

Optional block skip

Begin describing this feature in terms of how it allows a programmer to give the CNC operator a choice between two possible methods of program execution during the machining cycle. Stress the times when your own company uses this feature (with excess roughing stock, for trial machining, for moving the machine to its proper starting position, for multiple workpieces with one program, etc.). Then show the word involved (the slash code) and give examples of its use.

For CNC setup people and operators, one caveat I recommend giving is related to the on/off switch used with optional block skip. Commonly called *block delete*, this on/off switch, when on, will cause *block skip* commands in the program to be skipped. This is the opposite of what normally happens when a switch is turned on. With the optional block skip switch, turning it on will cause commands in the program to be ignored, meaning *less* will happen with this switch on. This can be confusing to beginning CNC people and you will need to confirm their understanding through practice.

Thread milling

As mentioned in Key Concept 2, even though thread milling is performed by a helical motion command (one of the motion types), I recommend delaying this presentation until Key Concept 6, since thread milling makes up only a small portion of CNC programming. Begin by reviewing helical motion. Point out that two axes will be moving in a circular manner (usually X and Y) while the third axis is moving in a linear manner (usually Z). Some people mistakenly call this a 3-axis circular move. It is not. Again, two axes are circular and one is linear. Stress that a helical motion is like making a spiraling movement with the radius of the spiral remaining constant. This is precisely the kind of movement needed when milling the helix required for a thread.

Next, introduce the kinds of thread milling cutters your company uses and discuss how they work. Discuss the pitch limitations related to a given thread milling cutter or insert (one pitch per cutter or insert on most cutters). If there are any limitations related to how much of the thread can be machined during one pass, be sure to point them out. Describe also how thread milling is done: how threads in blind holes, in through holes, and male threads are generated.

Then discuss how the helical motion for thread milling is programmed. Point out that the trick to thread milling is in determining the amount of Z-axis departure per individual helical command. This Z departure is related to the portion of a full circle being machined. If machining a full circle in X and Y, for example, the Z-axis departure must be one full pitch of the thread. If the machine is making a quarter circle approach or retraction, the Z departure must reflect this and be one quarter of the pitch. Finally, show examples of thread milling commands included within your own CNC programs.

Coordinate manipulation commands

Most CNC controls, and especially machining center controls, come with a series of coordinate manipulation commands. These commands, like scaling, axis rotation, and XY axis exchange are used to change the orientation of programmed movements. Commonly used with subprogramming techniques to repeat commands, coordinate manipulation commands can usually minimize the number of commands needed to machine a workpiece. If your company uses them heavily, you will need to include these presentations in Key Concept 6.

Multiple repetitive cycles

Also called *lapping cycles*, this special feature of programming applies only to turning centers. Like hole machining canned cycles, multiple repetitive cycles can dramatically simplify programming and reduce the program length. Cycles for rough machining on turning centers (rough turning, rough facing, and rough boring) allow the programmer to simply specify one command, telling the control how rough machining must be done. Based on a single command, and the finish pass definition, the control will completely rough-machine the workpiece. This ability to completely rough-turn in one command rivals what can be done by some of today's best CAM systems.

In similar fashion, cycles for finish machining (finish turning, finish facing, and finish boring) allow the entire finishing operation to be done by one command, based on the finish pass definition given previously. Other multiple repetitive cycles (for grooving, drilling, and threading) dramatically simplify the programming of other operations commonly performed on turning centers. Many turning center controls, for example, allow an entire thread to be machined by one simple command, regardless of how many passes are required to chase the thread.

In teaching multiple repetitive cycles, it is imperative, of course, that you reflect the particular turning center controls you use. Though there may be strict rules related to the various multiple repetitive cycles you use, most controls make it relatively simple to use them, which in turn makes your presentation easier.

Other programming features to discuss

As stated earlier, your particular CNC controls may have many more features you may wish to include in Key Concept 6. Many machining centers, for example, include a series of special milling cycles for face milling, pocket milling, and circle milling. Your machining centers may also include a series of hole pattern commands, making it easy to specify bolt hole patterns, grid patterns, window patterns, and line patterns of holes.

Additionally, your machine tools may come with certain accessories that require special programming. Machining centers may, for example, have pallet changers, rotary devices, probing systems, and postprocess gaging systems, among many other possible accessories. Turning centers may have tool touch-off probes, tailstocks, steady rests, live tooling, and other accessories your students must understand. Key Concept 6 provides the most opportune window for discussing any special programming functions your CNC controls require.

Questions for review of Key Concept 6

1. Why do control manufacturers include special programming features for their controls?
2. Name the most important special programming features for the CNC machines you will be working with.
3. Name the advantages of using hole machining canned cycles.
4. Name the most commonly used hole machining canned cycles and give their invoking G words.
5. Name three things all hole machining canned cycles have in common.
6. Name the words used with hole machining canned cycles and discuss their meaning.
7. How are hole machining canned cycles canceled?
8. When are subroutines required?
9. Name the programming words related to subprogramming.
10. When is optional block skip used? What is the programming word used with optional block skip?

Did we miss anything?

Given the wide variety of programming features related to CNC machining centers and turning centers, our discussions of CNC programming cannot possibly address every potential topic you may require for your special needs. At this point, we urge you to consider any programming functions important to you that are not included in this outline and discussion. With your knowledge of the key concepts for programming, you should be able to fit *any* programming topic into the structure of our six programming-related key concepts.

KEY CONCEPT 7: KNOW YOUR MACHINE (FROM THE OPERATOR'S VIEWPOINT)

The first operation-related key concept parallels Key Concept 1. The CNC setup person or operator must, of course, know the CNC machine tool. As discussed in Key Concept 1, the operator must know the machine's basic components, the directions of motion, and programmable functions. However, a CNC setup person or operator must be much more *intimate* with the CNC machine tool than a CNC programmer, understanding the specific tasks required to operate the machine tool. While this key concept applies more to CNC setup people and operators, the more programmers know about operation, the better programmers they can be.

Begin your instruction by relating the specific tasks your CNC operators and setup people must perform. Since my courses tend to be somewhat generic, I categorize these tasks into three levels of operation.

- Level 1. Many companies expect their CNC operators to perform only the most basic tasks related to running a CNC machine tool. In this level of operation, the CNC operator does little more than maintain production. Tasks include powering up and turning off the machine, loading and unloading workpieces, and updating tool offsets to maintain size during the production run.
- Level 2. The operator must do more than simply maintain production. An operator at this level may be responsible additionally for making setups and verifying the CNC program. This means the operator must also understand a higher level of programming functions in order to modify programs as mistakes are found during verification.
- Level 3. The CNC operator must do everything related to the use of CNC equipment. Additional tasks for this level include processing and programming.

Before you can teach CNC operation, you must isolate precisely those tasks your company expects of its CNC setup people and operators. This dramatically simplifies the effort required for making operation-related presentations.

Directions of motion from an operator's viewpoint

If the cutting tool does not move along with the direction of motion, it can be quite confusing for beginning operators to understand the polarity of axis motion (plus versus minus). You must ensure their understanding of this basic point through practice. For most turning centers, it is easy. Since the turret (holding the cutting tools) moves to form the axis of motion, the polarity is easy to visualize. Any motion in X growing smaller in diameter (toward the spindle centerline) is a minus motion on most turning centers. Any motion toward the face of the chuck is a minus motion. Additionally, a simple joystick (similar to the joystick of an arcade game) is commonly used to jog each axis. The operator can easily tell which way the turret is going to move when the joystick is used to cause manual motion.

On the other hand, many machining centers use *table motion* to form the direction of motion, making it harder to visualize plus and minus. The cutting tool is not actually moving for one or more of the machining center's axes. Many vertical machining centers, for example, commonly use table motion to formulate both the X- and Y-axis motions. As viewed from the front of the machine, X plus movement is when the table moves to the left. Y plus movement is when the table moves toward the operator. At least the Z axis does involve cutting tool motion on a vertical machining center. As the tool moves toward the table top, it is moving in a minus direction.

Keep in mind that most machining center operation panels involve two or more controls for manually jogging the axes. Typically a 3-position switch selects the axis to move (X, Y, or Z). Additionally, two push buttons (one for plus and one for minus), when pushed, actually cause the motion. This kind of operation panel complicates the plus versus minus problem for beginning operators, since if the wrong button (plus or minus) is pressed, the axis will move the wrong way.

Control panel buttons and switches

All CNC setup people and operators should know the purpose of every button and switch on the machine tool. There may be certain buttons and switches that are seldom, if ever, used; however, to be truly proficient, CNC people should know the function of all.

To simplify the learning of all buttons and switches, you can divide them into three basic categories, the first of which is the *control panel*. The control panel is made by the control manufacturer and includes all keys, buttons, and switches needed to interface with the CNC control. To simplify yet further, you can divide functions on the control panel into smaller categories.

Display screen functions. There are many display screen functions an operator or setup person must be familiar with. The most important functions include position, program, program check, offset, and possibly tool path display. Other, less important functions include setting, parameters, and alarm. Begin by introducing your students to these basic functions, stressing when they are required. Next go over the keys related to the use of the display screen.

Alphanumeric keys. This control panel function allows the CNC setup person or operator to enter letters and numbers. While some CNC controls allow only the letter addresses used with CNC programs, more and more accept all letters of the alphabet. Some controls even incorporate the same type of keyboard used with personal computers for alphanumeric entry.

Special CNC keys, buttons, and switches. Many CNC controls include a special set of buttons and switches on the control panel to allow many other CNC functions. Examples of special keys include program input and output functions, control reset for alarms and other problems, and keys used to display a directory of CNC programs currently residing in the control's memory.

Machine panel buttons and switches

The second panel type that all CNC people should understand is commonly made by the machine tool builder. Unfortunately, machine tool builders differ dramatically in their opinions as to what is important; even highly similar machine tools with identical controls can have very different machine panels. You must be prepared to address any differences in the design of machine panel buttons and switches, especially if you use several machines made by a variety of machine tool builders.

Mode switch. The mode switch is the most important switch on the CNC machine tool. Compare it to the function selector of a typical stereo sound system (to select from tuner, phono, tape, or CD), allowing the CNC operator or setup person to select from the most basic modes of the machine. Since almost all CNC functions require that this switch be correctly set *before* the function will be allowed, teach beginning CNC operators and setup people to look at this switch first, before attempting *any* function.

While I recommend introducing the actual positions commonly found on the mode switch, we devote an entire key concept to the understanding of CNC modes. For this reason, I recommend limiting your presentation during Key Concept 7 to simply explaining the importance of the mode switch.

Manual controls. You will also need to explain the buttons and switches related to manual axis motion (commonly jog and handwheel control). Since each machine tool builder determines just what it will allow operators to do manually, you must be prepared to present any differences from one machine type to the next. Common manually activated functions include axis motion, spindle control, coolant control, and tool changing.

Conditional switches. All CNC machine tools have a series of conditional switches (commonly on/off switches) to make the machine behave differently. While some of these functions are discussed in greater detail at other points in your instruction, you must at least introduce students to the function of each, including dry run, single block, machine lock, Z-axis feed neglect, optional stop, and optional block skip.

Accessory panels

If your machine incorporates any accessories like an automatic pallet changer, an automatic tool changer, a tailstock, a steady rest, or any other accessory, there will likely be additional buttons and switches for their use. The control of accessory devices is commonly handled with additional control panels. Be sure to cover the use of each.

Questions for review of Key Concept 7

1. Imagine you are standing directly in front of your machine and describe the directions of motion. Which way is plus and minus for each axis?
2. Name the three levels of operator skills.

3. For CNC operators, name the most important tasks you will be expected to perform.
4. Name the three most basic panels on your CNC machine tool.
5. Name the most basic display screen modes.
6. Name as many of the control panel keys, buttons, and switches as you can and discuss their function.
7. What is the function of the mode switch?
8. Name the buttons and switches related to manual controls.
9. Name the conditional switches and describe their function.
10. Does your machine have an accessory panel? If so, what buttons and switches are included on this panel?

KEY CONCEPT 8: THERE ARE THREE MODES FOR CNC MACHINE OPERATION

While there are many buttons and switches on the typical CNC machine tool, and while there are probably more than three positions on the actual mode switch, it is important that your students know that there are really only three basic modes of operation, and that all buttons and switches on the machine can be categorized into one of these three modes. Every machine operation task the CNC operator will perform will be in one of these three categories: a manual task, a manual data input (MDI) task, or a program operation task. Key Concept 8 focuses how and why these three modes are used.

Manual mode

In this mode, the CNC machine behaves like a conventional machine tool. A machining center used in the manual mode resembles a manually operated milling machine. A turning center used in the manual mode resembles an engine lathe. When the operator presses a button, some manual function corresponding to the button's function will be immediately activated.

Explain that there may actually be several mode switch positions that fit into the manual mode. Manual, jog, handwheel, and reference return are among the mode switch positions that are included in the manual mode.

Manual data input mode

You can easily explain the manual data input mode by relating that machine tool builders vary regarding how much control they give the CNC user in the *manual* mode. Some builders give excellent control, meaning very little (if anything) must be done in the MDI mode. Others give very little in the way of manual controls, forcing the CNC operator or setup person to do almost all manual tasks by manual data input.

Most machining center builders, for example, do not give the end user a way to cause tool changes manually. For these machines, the operator or setup person must go into MDI mode and give a CNC program-like command to cause the manual tool change (T01 M06, for example). Explain that the control executes the command and then forgets it. If another manual tool change is required, another MDI command must be entered and executed. On the mode switch itself, most machine tool builders designate the position used to command manual functions with the abbreviation MDI.

The operator or setup person may perform editing functions as part of manual data input mode. The machine is not activated in this mode, but the operator or setup person uses the keyboard and display screen of the CNC control to enter and modify programs. Most CNC people refer to the procedure of entering a CNC program in this manner as *manual data inputting* of a CNC program. The actual switch position used for this function is usually specified as EDIT mode.

Program operation mode

In this mode, the CNC operator or setup person is actually executing a CNC program. Newer controls may have only one mode switch position designated for running programs, commonly called MEMORY or AUTO. Older CNC controls may have a mode switch position labeled TAPE. When in the MEMORY or AUTO mode, the program is usually executed from within the control's memory. (The exception is when programs are so long they do not fit into the control memory.) In the TAPE mode, programs are being run from the machine's tape reader. The actual button that activates the CNC program is commonly named CYCLE START. To pause the program, most control manufacturers also include a button labeled FEED HOLD.

Questions for review of Key Concept 8

1. Name the three basic modes of CNC operation and discuss when each is used.
2. Name the mode switch positions included in the manual mode.
3. Name the mode switch positions included in the manual data input mode.
4. Name the mode switch positions included in the program operation mode.

KEY CONCEPT 9: YOU MUST UNDERSTAND THE KEY OPERATION PROCEDURES

Beginning CNC operators and setup people must understand that running a CNC machine tool is really nothing more than following a set of rather strict procedures. To turn the machine on or off, for example, the operator simply performs a series of steps in the correct order. Loading a program, adjusting an offset, measuring the program zero point, and manually starting the spindle are a few examples of critical tasks that involve simple step-by-step procedures. Experienced operators will agree that the real key to running a CNC machine lies in knowing *when* to perform a given procedure.

Though this is the case, it is amazing how many companies expect their beginning operators and setup people to totally memorize all procedures used on their

CNC equipment. Additionally, when they are first introduced to a CNC machine, many new operators get little more than a *demonstration* of how a given procedure is activated. From this single demonstration, they are expected to remember how the procedure is done. This is pitifully inadequate. While commonly performed procedures will eventually be memorized, no new operator, regardless of aptitude, should be expected to learn and remember in this manner.

You can get beginning operators and setup people productive much more quickly by preparing a set of the most common procedures. With this set of procedures, you will dramatically shorten the beginner's learning curve. The beginner will simply follow instructions instead of trying to commit the instructions to memory. Included here is an example procedure used on one popular CNC control for loading a program into memory. Notice how explicit each step is. Truly, even a beginning operator could easily follow this procedure without confusion.

To load a program into the control's memory:

1. Place the mode switch to *edit*.
2. Press the extreme left soft key until PROGRAM appears at the bottom of the display screen.
3. Press the soft key under PROGRAM until a program appears on the display screen.
4. Connect the computer and get it ready to send a program to the control.
5. At the control, type the letter address O and the program number to be loaded.
6. Press the extreme right soft key until READ appears at the bottom of the display screen.
7. Press the soft key under READ. (The word READ starts flashing on the display screen, indicating that the control is waiting to receive a program.)
8. Go to the computer and send the program.

I recommend developing a *procedure handbook* for each CNC machine your company uses. The small investment in time and effort it takes to develop this handbook will be repaid countless times over the course of a machine's use, since the handbook will make it easy for beginners to be productive. It will also help experienced personnel to perform an important, but rarely used, procedure that is not committed to memory. You can easily organize your handbook by type of procedure to make it easy for CNC operators to find a needed procedure. Here is a list of the most basic operation procedures we suggest you include. You may need more depending on your machine type and what accessories are supplied with your machine.

Manual procedures.

Turn on the machine.

Turn off the machine.

Do a manual reference return.

Manually start spindle.

Manually jog axes.

Use the handwheel.

Manually load tools into the spindle.

Manually load tools into the automatic tool changer (ATC) magazine.

Manually turn coolant on and off.

Manually make axis displays read any number.

Enter tool offset values.

Manually turn on mirror image.

Manually select inch or metric mode.

Manual data input procedures.

Change tools.

Turn on the spindle.

Do a reference return.

Move axes.

Program loading and saving procedures.

Load programs into memory from tape.

Load programs into memory from a computer.

Load programs into memory through the keyboard.

Save programs to a computer.

Program display and editing procedures.

Display a directory of programs.

Delete an entire program.

Search other programs.

Search for words within a program.

Alter words within a program.

Delete words and commands within a program.

Insert words in a program.

Setup procedures.

You can get as elaborate as you wish with this section of your procedure handbook. Since most companies prepare elaborate setup documentation for each job, we limit the procedures we show. However, you can include any specific task your setup people must perform when setting up your own company's machine tools.

Measure the program zero point.

Measure tool lengths (for machining centers).

Bore jaws (for turning centers).

Program running procedures.

Verify new CNC programs.

Verify previously run programs.

Pick up in the middle of the program and rerun a tool.

KEY CONCEPT 10: YOU MUST KNOW HOW TO SAFELY VERIFY, OPTIMIZE, AND RUN CNC PROGRAMS

CNC operators must know how to run CNC programs in production and rerun tools within the program. CNC setup people must in addition understand how to verify and optimize CNC programs. Though in many companies CNC programmers do not actually run the CNC machine tools, they should at least know how these important procedures are performed.

I like to begin my presentation of this important final concept by reviewing the three levels of safety related to running CNC machine tools (operator safety, machine safety, and workpiece safety).

Since new unproven CNC programs are the most dangerous and difficult to run (especially manually prepared programs), I recommend presenting four basic program verification procedures related to running new programs. You can include these procedures in your procedure handbook for each machine.

Explain that during a *machine-lock dry run*, the control will look for basic syntax mistakes in the program. If it encounters this kind of mistake, an alarm will sound and leave the program cursor close to the alarm-generating command, making it relatively easy to find and correct the mistake. At the completion of a successful machine-lock dry run, at least the setup person will know that the CNC control can understand the program.

During a *free-flowing dry run*, the setup person must be taught how to check for motion-related mistakes in the program. Let beginners know they will be checking for interference, spindle direction, and anything else about how the program runs during this procedure. They must be made aware of the switches available to help with this procedure, including dry run, single block, feed-rate override, and rapid override. Additionally, make sure the students understand the very helpful *distance-to-go* function commonly shown on the *program check* page of the display screen (if your controls have this page) and how it can help the setup person monitor for motion mistakes.

Since it is difficult to tell the difference between cutting commands and rapid motions during a free-flowing dry run, the *free-flowing normal run* will help the setup person confirm that the cutting commands (G01, G02, and G03) are included in the program as they should be. Explain that the program is simply run in its normal manner without a workpiece in position. Once each tool's cutting motions are confirmed, the dry run switch can be used to speed up the tool motions, shortening the time required for this procedure.

And finally, you must teach how to *run the first workpiece cautiously*. As discussed in Key Concept 4, you must first confirm that students understand how offsets may need adjustment prior to running each new tool for the purpose of trial machining (especially when you are teaching turning center operation). Also, be sure they understand how single block and dry run (and possibly rapid override) can be used to control each tool's first approach to the workpiece, the most dangerous move of every tool. Once the tool is in its proper cutting position (and the setup person breathes a sigh of relief), dry run must be turned off; feed-rate override can be used to control the rate of each cutting movement.

Keep in mind that the only way for students to truly master this key concept is through practice. Your beginners must practice these procedures with your supervision prior to being turned loose in production. This means your company *must* allow the use of CNC machines for practice during training courses (some companies even supply their training departments with practice machines). Be sure you give your students plenty of practice on verifying new CNC programs.

Make sure students understand that even previously run CNC programs will require verification. While they need not be overly cautious during actual cutting motions, be sure they understand that incorrectly set offsets and mistakes made during the program zero assignment can still cause disaster during the running of the first workpiece. Have them take control of each tool's first approach to the workpiece with single block and dry run. Once in position, they can let each tool run at its normal rate, as long as they are absolutely sure that the program has been successfully run before. Until operators feel very comfortable verifying previously run CNC programs, you may additionally wish to have them perform a free-flowing dry run prior to running the first workpiece. Again, practice is the key to success.

During this key concept, you will also need to relate how tools are rerun. As long as the programmer follows the recommendations made during Key Concept 5, rerunning tools should be relatively easy (especially if you have written a procedure for doing so as suggested in Key Concept 9).

Once the first workpiece is machined from a new program and passes inspection, further monitoring of the program will be necessary for optimizing purposes if high production quantities are to be run. Simple improvements like the elimination of unnecessary rapid motions can be commonly handled by the setup person or operator. Tool life and other cutting condition improvements may require the help of the CNC programmer.

OTHER CNC DEVICES REQUIRING TRAINING

As we hope you will agree by now, the key concepts approach allows you to incorporate every usage technique required for CNC machine tools in a highly organized and structured manner. However, your company may own certain CNC related support devices that are not discussed in our key concepts approach.

Keep in mind that the scope of training discussed to this point has included programming, operation, and setup of CNC machine tools. The devices we mention here may not be closely related to what has been presented thus far, yet they must be understood in order to get the most from your CNC machine tools.

How closely the device relates to the actual CNC machine tool determines how you handle the training. For devices attached directly to the CNC machine tool, you should incorporate your accessory device presentation during your presentation of the key concepts (discussing them during the point in the course that makes the most sense). Other, more complex devices that are not so closely related to the CNC machine tool may require a complete and separate training course. Here we list a few devices and offer suggestions as to how they can be handled during training.

MACHINE ACCESSORIES

Almost all CNC machine tools come with some set of accessories that CNC people must know how to handle. In most cases, understanding these accessories is as important as understanding the CNC machine tool itself.

Probing systems

If the probing systems you use are required on a daily basis by your CNC setup people and operators, I recommend incorporating them in the key concepts. However, since probes can be used for a variety of purposes, it may be a little difficult to determine where they should be placed in the key concepts.

If your probe is used primarily to help with setups, as many are, you can tailor your course instruction accordingly. For example, many spindle probes on machining centers are used to help with the assignment of program zero (as are tool touch-off probes on turning centers). In this case, you must introduce the probe during Key Concept 1, when you are describing how program zero can be assigned. You will also need to discuss the probe during Key Concept 4, when describing how fixture offsets (or geometry offsets) are used. You may also need to develop procedures for the probe's use to be presented in Key Concept 9.

Another application for probing is to measure the lengths of tools used on machining centers. Since tool-length-measuring probes on machining centers are used to simplify the tool length compensation measurements, you must discuss them in detail during your presentation of tool length compensation in Key Concept 4. During your discussion of key procedures in Key Concept 9, you can include a procedure to show how they are used.

Rotary axes

Many CNC machines, including machining centers and turning centers, are being equipped with rotary axes. If your machines have rotary axes, you can introduce the rotary axis as part of your presentation of motion directions during Key Concept 1, specifying the rotary axis name and showing how its motion relates to the other (linear) axes. Also relate how the decimal format differs from other axes designated in the inch or millimeter mode (3-place format instead of four). Mention that a program zero point is commonly assigned to the rotary axis, and, just like any other axis, the rotary axis has a reference position and can be programmed in absolute and incremental modes. During Key Concept 4, when discussing fixture offsets, you will need to describe how the program zero point is assigned for a rotary axis. During your presentation of how feed rate is specified during Key Concept 2, you will need to discuss how rotary axis feed rate is specified (in degrees per minute, not inches or millimeters per minute).

For a turning center rotary axis (commonly called the C axis), you will need to include a presentation on polar coordinate interpolation during your discussion of the motion types in Key Concept 2 if your company performs contour milling work on the end of your workpieces. As with the presentation of thread milling for machining centers, you may wish to postpone your lengthy presentation of the use of live tooling programming on turning centers until Key Concept 6, when you are discussing special programming features.

Pallet changers

Pallet changers can be introduced as part of your presentation of programmable machine functions during Key Concept 1. You may also need to include a presentation on how they are programmed during your discussion of Key Concept 6. Since most pallet changers require multiple programs (at least one per pallet), subprogramming techniques are commonly used in conjunction with pallet changes. During Key Concept 9, you will also need to include a procedure describing the activation of the pallet changer.

Bar feeders

Like pallet changers on machining centers, I recommend introducing turning center bar feeders during your discussion of programmable machine functions in Key Concept 1. Since bar feeders commonly utilize subprograms to perform the actual bar feed motion, I recommend showing how bar feeding is programmed as one of the applications for subprogramming in Key Concept 6.

DNC SYSTEM USE

The use of both types of DNC systems (*distributive* numerical control systems and *direct* numerical control systems) can be easily incorporated during your presentation of the key concepts. With distributive numerical systems, introduce the system during your discussion of key operation procedures in Key Concept 9, since most CNC operators and setup people need only be able to upload and download CNC programs. Simply include a procedure to upload and download in your procedure handbook. If using a direct numerical control system that runs the CNC machine tool from a host computer system, you may additionally wish to introduce the DNC system during your presentation of programmable machine functions in Key Concept 1.

CAM SYSTEMS

Most companies use some form of computer-aided manufacturing system to help with the preparation of CNC programs. Since most CNC operators and setup people do not actually use the company's CAM system, you may wish to include just a short discussion about your company's CAM system for them during your presentation of special features of programming in Key Concept 6. However, when it comes to teaching CNC programmers how to use the CAM system, you will have to present much more information than can be easily covered during Key Concept 6. For this reason, it may be necessary to develop a separate course on the use of your company's CAM system.

While we will not offer detailed discussions for the teaching of a course on CAM systems, most personal-computer-based, graphic-style CAM systems have enough in common to enable us to present a general outline for such a course. Note that our outline still incorporates the key concepts approach.

I.   Understand what your CAM system can do
   A.   Components
      1.   Hardware
         a.   Computer
         b.   Input tablet
         c.   Toolpath plotter
         d.   Printer
      2.   Software
         a.   Geometry definition module
         b.   Machining center module
         c.   Turning center module
         d.   Postprocessing modules
         e.   Text editor module
         f.   Toolpath simulation module
         g.   Communication module
   B.   CAD versus CAM
      1.   How drawings are created
         a.   In our company's CAD system
         b.   In the CAM system
      2.   Drawing import limitations
         a.   Program zero position
         b.   Drawing elements required
         c.   Scaling problems
         d.   Duplicated entities
         e.   How the CAM system expects drawings
            For threading
            For holes
            Other drawing limitations
II. Step 1 to programming—define geometry
   A.   How to import from CAD system
   B.   Draw within CAM system

        1. Understanding entity definition types
           a.   Points
           b.   Lines
           c.   Circles
           d.   Splines
           e.   Others
    C. Understanding definition order

III.   Step 2 to programming—Entity trimming
    A. Making geometry look like your drawing
    B. Trim and break types
        1. Line trimming
        2. Circle trimming
        3. Divide trimming
        4. Break lines and circles

IV.   Step 3 to programming—define cutting operations
    A. Generating available machining operations
        1. For machining centers
        2. For turning centers

V.   Checking your output program
    A. Toolpath verification
    B. Using the text editor
    C. Transferring CNC programs to the CNC machine tool

PART
III

IMPROVING UTILIZATION THROUGH SETUP AND CYCLE TIME REDUCTION

A CNC machine is productive only when it is machining workpieces. During our discussion of value-added principles in Part I, we emphasized the importance of keeping your CNC machining and turning centers' tools in operation. Unless they are cutting, there is no value being added to the product you produce.

CNC machines have come to be regarded as a type of production equipment well suited to machining small production quantities. When running small quantities, changeover time is extremely important. When you compare the time it takes to change over from one production run to the next on CNC equipment to that of more dedicated production machinery (transfer lines, screw machines, etc.), CNC machines win easily. This makes CNC an excellent choice for companies that incorporate just-in-time techniques.

However, as production quantities grow, cycle time becomes more and more important. Generally, dedicated equipment can still easily outperform CNC equipment for high production quantities. Though this is the case, many companies still elect to use CNC equipment even for higher production quantities. Though several CNC machines may have to be purchased to keep up with production, companies do so willingly to gain the many flexibility-related benefits CNC machines provide.

Many contradictions are exposed when you start comparing the utilization of CNC equipment for running small quantities to that for large production runs. What is extremely important to a company that runs predominantly small quantities may be unimportant to a company that runs high quantities, and vice versa. Our intention in Part III is to show as many specific techniques as possible for reducing setup and cycle time.

Generally speaking, companies that run small quantities should be most interested in reducing setup time, which is the subject of Chapters Seven and Eight. While these companies will also be interested in minimizing cycle time, their primary concern will be in minimizing the time it takes to go from one production run to the next. Since the time it takes to actually machine the small quantity of workpieces is usually very short, a high percentage of machine time may be required for the nonproductive on-line task of setup.

On the other hand, companies that run high production quantities will be much more interested in minimizing cycle time, which is the subject of Chapter Nine. While they will be interested in techniques that reduce setup time, their primary concern will be in producing as many workpieces as possible in the shortest time. The vast majority of machine time will be spent machining workpieces. Only a small percentage of machine time will be spent in setup.

Chapter Seven

Setup Time Reduction Principles

A s stated in Part I, making setups is a necessary support task, yet it does nothing to add value to your product. While all CNC people will agree that anything that can be done to reduce setup time should be done, companies differ dramatically in terms of the lengths they will go to do it.

Product-producing companies tend to heavily staff their CNC environments. Their goal is to keep the CNC machine tools running for the highest possible percentage of time. *Any* unnecessary downtime is seen as a waste of time, and these companies tend to do whatever it takes to eliminate as much downtime as possible. On the other hand, contract shops tend to minimize the number of people in their CNC environments. A contract shop may compromise on wasted machine time during setup for the ability to handle all CNC-related tasks with as few people as possible.

Regardless of the company type, setup time is commonly viewed as lost production time, and all companies should be highly interested in minimizing this lost time. In this chapter, we introduce the basic principles of setup time reduction. These principles can be applied to any form of manufacturing equipment, including CNC machine tools. In Chapter Eight, we offer many specific techniques that can be applied to reducing setup time on your CNC machining centers and turning centers. Some of these principles will even be helpful during our discussion of cycle time reduction in Chapter Nine.

THE IMPORTANCE OF REDUCING SETUP TIME

The number of workpieces you make per production run (quantity) is the primary factor that dictates the importance you should attach to reducing setup time. The higher your production quantities, the greater the percentage of time the machine will be in production and the fewer the number of needed setups. With very high production quantities, it may be possible that as little as five percent (or less) of the machine's overall production time will be spent in setup. There are CNC machines, in fact, that are *dedicated* to running only one workpiece. Once the job's setup has been made, the machine will *never* be in setup again.

For companies that spend less than 10 percent of a given CNC machine's time in setup, any reduction in setup time will have only a small impact on the machine's overall utilization. While setup time reduction may still be considered somewhat important, we recommend concentrating on reducing *cycle time* (the subject of Chapter Nine) as a way of improving the CNC machine's utilization.

The vast majority of CNC users run relatively low production quantities ranging from 1 to 500 workpieces. Indeed, one of the primary reasons for using CNC machine tools in the first place is their ability to efficiently run small lot sizes. In fact, many companies (especially in toolroom and prototype environments) run only one workpiece per production run. Figure 7-1 depicts the two extremes related to percentage of time spent in setup versus running production.

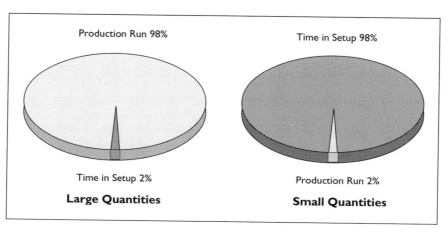

Figure 7-1. The two extremes of setup time versus production time.

It is likely that your company currently spends much more than five percent of a given machine's production time in setup. The greater the percentage of time spent in setup, the more important it is that you streamline your company's setup procedures to reduce setup time, and the more your company should be willing to invest to achieve that goal. Reaching the goal will result in increasing the machine's overall percentage of time spent machining workpieces, the value-added time.

Keep in mind that the extremes in workpiece quantities present special problems for incorporating setup time reduction techniques. For ultrahigh production quantities, as already mentioned, so little of the machine's production time is taken by setup that it would be wiser to concentrate on reducing cycle time or workpiece loading time. However, *very low* production quantities also present special problems for setup time reduction. Many of the setup time reduction techniques we show depend on having people perform certain setup-related tasks during the production run. If you are running only one or two workpieces with a short cycle time, it is likely that *nothing* can be done during the production run to get ready for the next setup.

Aside from production quantities, a second factor that contributes to the importance of reducing setup time is the throughput of workpieces. If your company incorporates just-in-time techniques, it is *very* important to ensure that workpieces arrive at each point along the manufacturing process when planned. Any enhancements that allow setup time reduction will help to ensure the flow of workpieces.

Regardless of how important it is that you reduce setup time, doing so can be quite challenging. Given the wide variety of manufacturing methods used today, each company will have its own special obstacles to overcome. Certain machine tools, for example, lend themselves to quick setups better than others. Bar-feed turning centers are generally quicker and easier to set up than horizontal machining centers. Diversity of work also presents special setup-related problems. Companies that produce families of parts will generally find it easier to minimize setup time than those that produce a wide variety of very different workpieces. Available personnel is still another variable in the setup time equation. The more people you have in your CNC environment, the easier it will be to incorporate setup time reduction techniques. Even something as basic as the age of your machine tools can affect how easy or difficult it will be to reduce setup time. The state of the art in CNC technology is constantly changing. Newer machines have functions designed to help minimize setup time.

JUSTIFICATION FOR SETUP TIME REDUCTION

Any setup time reduction program will cost something. Even if you or your people are simply applying a technique that does not require purchase of some new device, the time it takes to incorporate and test the technique must be considered. Depending on how your company operates, you may have to provide justification for the time, personnel, and money you must invest in the setup-time reduction program. Doing so requires an understanding of basic justification principles.

Given the high cost of CNC machine utilization, it should be relatively easy to justify most setup-time reduction techniques, especially those that require an investment in only time and effort. Since we are talking about justifying setup *time* reduction, your justification must be time-based, meaning you must know the shop rate of the machine tools involved.

When you have a setup-time reduction technique that requires justification, first calculate how much your setups currently cost. Knowing the machine's shop rate and how long it takes to make a given setup, you can easily calculate this cost. Knowing how many times the setup is made per week, month, year, etc., you can determine how much your company spends on a given setup. If the improvement you intend to make will be applied to several setups, be sure to repeat this process for each.

Second, approximate the savings in time your improvement will provide per setup. You should be able to easily calculate an expected savings from the time saved, the machine's shop rate, and the total number of setups involved. Compare the potential savings to the cost for implementation to determine if your setup-time reduction technique is justifiable.

Remember that a savings in setup time provides an equivalent increase in production time; therefore, reducing setup time provides a double benefit. Not only are you saving machine time, you are increasing the machine's productive time. Remember to factor this added savings into your setup-time justification equation.

We cannot stress enough the importance of knowing the current costs of making setups as you begin any setup-time reduction program. Only by knowing your current costs can you justify any investment in personnel, time, effort, or equipment needed to reduce them.

THE RELATIONSHIP BETWEEN PRODUCTION QUANTITIES, PROCESS, AND SETUP TIME

As stated, the number of workpieces to be produced per production run has the greatest affect on how important it is that you reduce setup time. Indeed, production quantities dramatically affect all important decisions made for machining of a given workpiece, including the *process* by which the workpiece is produced.

Production quantities, process, and setup design are very closely related. The number of workpieces to be produced is the largest single factor that determines how the process should be engineered. Generally speaking, the more workpieces to be machined, the more important it is that workpieces be machined efficiently, resulting in a more elaborate process.

If, for example, only 25 workpieces are to be machined, the only real priority may be machining *acceptable* workpieces. The process engineer will use standard cutting tools, and unless it is unavoidable, nothing special will be purchased for such a low production quantity. On the other hand, if 1000 workpieces are to be machined in one production run, the process engineer will likely design a completely different process. Instead of simply having to produce acceptable workpieces, acceptable workpieces must be produced *efficiently*. The higher the production quantities, the higher the emphasis on efficiency. Much more engineering will go into the process to minimize machining time. Special components related to the setup, including special cutting tools, fixtures, and possibly even machinery, can be justified if production quantities are high enough.

Just as production quantities determine how elaborate the process must be, so does the process determine how elaborate the *setup* must be. Because quantities are very high for a given job, for example, the process may call for multiple workpieces to be machined in one machine cycle. This, of course, requires the workholding setup to be capable of holding multiple workpieces.

In similar fashion, just as the quality of the process dictates the resulting cycle time, so is setup time a slave to the process. Generally speaking, the overall quality and sophistication of the setup will be a simple reflection of the process, which is in turn a result of the number of workpieces to be machined.

As a simple analogy, compare the relationship of production quantities, process engineering, and setup design to matching tires to your automobile. Your tires should be chosen to match the kind of car you drive. If you drive a sports car, the quality of your tires must reflect the high performance capabilities of your car. If you buy tires that were designed for a family car, the tires will be the weak link in your sports car's performance. If, on the other hand, you drive a family car, purchasing high performance tires would be overkill. The family car now becomes the weak link. Just as tires must be correctly matched to an automobile to ensure that there are no weak links in the car's performance, so must production quantities, process, and setup be matched to ensure that there are no weak links in the manufacturing process.

The engineering effort that goes into the design of the setup plays the biggest role in determining how quickly the setup can be made. Just as every process could be engineered to allow the absolute minimum cycle time, so can every setup be engineered to allow the absolute minimum setup time. However, feasibility, based on production quantities, limits how much engineering goes into the process. In similar fashion, feasibility, based on the process, limits the engineering that goes into the design of the setup.

Though the design of the setup plays a major role in determining setup time, cutting tool and fixture design lies beyond the scope of this text. We freely acknowledge its importance in reducing setup time, but we limit our scope to specific CNC techniques to reduce setup time.

SETUP TIME DEFINED

Our broad definition of setup time is: *The time it takes to go from making the last workpiece in the previous setup to efficiently making the first good workpiece in the next setup.* Since the machine tool is down (not running production), *anything* that happens between production runs must be considered as setup time.

Here are some tasks commonly associated with setting up CNC machine tools:

- Tear-down of old setup.
- Making of workholding setup.
- Tool assembly.
- Tool measurement (if required).
- Tool loading.
- Program zero measurement.
- Offset entry for tooling information.
- Offset entry for program zero information.

- Program loading.
- Program verification.
- First workpiece inspection.
- Modifications based on first-piece inspection.
- Optimizing program (for higher production quantities).

By our definition, several tasks affect setup time that you may not feel are truly part of a CNC machine setup. Program verification, for example, though it is not actually part of the workholding or cutting tool setup, keeps production from running and must be considered as setup time. First-workpiece inspection, as well as the time it takes to make adjustments to the CNC program in order to get a workpiece to pass inspection, holds up production and must be considered as setup time. Program optimizing time, though it may dramatically improve *cycle time*, also holds up production and must be considered as part of the setup. Time spent searching for tools, inserts, gages, fixtures, and anything else needed during the setup must be considered setup time. Even lunch, breaks, and all forms of personal time, if taken while a machine is down in setup, must be considered as setup time—and companies pay an extra penalty for personal time taken during the setup of CNC machines that can run unattended, like bar-feed turning centers. These machines can normally be running production even during breaks, lunch, and other personal time.

Though some of these functions have nothing to do with setting up a CNC machine tool, most are necessary support tasks that hold up production—and anything that can be done to minimize these tasks will effectively reduce setup time.

FINDING A PLACE TO START

As you consider setup time reduction for any kind of machine tool, you should begin by analyzing your *current* setup procedures. This will provide a point of reference for any changes you make and, if done correctly, will usually help you set priorities on which changes will have the biggest impact on setup time.

In your analysis, enlist everyone involved with designing and making setups. The more experts you involve, the more potential you have for finding the best, most feasible suggestions and solutions. To provide your experts with something to analyze, videotape the current setups in question. Again, be sure to invite everyone to analyze and critique the videos.

You will find that some tasks during setup take longer to perform than others. The first step will be to isolate each task in the process to determine what percentage of the overall setup each task takes. When this is finished, you can easily illustrate the tasks related to the setup by making a graph similar to the one shown in Figure 7-2. This makes it very clear to everyone involved in your analysis just how much time each task takes. It also makes an excellent starting point for your setup time reduction effort. Those tasks that take the most time generally offer the most potential for improvement.

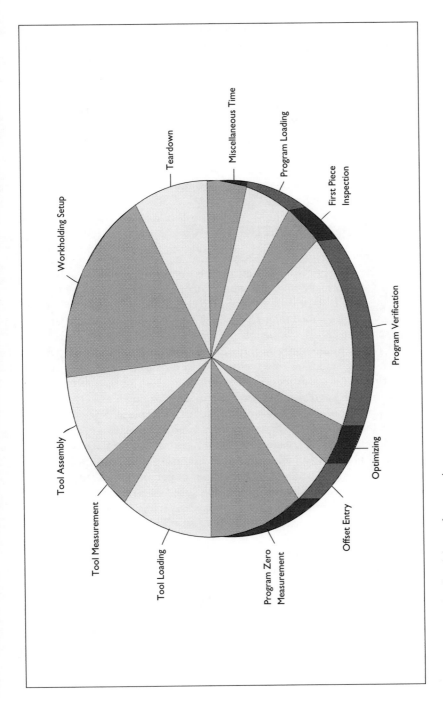

Figure 7-2. Time-slice breakdown of setup tasks.

Since it takes time to videotape the setup and organize the results, you may be tempted to bypass these steps and proceed directly to incorporating the setup time reduction technique shown in the next chapter. Or you may be tempted to bypass these steps because you think you know what your setup people are currently doing. We urge you to avoid this temptation for three reasons. First, with no point of reference, you will have no way of gaging the impact of the setup time reduction effort. Second, without this analysis, it is likely that you will not come up with the best solutions to *your* particular setup problems. Third, and most important, analyzing actual setups is the only way of truly seeing exactly how long setup-related tasks truly take. If you have never incorporated setup-time reduction techniques, you will probably be very (unpleasantly) surprised at just how much time is wasted in your setups.

THE TWO TYPES OF SETUP TASKS

There are only two types of tasks related to making setups: those that are performed while the machine is down between production runs and those that are performed up front, prior to the machining of the last workpiece in the current production run. Tasks that are done while the CNC machine is down are called *on-line tasks* (they are also called *internal tasks*). Tasks that are done in preparation for the next setup are called *off-line tasks* (they are also called *external tasks*).

Some tasks that are commonly performed as on-line tasks include tearing down the old workholding setup, making the new workholding setup, loading cutting tools, measuring cutting tool lengths, measuring program zero, entering offset values for cutting tools and program zero assignment, loading the CNC program, verifying the CNC program, and inspecting the first workpiece. Some tasks that are commonly thought of as off-line tasks include programming; cutting tool assembly; and locating fixtures, gages, and other tools needed for the next setup.

Companies differ greatly with regard to which tasks they perform on or off line. Indeed, the size of the company, the number of people in the CNC environment, and the engineering that goes into the design of the setup are just a few of the factors that help determine which tasks *should* be on line and which should be off line. Though this is the case, you must understand that *the actual time your CNC equipment is down between production runs (setup time) is the sum total of the on-line tasks.* Your CNC machine must be down (not productive) during on-line tasks. It can be running production during off-line tasks.

THE THREE WAYS TO REDUCE SETUP TIME

There are only three general ways to reduce the time required to go from one production run to the next. *Every* setup time reduction technique we show will fit into one of these three categories:

  1. Eliminate on-line tasks.

2. Move on-line tasks off line.

3. Facilitate on-line tasks.

Since setup time is the sum total of on-line tasks, whenever you eliminate an on-line task or move it off line, you effectively reduce setup time by the length of time it was taking to perform the tasks on line. For this reason, minimizing the number of on-line tasks that must be performed during setup should be the highest priority of any setup-time reduction program. By *facilitating the on-line tasks*, we mean that you must make it as easy (and efficient) as possible for the setup person to perform on-line tasks. In this chapter, remember that we are simply introducing setup-time reduction concepts. In Chapter Eight, we discuss in much greater detail how they apply to CNC equipment.

Eliminating on-line tasks

By *eliminating* on-line tasks, we mean finding ways to make a setup without performing certain on-line tasks. Though your solutions to eliminate on-line tasks can sometimes involve spending money, it usually takes little more than ingenuity and the determination not to give up until an answer is found.

The evolution of CNC technology in recent years has a lot to do with your ability to eliminate tasks that have been traditionally done on line. Newer CNC controls offer features that eliminate the need to perform certain tasks. One example of eliminating an on-line task has to do with measuring program zero positions for machining center programs (this is shown in much greater detail in Chapter Eight). Many companies require their setup person to measure the program zero positions for each setup. This on-line task involves taking three measurements (one for X, one for Y, and one for Z) for each program zero position used within the program. For horizontal machining center applications in which several surfaces are machined, this can require several measurements, resulting in a great deal of wasted setup time.

By using the fixture offset techniques enabled by today's CNC controls, it is likely that these redundant measurements during setup can be eliminated. If some critical machine measurements are taken once, early on in the machine's use, the results of these measurements can be used for every setup made. For example, the location of the machine table's key slots and critical location surfaces can be measured one time. If your workholding tooling is keyed to the table using these key slots, the programmer can easily calculate the location of each program zero position, eliminating the on-line task of measuring program zero for every setup made. The programmer can even include fixture offset setting commands in the program. This frees the setup person from having to enter fixture offset values manually, eliminating another on-line task. Programming fixture offset values in this manner also eliminates the possibility of entry errors in fixture offset values.

Ingenuity and a thorough understanding of your CNC machine's functions and features are the keys to eliminating on-line tasks. By scouring your CNC control's programming manual, and with the assistance of your machine tool builder's application engineers, you may be surprised at how easy and inexpensive it can be to eliminate many on-line tasks. Given unlimited financial resources, *any on-line task can be eliminated.*

Moving on-line tasks off line

The second way to reduce setup time is to move the tasks you are currently performing on line to off line. Instead of performing setup tasks while the machine is down between production runs, do as much as you can up front, before the time comes to actually make the next setup. While the task still has to be performed, at least the CNC machine tool can still be producing workpieces. It is not down, waiting for the task to be accomplished.

One example of moving an on-line task off line (many more will be shown in Chapter Eight) is tool length compensation on a machining center. If you are currently measuring tool length values on line (regardless of how efficient your tool-length measuring techniques), you can reduce setup time by moving the task of tool length measurement off line. A tool-length measuring gage (a height gage, for example) can be used to determine the length of each tool. Each tool's length can be written down so the operator can easily enter the tool length value (offset) during the setup. Or better yet, the tool-length measuring gage can be attached (via a serial port) to a personal computer. As soon as a tool length is measured, it can be recorded by the computer and entered (automatically) into a program as part of a tool offset setting command. Using distributive numerical control (DNC) techniques, the offset setting program can be quickly loaded and run during the setup. This technique not only moves the tool-length measuring task off line, it also eliminates the on-line task of manually entering tool offsets.

Two important factors contribute to how feasible it will be for your company to move on-line tasks off line. First, your company must have the available personnel. If you expect one person in your company to do all CNC-related tasks (write the program, make the setup, verify the program, inspect the first workpiece, and actually run the machine during the production run), as is the case in many smaller companies, it may be difficult for this one person to handle any additional work. From time to time there may be those jobs when off-line work is possible, but most of the time your programmer/operator/setup person/inspector will be so busy performing normal responsibilities that there will be no time for additional off-line setup-related work. If you work in this kind of CNC environment, keep in mind that where several CNC machines are in use in production, a great deal of production time may be wasted while your operator/setup people perform tasks on line. The more machines your company owns, the easier it will be to justify hiring a person whose sole responsibility is to perform setup-related tasks off line.

Second, your CNC machines must be in production long enough to allow off-line tasks to be performed. If you run very small production quantities with very short cycle times, it is likely that *nothing* can be done to get ready for the next setup while the machine is in production. The suggestions we offer in Chapter Eight for moving on-line tasks off line assume you have large enough production quantities to allow the time needed to perform certain setup tasks off line. We also assume you have the personnel in your CNC environment needed to perform tasks off line. Companies that currently do *not* have adequate staffing to do this can use our suggestions to help determine whether hiring a separate person to perform off-line tasks can be justified.

Facilitating the on-line tasks

The third way to reduce setup time is to make it as easy as possible for your setup people to make setups. In essence, you will be attempting to shorten the length of time it takes for your setup people to perform on-line tasks. Since our definition of setup is so very broad, there is a wide variety of potential techniques you can use to help your people. Any task you can facilitate—from tearing down the previous setup to getting the first workpiece to pass inspection—is fair game.

One area that usually offers particularly good opportunity to help setup people is *program verification.* By program verification, we mean not only confirming that the program is correct, but also ensuring that the first workpiece comes out to size. In many cases, a setup person must trial-machine and rerun sections of the program in order to get tools to machine the workpiece properly. If no consideration is given to the difficulty of trial-machining, program verification can be wasted. Later in this text, we show several specific examples of how a programmer can facilitate the running of the first workpiece with special programming techniques.

SPECIAL CONSIDERATIONS FOR REDUCING SETUP TIME FOR CNC EQUIPMENT

To this point, the presentation has been general; all points made can be applied to reducing setup time for just about any kind of production machinery in your company. For the balance of this chapter and all of the next, we make points that apply primarily to setup reduction for CNC machining centers and turning centers.

CONVENIENCE FEATURES VERSUS SETUP TIME

We have been pointing out contradictions in CNC technology throughout this text. Here we expose yet another—one related to setup time.

CNC control manufacturers are constantly striving to make their controls (as well as the machine tools they drive) easier and more convenient to use. Hundreds, if not thousands, of special features have been developed over the years to make CNC machines easier to program and operate, safer to run, and in general, more convenient to use. Indeed, today's CNC user has a much easier time of utilizing current CNC equipment than machines made even as recently as five years ago.

While many control enhancements provide benefits with no adverse effect on setup or cycle time, some require a conscious effort on the user's part to avoid costly setup or cycle time penalties. Said another way, if setup time reduction is of the utmost priority, it may mean you must modify the ways you utilize certain CNC features. In extreme cases, it may even mean you cannot take advantage of some of these very helpful features without paying a penalty in longer setup or cycle time.

An example serves to emphasize the point. All CNC machining center users are taught to use the *tool-length compensation feature*. This very important feature allows the programmer to forget about tool lengths as a program is written. Regardless of each tool's length, the program will behave correctly according to the *tool-length offset value* the operator enters at the time of setup. In essence, tool length compensation allows the CNC user to separate the tool length entry from the CNC program. While this is an extremely valuable feature, one that *all* CNC machining center users should use, if no concern is given to the effect tool length compensation can have on setup time, a great deal of production time can be wasted during the life of a CNC machining center.

Those of us who are old enough to remember the days of NC (before computers were incorporated into the machine control) know that tool length compensation was not always available. In the early days, machining center users were forced to *preset* tools to a previously determined length. The NC programmer had to know the exact length of all tools before a program could even be written, making it quite difficult and cumbersome to program Z-axis movements on machining centers.

Though these old machines were quite difficult to work with because of the problems created by tool length presetting, they were actually quite efficient when it came to minimizing setup time. Since tools had to be preset to a given length *before* they were loaded into the machine tool, and since absolutely no on-line setup time was taken to measure tool lengths, setup time was kept to a minimum. In essence, old NC machining centers *forced* their users to perform the task of presetting off line.

When tool length compensation was first introduced, most versions required that the tool-length compensation value be measured right on the CNC machine tool (on line). Many CNC users were so thrilled that cutting tools no longer required presetting, they willingly compromised on this seemingly minor point to gain the *convenience* of tool length compensation. Though all current CNC controls allow tool-length compensation values to be measured off line, there are still CNC users who measure tool-length compensation values on line, and still pay the setup time penalty for having the convenience of tool length compensation.

This exemplifies how a convenience-oriented feature can affect setup or cycle time. Today's CNC controls have many such features. Others include spindle probes, tool-length measuring systems, tool touch-off probes, and conversational (shop-floor-programmed) controls. While the benefits offered by these functions may be substantial and right for given applications, remember that they can have an adverse effect on setup time if no consideration is given to their method of use.

Companies that have one person performing all CNC tasks tend to take great advantage of the convenience-oriented features, while companies that heavily staff their CNC environments should exercise caution whenever a convenience-oriented feature is used. Remember that the shop rate for any CNC machine in your shop is probably at least three or four times the wage of any worker in the CNC environment. If optimum CNC machine utilization is your priority, having your setup people perform certain tasks off line in a time-consuming manner will often be preferable to having them perform an on-line task quickly and in a very convenient manner.

GETTING HELP

If you have been associated with CNC for any length of time, you know that almost every CNC function can be handled in several ways. This is evidenced by the fact that no two CNC programmers, if left to their own devices, will ever develop identical CNC programs. There are at least two ways of handling almost every CNC function.

Setup-related CNC features are no exception. In almost all cases, there will be more than one way to deal with any given problem or situation. However, control manufacturers and machine tool builders tend to teach only one way of handling setup-related features. They tend to pick the easiest or safest techniques to teach in their machine courses.

It is likely that your special setup needs are not a perfect match for the setup-related procedures your machine tool builder teaches. Your machine tool builder's generic methods may get setups made, but they may not be made as efficiently as possible. Worse, you may not even realize that another method exists for minimizing your setup time.

Many of the specific techniques we offer in the next chapter stray from those commonly taught by machine tool builders and control manufacturers in their standard machine courses. Yet all of the techniques we show can be adapted to almost any CNC control on the market today. Some of the techniques require significant changes in the way you do things. Some even require parameter changes that modify the way your CNC control functions.

In all cases, when you are in doubt about how to incorporate the changes we suggest, be sure to seek the help of your machine tool builder or control manufacturer. You may find a willing, knowledgeable, and experienced applications engineer (although your first telephone call may *not* locate such a person) who will help you incorporate your desired changes. Once your particular setup problems are understood, the applications engineer may even be able to improve on the suggestions made in this text.

Chapter Eight

Setup Time Reduction Techniques

W ith a firm understanding of setup time reduction principles, you should be able to easily grasp the specific techniques described in this chapter. Keep in mind that no matter how lengthy this chapter may be, there is no way that we can cover every potential technique that can reduce setup time on CNC equipment. While our primary objective is to describe as many CNC machine setup time reducing techniques as possible, our secondary objective is to get you thinking about how you can overcome your own specific setup-related problems. If you follow our recommendation to videotape your own people making setups, we can almost guarantee that you will come up with several ideas on your own as to how your setups can be improved. This experience, combined with the knowledge of our techniques, should get you started with your company's setup time reduction program.

Our presentation is organized in the same general order in which setups are commonly made. At each step of the setup process, we discuss common pitfalls and offer suggestions as to how improvements can be made. For each suggestion, we make it clear whether the technique eliminates an on-line task, moves a task from on line to off line, or facilitates an on-line task. Further, at the end of this chapter we arrange all key suggestions into three lists corresponding to the three categories of setup time reduction.

PREPARATION AND ORGANIZATION FOR SETUP

We again urge you to videotape setups being made. If you do, your video will surely document several wasteful periods during your setups caused by something as basic as lack of preparation and organization. If the setup person does not have *everything* needed to complete the setup *before* the last workpiece is machined in the most recent production run, precious setup time will be wasted in a search for needed items. Here we identify several easily avoidable trouble spots.

DOCUMENTATION

Documentation is extremely important in the CNC environment. In fact, Chapter Ten is devoted to showing you important documentation techniques. The better the setup documentation, the easier it will be for the setup person to make the setup and the less chance for confusion and mistakes that waste time. By making setup instructions as clear as possible, you can eliminate (or at least minimize) the on-line time a setup person takes to study and interpret the documentation.

You must ensure that the documentation is brought to the setup person *before* the last workpiece is machined in the current production run, eliminating the on-line time a setup person wastes waiting or searching for setup documentation. Ideally, the setup person should have time to study and interpret the documentation before actually making the setup, especially for complicated setups. This will ensure that he or she is ready to start as soon as the current production run is completed.

Keep in mind that the setup person will require certain production control documentation at the time the setup is made. This documentation includes the workpiece print, routing sheet, and special notes and comments about operations performed prior to the CNC operation.

TOOLING

As you watch your setup videotape, you may find that the operator disappears at times to look for a hand tool, cutting tool, vise, fixture, or gage. Possibly the setup person even lacks something very basic like a wrench or screwdriver. Consider for a moment how wasteful this kind of holdup is. Production on a $100,000 (or more) CNC machine tool is being delayed because the setup person does not have an $8 wrench. Considering the high shop rates for CNC equipment, there is absolutely no excuse for the holdup.

The same problem may exist with cutting tools, fixtures, and gages. In many shops, tools are stored in a tool *crib*. If the CNC machine must be down while the setup person walks to the tool crib and gets the needed tools, a lot of setup time can be wasted, especially if the trip must be repeated for several tools.

One simple and effective way to move the task of locating tools from on line to off line is to prepare a tool *cart* for every setup to be made. All tools related to the setup can be easily listed in the setup documentation. The setup person gathers all tools, including cutting tools, fixtures, gages, and hand tools, and places them on the tool cart while the machine is still in the current production run. Everything will be readily available to the setup person from the moment the last workpiece in the current production run is finished. While some duplicate tooling (especially hand tools) will be needed to implement the tool cart method, this relatively low investment should be very easy to justify.

PERSONAL TIME

As human beings, we all value our own time. CNC setup people are no exception. However, if we compare the cost of a setup person's time to the shop rate of any CNC machine tool, it is quite likely that the CNC machine's rate will be at least three to four times the hourly rate of even your highest paid setup person.

The knowledge that the CNC machine tool's time is more valuable than their own (while not very palatable) should inspire concerned setup people to place a very high priority on getting setups made with a minimum of unnecessary personal time. Though it may be difficult to inspire this behavior in all but the most advanced CNC shops, a great deal of production time can be saved if the setup person postpones all forms of personal business—including phone calls, breaks, personal conversations with fellow workers, and even lunch or dinner—until *after* the setup is finished and production is running.

Keep in mind that an extra penalty is paid while the setup person takes personal time during the setup of a CNC machine that can run unattended. A CNC bar-feed turning center, for example, can be running production even during breaks, lunch, and other employee personal time. This production time is lost if the machine is down for setup during any form of personal time.

PRIORITIZING SETUPS BASED ON MACHINE TYPE

Regardless of how well organized your company is, there will be those unfortunate times when two or more machines are down at the same time, waiting for setup. You can facilitate the entire setup procedure for your setup people by having a firm *pecking order* in place that makes it clear as to the order in which setups should be made. Your criteria for determining which setup gets made first might be based on the machine's shop rate. The most expensive machine gets the first setup. If machines have similar shop rates, your criteria for determining the pecking order may be based on which setup is quickest or easiest, or which workpiece is needed most, or which machines can run unattended.

TAKING ADVANTAGE OF NONPRODUCTION TIME

The best time to be performing setups is when CNC machines would normally be down for other reasons. I call this *nonproduction time*. One example of nonproduction time is when CNC operators are on their own time (during breaks or lunch, for example). Another nonproduction time is off-shift hours when there are no CNC operators to run the machines (overnight or weekends, for example). Tasks normally classified as on line automatically switch to off line if they are performed during nonproduction times. For this reason, many companies stagger the working hours of their setup people to offer the *potential* to make setups during nonproduction times. Some even schedule a special (usually third) shift during which setups can be made with no effect on production.

One company, for example, works its CNC machines for two 8-hour shifts with a half-hour break for lunch each day. Day shift starts at 7:00 a.m. and ends at 3:30 p.m. Evening shift starts at 3:30 p.m. and ends at midnight. This company has its setup people work 9-hour days. The day-shift setup people come in at 6:00 a.m. and leave at 3:00 p.m. Night-shift setup people come in at 3:30 p.m. and leave at 1:00 a.m. Lunch periods are also staggered, giving the potential for another half-hour of nonproduction time for setup per shift. This simple organization technique allows three hours of potential setup time per day during nonproduction times.

Admittedly, there will be days when there are no setups to make, or when the end of a production run does not nicely correspond to nonproduction time setups. Still, this company maintains that any loss in wasted pay to setup people is far outweighed by the benefits reaped from having at least *some* of the setups made during nonproduction time.

TEARING DOWN THE PREVIOUS SETUP

The on-line part of making a setup begins the instant the last workpiece is machined in the previous setup. One of the first things the setup person will have to do is tear down the setup for the previous production run. Almost all CNC machine tools must be down for the entire time it takes to tear down the old setup.

Tasks related to tearing down setups include removing the workholding devices, erasing program zero setting offsets, removing (and disassembling) cutting tools, clearing tool offsets related to cutting tools being removed, downloading the CNC program to a distributive numerical control device (if changes have been made to the CNC program during production), erasing the CNC programs from the control's memory, and cleaning the machine for the new workholding setup.

JOB ORDER PLANNING CAN MINIMIZE SETUP TIME

Each company sets its own priorities for the order in which workpieces are machined. The production control department normally sets the priorities, usually according to just-in-time principles. Those workpieces required at assembly first are commonly the first workpieces to be machined.

Though JIT principles are extremely important, keep in mind that setup time is highly dependent on the order in which jobs are run. For example, if you run similar workpieces, your machining center fixtures may be designed to hold different workpieces in a family of parts. If all jobs related to a given family that require a particular fixture are run consecutively, the time required to remove and replace the fixture between setups can be eliminated. In similar fashion, the need to assemble, measure, load, and input offsets for cutting tools from one setup to the next can be eliminated for jobs that use the same cutting tools. With very close families, it may even be possible to totally eliminate work related to setup.

Even if you do not consider your workpieces to be in a family of parts, you may still be able to minimize the teardown of setups in certain situations with regard to workholding devices and cutting tools. If you use CNC bar-feed turning centers, for example, you can eliminate any adjustments to the bar feeder (changing collets, changing spacers, etc.) if you consecutively run all workpieces required from a particular bar material and size. Similar techniques can be applied to vertical machining center vise setups. If all workpieces that use a particular vise are run consecutively, there will be no need to tear down or set up the work-holding device. These are but two examples of the many times when setup can be minimized with nothing more than careful job order planning.

By working closely with your production control people, you may be able to dramatically reduce setup time with little adverse effect on your company's just-in-time efforts. As you study the workpieces you machine on a given CNC machine, group the workpieces that require the same cutting tools and workholding devices. Supply this information to your production control people. Let them know that if these workpieces in each group can be run consecutively, turnaround time for the workpieces can be dramatically reduced.

MAKING THE WORKHOLDING SETUP

There are many forms of workholding devices used with CNC equipment, and it is not within the scope of this text to describe the pros and cons of each. Our intention in this section is simply to help you shorten the time it takes to make workholding setups requiring only minor changes in your current workholding methods. While we speak in rather general terms, we hope to reveal facets of your workholding setup methods that allow room for improvement.

MOVING THE WORKHOLDING SETUP OFF LINE FOR MACHINING CENTERS

It is quite common for workholding setups to be made right on a CNC machine while it is down between production runs. On a vertical machining center, for example, the machine may be down while the setup person mounts a vise, in-dexer, or fixture to the table. Making on-line workholding setups for devices that can be simply mounted to the table is bad enough. In worse situations, your setup person may have to make very tedious and time-consuming block setups.

Depending on the complexity of the setup, a great deal of time may be saved if you can move the task of making workholding setups off line. For complicated vertical machining center setups, one way to accomplish off-line workholding setups is to utilize subplates. At least two subplates will be required to take advantage of this technique. The subplates, usually about 1 to 3 inches (25 to 76 mm) thick and slightly smaller than the machine table, will be machined in such a way that the subplate simulates the machine's table top. Either key slots or a combination of dowel holes and tapped holes will be machined, to work as location and clamping sites for workholding devices. The subplate will be keyed so that it can be quickly and accurately mounted to the table top.

The workholding setup will be made on the subplate instead of on the machine's table. This can, of course, be done off line, while the machine tool is currently running workpieces. Once made, the subplate will be mounted to the machine table for production. While production is being run, the setup person can be making the workholding setup for the *next* production run. Between production runs, on-line workholding setup time will be limited to the length of time it takes to remove the old subplate, clean the table top, and mount the new subplate.

The amount of time saved by incorporating subplates will vary with the complexity of the setup and the size of the subplate. Generally speaking, subplates are most helpful when you are making rather complicated setups. If, for example, the workholding setup requires holding 10 workpieces (in 10 vises), it is quite likely that the subplate can be mounted to the machine table much faster than can 10 individual vises. On the other hand, there will be times (for simple setups) when the time it takes to change subplates may be as long or longer than making the setup right on the machine's table. Additionally, obstructions, like the headstock and coolant guarding, may make it difficult to remove and load subplates directly from the machine's table.

Another similar method to move the workholding setup off line is to use a *pallet changer*. Pallet changers can be manual or automatic and shuttle the entire workholding setup in and out of the machining area. While one workpiece is being machined, the machine operator can safely load the next. While pallet changers are commonly purchased to move the task of workpiece loading off line (make workpiece loading internal to an individual machining cycle), they also allow workholding setups to be moved off line.

To reap the setup-time-reducing benefits of pallet changers, additional pallets are purchased. During a production run, the setup person can make the setup required for the next workholding setup on an open pallet, moving the task off line. At the completion of a production run, the pallet in the loading station can be safely and easily removed from the loading station of the pallet changer. Since there will be no obstructions, removal of the pallet will be very easy compared to removing a subplate from the table top (on most vertical machining centers).

The pallet containing the new workholding setup can be placed in the loading station with the same relative ease. And of course, all of this can be done even while a workpiece is still being machined, meaning the on-line task of making the workholding setup can be completely moved off line.

If your CNC machining centers do not currently have pallet changers, keep in mind that there are a number of aftermarket suppliers that can provide them. The more setups you make, the easier it will be to justify their cost. Most aftermarket pallet changers are manually activated; that is, the operator or setup person must manually perform the pallet change. Though not as convenient as totally automatic pallet changers, most are still quite efficient and easy to operate.

Other machining center suggestions

As you study your setup people making workholding setups, be alert for ways you can help them. Here are just a few suggestions to get you thinking.

All workholding devices should be keyed to the table, to eliminate the need to square the device to the table with a dial indicator or other aligning tool. Squaring devices to the table takes time. Along the same lines, but not actually part of the workholding setup, if workholding devices are not keyed in some manner, time-consuming measurements must be taken to find the position of the workholding device for the purpose of assigning program zero.

Block setups on machining centers should be avoided unless you can make setups off line. Block setups are the most difficult and should be made only as a last resort. Unfortunately, many contract shops must make block setups because of the expenses related to purchasing more dedicated tooling. For any repeating jobs, however, strong consideration must be given to making a fixture, easing the on-line task of making the workholding setup. In similar fashion, possibly a standard vise can be made to hold the workpiece if machined soft jaws are used.

Manual clamping should be done using table slots with socket-head bolts. Keys in key slots are easier for the setup person to work with than dowel holes and pins. Socket-head bolts (Allen type) are easier and faster to deal with than hex-head bolts.

FACILITATING TURNING CENTER SETUPS

By the very nature of most turning centers, it may be impossible to move the on-line task of making the workholding setup off line. For this reason, the suggestions we offer for turning center holding setups fall into the category of facilitating the on-line tasks, making it as easy and efficient as possible for the setup person to make setups.

Turning center workholding setup usually requires work on some form of *chucking device*. The most common forms of chucks include 2-jaw and 3-jaw chucks (usually pneumatic or hydraulic), collet chucks (commonly used for bar-feed applications), and index chucks (used to rotate the workpiece during the machining cycle). Additionally, there may be some form of work *support* device that works in conjunction with the workholding setup (bar feeder, tailstock, or steady rest).

When the CNC turning center is originally purchased, one of the most important choices made will be which form of workholding devices (3-jaw chuck, collet chuck, etc.) best suits your company's needs. Once the machine is in place, the workholding device is commonly left on the machine permanently. While there are companies that incorporate more than one workholding device on their turning centers, most companies do not require the setup person to change the workholding device from setup to setup.

There are some exceptions to this general statement. A company that splits the application of the turning center evenly between bar-feed work and chucking work will need to have the machine set up to run as efficiently as possible for both types of work. It will commonly use a collet chuck for bar-feed work and a 3-jaw chuck for chucking work. This is another case where wise job order planning can minimize setup time. If production control can schedule a large group of bar-feed or chucking jobs to run consecutively, much chuck changing time can be saved.

Tooling manufacturing designs vary when it comes to the ease of manipulating the workholding device. Collet chucks, for example, are usually very easy to manipulate. The collets within collet chucks can usually be changed in less than two minutes (with the proper hand tools).

Three-jaw chucks, on the other hand, can be rather difficult to work with. Some form of *top tooling*, usually in the form of hard jaws or soft jaws, must be mounted to the chuck from one setup to the next. Depending on the 3-jaw chuck manufacturer, mounting top tooling can range from quick and easy to difficult and time-consuming. Most 3-jaw chucks require each jaw of the top tooling to be located on the chuck in exactly the same manner. Serrations in the top tooling must be matched to those on the master jaw of the chuck. With some chucks, it can be very time-consuming for beginning setup people to match serrations.

One form of 3-jaw chuck that dramatically reduces the amount of time it takes to manipulate top tooling is the *quick-change 3-jaw chuck*. Offered by several chuck manufacturers, this form of chuck allows top tooling to be easily changed. Most such chucks allow each jaw to be positioned in less than 30 seconds. While quick-change chucks tend to be more expensive than conventional chucks, you can dramatically facilitate the setup person's ability to mount top tooling by utilizing this form of chuck.

Keep in mind that mounting the top tooling to the chuck may be only part of the top tooling preparation. While some top tooling (like hard jaws) can be used as soon as it is mounted, other top tooling (like soft jaws) must be machined to the diameter it will be gripping every time it is used. Machining soft jaws must be done on line, since the concentricity of the workpiece to be machined depends on the quality of the soft-jaw machining operation. Since the machining of soft jaws can be quite time-consuming, anything that can be done to facilitate this task will effectively reduce setup time.

Before soft jaws can be machined, they must be properly mounted to the chuck. In addition to being placed in the proper serrations, soft jaws will be closed on a chucking ring (as shown in Figure 8-1) at about the halfway point of the jaw stroke. In order to ensure good concentricity, the clamping pressure should be set the same when boring jaws as when actually clamping on the workpiece.

At this point, many setup people machine soft jaws completely by hand. They manually start the spindle, position the jaw machining tool close to the soft jaws, and use the handwheel to machine the jaws. Depending on how much material must be machined to bring the soft jaws to the desired diameter, this can take a great deal of time.

This task can be done much faster and made much easier with some simple programming. Instead of manually boring jaws, your setup people can take advantage of certain CNC features (including canned cycles, multiple repetitive cycles, and even parametric programming functions) to make it very quick and easy to machine soft jaws.

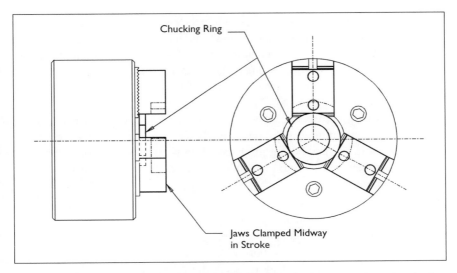

Figure 8-1. Detail showing how soft jaws should clamp on chucking ring prior to jaw boring.

Here is an example program that bores soft jaws 1.000 inch (25 mm) deep from a diameter of 3.000 to 4.000 inches (76 to 102 mm). While it is written for one popular type of CNC control, you should be able to modify it easily for any control type. The program zero point for this program is the center of the spindle in the X axis and the face of the jaws in the Z axis. Prior to running this program, the setup person places the boring bar in turret station 12 and confirms that the program zero assigning numbers are properly specified for the tool (commonly in *geometry offsets*). Then the operator manually positions the boring bar tip within about 0.5 inch (13 mm) of the jaws to be machined.

Program:

O0001 (Program number)
N005 T1212 (Index to boring bar station and instate offset)
N010 G96 S300 M03 (Start spindle at desired speed in surface feet per minute [sfm])
N015 G00 X3.0 Z0.15 (Rapid to clearance position)
N020 G71 P025 Q035 U-0.01 W0.005 D0.100 F0.010 (Rough bore)
N025 G00 X4.0 (Finish pass definition begins)
N030 G01 Z-1.0 (Feed to Z depth)
N035 X3.0 (Feed back to beginning diameter)
N040 G70 P025 Q035 F0.005 (Finish bore)
N045 M30

This program uses a feature called *multiple repetitive cycles* to take 0.100-inch (2.54-mm) deep passes to completely rough bore the jaws (based on line N020).

The jaws are finish-bored in line N040. While the programming format will likely change for your particular machines, you should be able to see how easy it can be to program the soft-jaw boring operation. Also, if this program is left in the control, it can be easily modified whenever jaws must be bored. Only four words in the program need be changed (the X of line N015, the X of line N025, the Z of line N030, and the X of line N035).

Your jaw-boring program can be made even easier to use if your machine has parametric programming capabilities. This function allows variables to specify certain words in the jaw-boring program. Here is another jaw-boring program written using Custom Macro B techniques (the most popular version of parametric programming).

```
O0002 (Program number)
#101 = 3.000 (Starting diameter to bore)
#102 = 4.000 (Ending diameter to bore)
#103 = 1.000 (Z depth of jaw boring)
#104 = 0.100 (Depth of cut for rough boring)
#105 = 300 (Speed for boring in sfm)
N005 T1212 (Index to boring bar station and instate offset)
N010 G96 S#105 M03 (Start spindle at desired speed in sfm)
N015 G00 X#101 Z.15 (Rapid to clearance position)
N020 G71 P025 Q035 U-0.01 W0.005 D#104 F0.010 (Rough bore)
N025 G00 X#102 (Finish pass definition begins)
N030 G01 Z-#103 (Feed to Z depth)
N035 X#101 (Feed back to beginning diameter)
N040 G70 P025 Q035 F0.005 (Finish bore)
N045 M30
```

With this program, the setup person simply changes the values of variables #101 through #105 at the beginning of the program in order to specify the new criteria for jaw boring and runs the program. Jaws will be bored to the diameter specified from the starting diameter using the specified depth of cut.

Two more points about using programmed jaw boring as opposed to manually boring jaws:

1. Jaws will be bored under optimum (and consistent) cutting conditions if a program is used. When jaws are bored completely manually, many times the setup person will barely *tickle* the jaws, making for very inefficient cutting conditions.

2. With programmed jaw boring, you are able to make movements not easily made when manually boring jaws. For example, boring jaws to hold on a tapered surface is easily possible through programming techniques. Boring taper jaws completely manually is next to impossible.

Work support devices on turning centers

The device that actually clamps the workpiece (3-jaw chuck, collet chuck, etc.) may not be the only device in the turning center workholding setup. In shaft applications, for example, a tailstock provides support for the end of the workpiece opposite the chuck. Depending on the length of the workpiece, one or more steady rests may also be required to provide support in the middle of the workpiece.

As with any lathe that uses a tailstock, the tailstock center must be aligned with the spindle centerline to prevent taper in the workpiece. Some form of manual adjustment within the tailstock will always be available to adjust for taper.

Unfortunately, once a turning center is purchased there is not much that can be done to facilitate the on-line task of tailstock adjustment to eliminate taper. Since turning center suppliers differ greatly on how easy it is to adjust the tailstock (and how often it must be adjusted), you must confirm what will be involved with the tailstock's taper adjustment *before* the machine tool is purchased.

With some machines, once the tailstock is placed in one position along the Z axis and adjusted, it will remain perfectly centered throughout the entire length of its travel. For this kind of machine, the setup person can easily go from one setup to another with absolutely no concern for the tailstock's alignment. This type of machine eliminates the on-line task of taper adjustment (until a mishap, like the turret's crashing into the tailstock center). On the other hand, there are turning centers that require the setup person to check and adjust for taper every time the body of the tailstock is repositioned.

Adjusting the tailstock for taper can be a tedious and time-consuming procedure. It usually involves trial turning a shaft and using a dial indicator to gage the position of the tailstock center. The actual movement of the center is commonly done with a rather crude set-screw adjustment.

Unfortunately, if a turning center is purchased without concern for the ease of tailstock usage, a great deal of setup time may be wasted over the life of the machine tool. If you will be performing shaft work requiring tailstock support, you *must* confirm that your intended machine's tailstock will remain in alignment throughout its travel.

When the length of the shaft exceeds about five to six times its diameter, chuck and tailstock support will not be enough to ensure rigid machining, especially in the middle of the workpiece. Steady rests are used to provide additional support for very lengthy workpieces. As with tailstocks, steady rests vary with regard to how easy their position and diameter can be changed. This is another function you will need to confirm before the machine and steady rest combination is purchased, as there is little that can be done to facilitate on-line tasks related to steady rest adjustment after the machine is in place.

We also categorize bar feeders as work-support devices. Used to provide unattended operation for the entire length of the bar (and possibly even several bars), bar feeders must be considered part of the workholding setup. As with tailstocks and steady rests, bar feeders vary dramatically in ease of going from one bar size to another.

Bar feeders that allow only one bar size at a time require on-line setup tasks. At the very least, some form of spacers and/or retainers must be changed when changing bar sizes. This, of course, is an on-line task that must be done while the machine is down between production runs.

The on-line tasks related to bar diameter changes can be either moved off line or eliminated by purchasing more expensive multiple bar feeders. This kind of bar feeder incorporates several bar tubes and allows several different bar diameters to be held. If you machine a limited number of different bar diameters (no more than five different diameters), you can completely eliminate the on-line task of bar feeder changeover with this kind of bar feeder. When going from one bar size to another, the setup person simply rotates the bar feeder's turret to position the proper bar tube into the feeding position. If you machine more bar diameters than are allowed by the multiple bar feeder, you will at least be able to move the task of bar feeder changeover off line. Most multiple bar feeders allow the setup person to change spacers within the bar feeder for one tube while the machine is in production.

CUTTING TOOL ISSUES

Cutting tools are usually dealt with early in the setup procedure. Here we offer suggestions on how the proper handling of cutting tools can reduce setup time.

ORGANIZING TECHNIQUES

Just as getting organized can help the setup person make the workholding setup, cutting tool organization goes a long way toward minimizing cutting tool setup time. Several common-sense organization techniques can help the setup person avoid performing redundant tooling-related tasks from setup to setup.

For example, if you utilize standard tooling stations for your most often-used cutting tools, you can minimize the on-line task of removing and reloading tools from one setup to the next. For tools that remain in the machine from setup to setup, you also eliminate the related tool measuring and offset-entering tasks.

Because of the similarity of operations required for turned workpieces, turning centers usually allow users to easily incorporate standard tool stations. For example, say station 1 is designated as the rough turn-and-face standard tool station and station 2 is designated as the finish turn-and-face standard tool station. Since almost all turned workpieces require rough-and-finish turning, these tools will probably *never* be removed from the turret. In similar fashion, station 3 can be designated as the grooving tool and station 4, the threading tool. While not all workpieces require grooving and threading, at least you can minimize the *potential* for having to remove and reload tools. Any drill might always be placed in station 5, the rough boring bar in station 6, and the finish boring bar in station 7. Stations 8 through 12 may be reserved for special tools that change often. With this utilization method, tool loading and the related adjustments will be minimized (in some cases eliminated) at least for stations 1 through 7.

Similar techniques can be used for machining centers, and especially for machines that have large tool magazines. Commonly used tools, like center drills, drill and tap combinations, end mills, thread mills, and face mills can be assigned a permanent tool station number. This minimizes the number of tools that must be loaded (as well as the related tool length and cutter-radius compensation offset entry) during setup.

Clearly, you can eliminate a great amount of setup time with standard tool stations. Since contract shops may not be able to predict what tooling will be required for future jobs, the benefits they gain from utilizing standard tool stations may be limited to what has been presented thus far. However, product-producing companies having a set number of different workpieces to machine can take this standard tool station concept further.

By analyzing the group of workpieces to be machined by a given CNC machine tool, you can come up with all the cutting tools required by each CNC machine. You can then determine if you are using similar tools that can be combined to minimize the number of different cutting tools required to machine your set of workpieces.

You may, for example, find that one of your machining centers uses many different end mills, when just a few can handle the machining of all workpieces. Or you may find that a given turning center is using three different finish turning tools when just one would function properly in all cases. While you must consider potential problems, like tool interference or lack of rigidity, the fewer tools used, the better the chances will be that cutting tools will not have to be changed from one setup to the next. In extreme cases, you may even find that minimizing the number of cutting tools needed will allow you to fit *all* tools required to machine your workpieces into the machine's magazine or turret.

CUTTING TOOL DOCUMENTATION

For most tooling, the actual cutting tool itself is usually only part of the complete cutting tool assembly. With machining center cutting tools, for example, the cutting tool is commonly made up of a cutting tool, a tool holder, an adapter, an extension, and possibly other components. Some turning center tooling also requires this kind of assembly prior to placement into the turret. The method by which cutting tools are assembled can have a dramatic impact on the machining operations they perform.

Many CNC programmers limit the amount of cutting tool documentation they provide to the cutting tool's description and the tool station number in which the tool is to be placed. Figure 8-2 shows an example of a typical tool list as supplied by the programmer.

Tool List For Machining Centers			
Station	Tool Description	CLC	CRC
1	#3 Center Drill	1	
2	13/64 Drill	2	
3	1/4-20 Tap	3	
4	3/4 End Mill	4	34
5	3/4 End Mill	5	35
6	4" Face Mill	6	
7	4" Face Mill	7	
8	27/64 Drill	8	
9	1/2-13 Tap	9	
10	31/32 Drill	10	
11	.5005 Reamer	11	
12	1", .0833 Pitch Thread Mill	12	42
13	3/4 Drill	13	
14	1.25 Drill	14	
15	2.0005 Boring Bar	15	
16	1" Counter Bore	16	
17			
18			
19			
20			
21			
22			
23			
24			
25			

Figure 8-2. Tool list commonly used for machining center setups. Note that a simple tool list is not very informative and assumes the setup person knows what tool holders and other tool components to use.

While this list may be sufficient to relate which tools belong in the various machining center stations, and while it does specify the related offsets for tool length compensation and cutter radius compensation, it does nothing to specify the components required to assemble the tool. Additionally, it does not allow for any special considerations required of a particular control. For example, most setup people, when seeing this list, will try to assemble cutting tools in a way to make the tools as short as possible. Short tools are rigid tools, capable of more powerful machining. However, one or more of the tools on this list may be expected to reach into a cavity in order to machine a given surface. Depending on the depth of the cavity, there may be an interference problem if the tool is kept *too* short.

If this problem goes undetected, it will cause wasted program verification time (on line) as the first workpiece is being machined. The machine will be down during the time it takes the setup person to extend the tool to a sufficient length.

This problem can be avoided only through good documentation of the cutting tools. One way to improve a tool list is to include all components that make up each tool right on the tool list. Figure 8-3 shows an example of this kind of list. Others make a special tool drawing for each tool that graphically illustrates each component. Figure 8-4 is an example of this kind of drawing. Regardless of the

Tool List For Machining Centers

Station	Tool Description	CLC	CRC	Min Length	Tool Assembly Components
1	#3 Center Drill	1		7.500	H1248, A2387. E2355, C2312
2	13/64 Drill	2			H1248, A2334, C3422
3	1/4-20 Tap	3			H1250, A2339, TC2355
4	3/4 End Mill	4	34	7.500	H2323
5	3/4 End Mill	5	35	7.500	H2323
6	4" Face Mill	6			H3433
7	4" Face Mill	7			H3433
8	27/64 Drill	8			H1248, A2332, C2344
9	1/2-13 Tap	9			H1250, A2335, C3425
10	31/32 Drill	10			H1248, A2397, C3233
11	.5005 Reamer	11			H1223, A2332
12	1", .0833 Pitch Thread Mill	12	42		H2344
13	3/4 Drill	13		6.000	H1249, A2390, C1000
14	1.25 Drill	14		6.000	H1251, A2392, C1250
15	2.0005 Boring Bar	15			H1266
16	1" Counter Bore	16			H1224, A2335
17					
18					
19					
20					
21					
22					
23					
24					
25					

Note: Minimum length is not critical unless specified.

Figure 8-3. In this more complete tool list, component identification numbers are specified to ensure the setup person's understanding of each tool's setup.

method of documentation, by one means or another the programmer must make it very clear to the setup person how each tool must be assembled, to avoid confusion at the time of setup and downtime at the time of program verification.

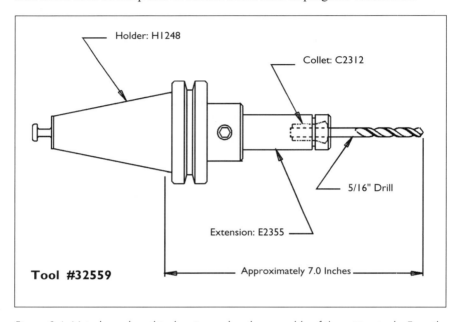

Holder: HI248

Collet: C2312

5/16" Drill

Extension: E2355

Tool #32559

Approximately 7.0 Inches

Figure 8-4. Note how clear this drawing makes the assembly of the cutting tool. Even the approximate overall tool length after assembly is shown.

MOVING CUTTING-TOOL TASKS OFF LINE

A great potential for reducing cutting-tool setup time rests in doing as much as possible *off line*. Ideally, the only on-line task should be the act of loading the cutting tools into the machine's magazine, carousel, or turret. The goal will be to assemble, take any necessary measurements, and even enter offset data for cutting tools completely off line. In the real world, this goal may be somewhat difficult to achieve, though not impossible.

Assembly

The benefits of performing certain cutting tool tasks off line should be obvious. The cutting tool's assembly, for example, if done off line, will eliminate a great deal of setup time. Though this is the case, there are still those companies that have their setup people assemble cutting tools on line. They do not *begin* assembling cutting tools until the last workpiece in the previous production run has been machined. There are three reasons commonly given as to why companies do not assemble cutting tools off line:

1. Many companies do not own enough tool holders and other cutting tool components to assemble tools off line. Admittedly, assembling tools off line can require a duplication of many cutting-tool components, since it is likely that at least some tool components needed for the next setup will be in use during the current production run. However, when you consider how much production time can be saved by assembling cutting tools off line, and if you accept the fact that your machine's downtime can be directly tied to a loss in money, it should not be difficult to justify the purchase of duplicate cutting-tool components.

2. As mentioned in the previous chapter, all off-line tasks require that personnel be available to perform the off-line tasks. If your company utilizes only one person to perform all CNC-related functions, it is possible that this person will not have time to be assembling tools during a given current production run. Also as stated in the previous chapter, if your company has several CNC machine tools, it may not be difficult to justify the hiring of a person whose sole responsibility will be to perform off-line setup-related tasks.

3. It is possible that the production run is so short that there will be no time to assemble cutting tools for the next setup while running production. During a run of only three workpieces with a 2-minute cycle time, for example, the machine will be in production for only six minutes. In this case, it is unlikely that anything substantial can be done to get ready for the next setup; therefore, assembling tools on line may be the only alternative.

While it is commonly thought of as a machining center function, keep in mind that assembling tools for turning centers off line can also minimize setup time. Inserts can be placed into insert holders. Insert holders, in turn, can be placed into tool holders. Drills can be placed into bushings, and bushings mounted into tool holders. There is no turning center cutting tool that does not require at least some assembly. If assembly is done off line, the only on-line task left to be done will be mounting the tool holder into the turret.

Additionally, more and more turning centers are being equipped with *quick-change* tooling. Quick-change tooling for turning centers commonly has shanks that resemble the tapered shanks in machining center cutting tools. To replace the entire tool holder in the turret of a turning center that has quick-change tooling, the setup person simply presses a button to release the clamp holding the tool holder in the turret. The tool holder is easily removed and the new holder placed in the turret station. The button is pressed again to clamp the tool holder into the turret. This method resembles the method by which tools are manually placed in and removed from the spindle of any CNC machining center.

Quick-change tooling for turning centers provides many benefits over conventional tool holders in regard to setup time. Each conventional tool holder may require one to five minutes or more for replacement (depending on the turret design and the skill of the setup person), while quick-change tools can be easily changed in less than 15 seconds. For most applications, when you consider how many times a setup person will have to change tool holders, the additional cost of quick-change tooling should be very easy to justify.

Another turning center feature that is becoming more and more popular among turning center manufacturers is an automatic tool changer that resembles the automatic tool changer of a machining center. A magazine holding any number of (quick-change) tools supplies tools for the automatic tool-changing system. At any time, a tool in the magazine can be exchanged with a tool in the turret. While this kind of automatic tool-changing system is usually purchased to allow more tools than the machine can hold in its turret, the automatic tool changer also offers a major benefit when it comes to setup time reduction.

With an automatic tool-changing system on your turning center, almost all tasks related to tool assembly and loading can be moved off line. Since many applications for automatic tool changers on turning centers do not actually require that tool changes be made during the machining cycle, a setup person can be loading tools into the magazine while production is running. The only on-line time for tool loading will be the time it takes for the automatic tool changer to transfer the cutting tools into the turret (usually less than 10 seconds per tool). Moreover, because it expands the number of tools the turning center can accommodate, this kind of tool changer also improves the benefits related to *standard tool station* techniques discussed earlier in this chapter.

When it comes to loading tools into the magazine of a machining center, keep in mind that most machines are interlocked in such a way that loading tools during production is not possible. For safety reasons, most machine tool builders do not allow the setup person to rotate the tool magazine while the machine is in cycle. For machines that have this kind of interlock, the physical act of tool loading must be done on line.

Measuring and determining cutting-tool offsets

As you look for ways of moving on-line tasks off line, one of the first things you should consider is how *any* measurements taken during setup (on line) can be done up front. Measurement of cutting-tool offset is no exception. Given an adequate number of people and production quantities that allow the time required to perform on-line tasks off line, you should be able to eliminate *all* on-line measurements associated with cutting-tool offsets.

The degree of difficulty related to attaining this goal varies with the machine type. For machining centers, it is relatively easy and inexpensive. In fact, if you have the people and time during production runs to measure tool offset values for machining centers off line, there is *no excuse* for not doing so. When it comes to measuring cutting-tool lengths, for example, a CNC machining center makes a very expensive height gage. Turning centers tend to present more of a challenge. It will be next to impossible to determine wear offset values up front (offsets associated with sizing the workpiece and allowing for tool wear). However, tool-nose radius compensation offsets are commonly known prior to making the setup. Additionally, the program zero setting offsets (commonly called *geometry offsets*) can be determined up front with a little effort.

Measuring tool-length compensation offsets off line. There are two popular ways to utilize tool length compensation. With one method, the tool's length (a positive value) is used as the tool offset value. With the other, the distance from the tip of the tool (while the machine is resting at its reference position) to the program zero point is used as the tool-length compensation offset value. Figure 8-5 shows a comparison of the two methods.

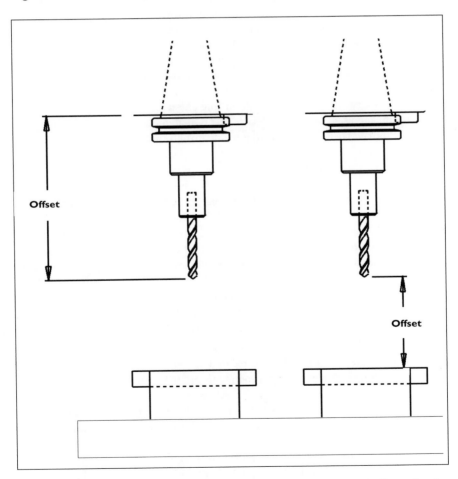

Figure 8-5. This illustration details the two most popular ways of using tool offsets related to tool length compensation.

Using the tool's length as the offset value offers several advantages, including three related to setup time reduction:

1. For times when a tool is used from one production run to the next, the tool-length compensation value will not change, whereas if the distance from the tip of the tool to program zero is used as the offset, the offset value will change for every production run (based on the new position of the program zero point in Z). Using the tool length is especially helpful when you utilize standard tooling stations; you can eliminate the on-line task of measuring tool-length compensation offsets for tools used from one production run to the next.

2. Offset values will remain the same from one machining center to another. If you have two or more machining centers that accept the same tool shank configuration (CAT-40 V flange, CAT-50 V flange, BT-40, BT-50, etc.), tools can be easily exchanged from one machine to another without requiring new tool-length compensation measurements. This can be very helpful with cutting tools used on several similar machines that tend to wear out quickly (rough milling cutters, rough boring bars, etc.). If you keep several identical tools assembled, measured, and ready to go, any tool can be quickly loaded into any machining center. The tool-length compensation value, measured off line, will be correct regardless of which machine receives the tool.

3. Most important, using the length of the tool as the offset value makes it much easier to move the task of measuring tool-length compensation values off line. It is much easier to determine the length of the tool than some arbitrary distance that is based on the machine tool and the workholding device. The length of the tool can be easily measured with a height gage or a special measuring tool specifically designed to measure tool lengths. Figure 8-6 shows a simple height gage being used to measure a tool's length off line. Determining the distance from the tip of the tool to program zero would require knowing some critical machine dimensions related to the setup as well as each tool's length.

Once the tool-length compensation value for each tool is determined, it must eventually be entered into the CNC control. One obvious way to do this is to have the setup person simply write down each tool-length compensation value for each tool at the time of measuring. This information can be used at the time the setup is actually made (on line). However, the act of actually entering tool-length compensation values is tedious, error-prone, time-consuming, and wasteful (since it is being done on line). Additionally, mistakes made in writing and entering tool-length compensation offset values can be disastrous.

Since most current-model CNC controls allow offsets to be entered through programmed commands, you can eliminate the on-line task of manually entering tool offsets into the control altogether if you develop a CNC program to enter

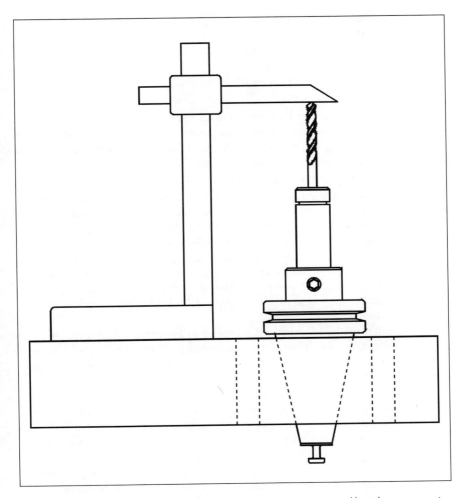

Figure 8-6. Even a simple height gage can be used to measure tool-length compensation values off line if the length of the tool is used as the offset value.

offsets. Many controls use G10 for the purpose of entering offset values. Here is an example program written in the correct format for one popular CNC control that uses G10 to set the tool-length compensation values for offsets 1 through 10.

O0001 (Program number)
G10 L1 P1 R5.3664 (Set offset 1)
G10 L1 P2 R3.4532 (Set offset 2)
G10 L1 P3 R8.7653 (Set offset 3)
G10 L1 P4 R6.8655 (Set offset 4)
G10 L1 P5 R6.4556 (Set offset 5)

G10 L1 P6 R6.5455 (Set offset 6)
G10 L1 P7 R4.6655 (Set offset 7)
G10 L1 P8 R8.7655 (Set offset 8)
G10 L1 P9 R6.7786 (Set offset 9)
G10 L1 P10 R4.5677 (Set offset 10)
M30

For this particular control's format, the L1 specifies that you are setting tool offsets (as opposed to fixture offsets or parameters). The P word specifies which offset you are entering (P1 for offset 1, P2 for offset 2, and so on). The R word specifies the actual value being placed into the tool offset.

The tool offset program can be created in two ways.

- One is to have the setup person who measures tool lengths actually type it in the text editor of a personal computer close to the tool setting gage. With today's text editors, after one offset setting program is created, the setup person will be able to use it to easily make others; therefore, creating this program is easier than it may first appear. It will actually take less time than it would to enter the offsets into the CNC control. However, this manual method of creating the offsetting program does not overcome the potential for human error. A tool offset value could still be entered incorrectly.

- The other method (which is also easier to implement than it may first sound) is to have your tool-length measuring gage output the tool-length compensation value directly to the personal computer (via the serial port of each device). Many current inspection gages, including height gages and gages designed specifically for tool length measurement, have a serial port specifically for this purpose. (By the way, it is this same serial port that allows easy data gathering for statistical process control [SPC] systems). A software program within the computer will take the tool length value coming from the gage and automatically place it in a G10 command of a CNC program. Some tool-length measuring gage manufacturers can supply the software program to collect tool-length compensation values and create the corresponding CNC program. Unfortunately, unless you purchase this kind of tool-length measuring gage, you will have to develop the computer program that creates the tool offset setting CNC program on your own.

Regardless of how the CNC program that enters tool-length compensation offset values is created, it can be easily transferred into the control through any distributive numerical control device. The procedures to do so are usually identical to those used for normal CNC program transfers. Once loaded, the tool offset setting program is run once to set the offsets.

With these techniques, the on-line time spent getting tool-length compensation offsets entered will be reduced to the time it takes to load the offset setting program and execute it once. As we discuss the task of program loading later in this chapter, you will see that today's automatic DNC systems allow CNC programs to be loaded in well under 30 seconds. Executing the offset setting program will require a few more seconds. Unless there are only a few cutting tools needed in the setup, this time will usually be much less than that required to enter offsets manually, meaning you can dramatically facilitate the on-line task of entering tool offsets with these techniques. (We have not mentioned the fact that manual intervention, including the related potential for mistakes, has been eliminated from the tool-length compensation offset setting task.)

Determining cutter-radius compensation offset values. Since a tool-length compensation offset must be used for every tool in every program, it is extremely important to make it easy for the setup person to enter them quickly. Though they are important, cutter-radius compensation offsets are used only for milling cutters, and only when milling on the periphery of the cutter. If you do very little contour milling, you may have little interest in the task of entering cutter-radius compensation offsets into the control.

The more contour milling you do, however, the more important it is to streamline the setup person's procedure for entering cutter-radius compensation offsets. Keep in mind that the same techniques shown for programming offset setting commands can be used to enter cutter-radius compensation offsets. If the setup person simply adds a few commands to the tool-length compensation offset setting program to specify the radius of each contour milling cutter, the on-line task of entering cutter-radius compensation offsets can also be eliminated.

Wear offsets on turning centers. Turning centers also require the entry of offsets. As with tool length compensation on machining centers, every tool in every program uses a wear offset. Wear offsets are needed for three reasons:

1. They allow the turning center operator to adjust for sizing problems caused by imperfections in tool placement. Wear offsets help the operator avoid physically moving the tool in the turret to allow for an imperfectly positioned tool.
2. Wear offsets help the operator allow for tool wear, keeping the surfaces machined by the tool to the proper size.
3. Wear offsets used in conjunction with trial machining techniques ensure that the first workpiece in the production run will be machined to size.

Unfortunately, it is nearly impossible to predict the values needed for wear offsets until the first workpiece is machined. Theoretically, if the tool is perfectly positioned in the turret, and if the program zero setting values for the tool are perfectly correct, the tool *should* machine the workpiece perfectly the very first time. However, the dynamics involved with machining—including tool pressure, amount of material being machined, and the rigidity of the tool and workholding setup—all influence the size to which a tool will machine the workpiece. Even tool touch-off probes used to help with the setting of program zero take static measurements that do not take into account the dynamics of machining.

When it comes to adjusting wear offsets for cutting tools, the best advice we can give is to utilize the standard tool station techniques discussed earlier. If a tool used from one production run to the next is not removed from the turret, the wear offset for that tool should not need to be changed. There are two possible exceptions. One is when there is a dramatic difference in workpiece material from one production run to the next. Possibly one production run machines an aluminum workpiece and the next machines tool steel. Material characteristics will likely require changes in wear offset settings (and probably insert material grade) because of differences in tool pressure. The second exception is when there is a major difference in the rigidity of the workholding setup. Again, tool pressure plays an important role in the values placed in wear offsets and changes with the rigidity of cutting tools and workholding setups.

More setup time reduction techniques related to wear offsets are given during the discussion of program verification. You will see that there are many things a CNC programmer can do to facilitate the setup person's ability to correctly set wear offsets, while ensuring that the first workpiece in the production run comes out to size.

Tool-nose radius compensation offsets. When using tool nose radius compensation in a turning center program (usually specified by G41 and G42), the setup person must at least enter the tool-nose radius value. Depending on the control model, the setup person may also enter a special code number specifying the type of tool being used (turning tool, boring bar, etc.).

Remember that most turning center programmers *specify* the size of the tool nose radius to be used. In this regard, the reason for using tool-nose radius compensation differs from that for using cutter radius compensation on a machining center. A machining center programmer may not be able to predict the exact size milling cutter to be used, and allows for a range of cutter sizes. While different tool-nose radius values can be used with turning center tool-nose radius compensation, cutting conditions (spindle speed, feed rate, depth of cut, etc.) will also change when the tool-nose radius value is changed. For this reason, when a given tool becomes dull and is replaced, it will be replaced with another tool having *exactly* the same radius. This, of course, means the programmer knows the values to be placed in the tool-nose radius compensation offsets as the program is being written, long before the setup is made.

The on-line task of entering tool-nose radius compensation offsets can be eliminated if the programmer includes offset setting commands within the program. Here are two commands that set the tool-nose radius compensation values of offsets 2 and 7 shown in the format for a popular CNC control. These commands need be executed only once to set the offsets correctly.

G10 P2 R0.0316 T3 (Set TNR compensation offset 2 to 0.0316 for a turning tool)

G10 P7 R0.0156 T2 (Set TNR compensation offset 7 to 0.0156 for a boring bar)

Another way to help the setup person avoid having to enter tool-nose radius compensation offsets, as well as to eliminate all other problems associated with tool-nose radius compensation, is to generate the CNC program's movements on the basis of the radius of the tool being used. While this may not be feasible for manual programmers because of the difficulty associated with calculating program coordinates, it is quite easy to do if a CAM system is used to prepare the CNC program. A CAM system can calculate movements based on the size of the tool nose radius just as easily as it can for the actual work surface being machined. If this is done, you completely eliminate the need for CNC-control based tool-nose radius compensation and the associated on-line task of entering the related offsets.

FACILITATING ON-LINE CUTTING-TOOL TASKS

Most of what we have been trying to accomplish to this point has been related to moving as many cutting-tool tasks as possible off line. However, as stated in the previous chapter, production quantities have a great deal to do with how much can be done off line. With low production quantities, it may be impractical to do *anything* off line.

Tool length measurement

If production quantities are so small for a machining center that tool lengths cannot be measured off line, they must of course be measured when the machine is down between production runs. However, just because tool-length measurement must be done on line does not mean the CNC machine must be used to take the measurements. The same kind of tool-length measuring gage described earlier can still be used, and will make it faster and easier for the setup person to measure tool lengths than using the CNC machine tool itself. Additionally, the same kind of program-generating, personal-computer-based software can still be used to create a CNC program to enter tool-length compensation offsets. Facilitating the on-line task of tool length measurement in this way not only reduces setup time, it saves wear and tear on the machining center and minimizes the danger related to measuring tool-length compensation values. Pressing the wrong button on the machine's control panel while measuring tools could result in disaster.

Tool-length measuring probes. One of the best ways to measure tool lengths on line is to utilize a tool-length measurement probe. This form of probe, which works best on *vertical* machining centers, is usually mounted right to the machine's table top. It facilitates the on-line task of measuring tool lengths and entering tool-length compensation values. Some of these probes can also measure the radius of milling cutters and enter cutter-radius compensation values.

Most tool-length measuring probes allow the setup person to simply specify the numbers of the tool stations holding tools to be measured. From this point, the operator simply activates a tool-length measuring program. For each tool station, the tool is loaded into the spindle and brought into contact with the tool-length measuring probe. After contact, the tool-length compensation value is automatically entered into the corresponding tool offset. Then the machine will automatically change to the next tool to be measured and repeat the measurement. This process is repeated for every tool. Since the machine is in an automatic tool-length measuring cycle, the setup person will even be freed to perform other on-line tasks while the machine tool measures tool lengths.

In the strictest setup time reduction sense, we must caution you that this form of probe is beneficial only in those applications where it is not possible to perform the tool length measurements off line. Though tool-length measurement probes may *appear* to be an attractive and convenient way to measure tool lengths, in reality, they will actually *add* to setup time as compared to measuring tool lengths off line.

Helping the setup person measure tool lengths. For machines that do not have the tool-length measurement probe, there may still be some things you can do to make it easier for the setup person to measure tool lengths and enter the corresponding tool offsets. If your CNC control is equipped with parametric programming (as many current controls are), you may have access to certain machine functions not accessible through normal CNC programming techniques. To help the operator measure tool lengths on the machining center, you will need access to the tool offset table as well as the machine's current position, which are accessible with many versions of parametric programming.

Here is an example that is written in the most popular form of parametric programming, Custom Macro version B. While at first glance it may be somewhat difficult to understand, it will dramatically minimize the amount of work your setup person must do to measure tool lengths.

With our example, the setup person will use the 3-inch (76-mm) side of a 1-2-3 block to touch tools (Figure 8-7). One time only, the setup person will measure the distance from the table top to the spindle nose while the machine is resting at its reference position in the Z axis. This value (which will be referenced by our custom macro) is placed into permanent common variable 500. The setup person will first (manually) move the X and Y axes so that the spindle is above the 1-2-3 block as shown in Figure 8-7. At this point, a special manual data input (MDI) command will be given to specify the numbers of the first and last stations that include tools to be measured.

If, for example, your setup person wishes to measure the lengths of tools in stations 1 through 15, he or she would first move the block under the spindle. Then this program would be called up for modification:

O9050 (Program number)
#100 = 1 (First tool)
#101 = 15 (Last tool)
#103 = #100 (Initialize counter to first tool)
N1 IF [#103 GT #101] GOTO 99 (If #103 is greater than #101, exit)
G91 G28 Z0 M19 (Return to tool-change position, orient spindle)

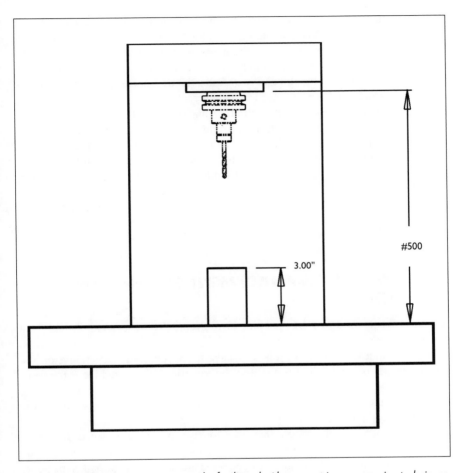

Figure 8-7. Tool-length measurement can be facilitated with parametric programming techniques.

T#103 M06 (Put tool in spindle)
#3006 = 101 (TOUCH TOOL TIP TO BLOCK)
G90 G10 P#103 R[#500 - 3.0 - ABS[#5023]] (Set offset)
G91 G01 Z.5 F30. (Move away from block)
#103 = #103 + 1 (Step counter)
GOTO 1 (Go back to test)
N99 M99

The first two commands set the first and last tool stations containing tools to be measured (#100 is the first tool station and #101 is the last tool station). When this program is executed, each tool will be automatically placed into the spindle and then the machine will come to a stop. The #3006 command is a program stop (like that caused by M00), but it additionally places the message contained in parentheses on the screen for the operator to see. During the program stop, the setup person places the mode switch to a manual mode to jog the tool down to the 3-inch (76-mm) side of the 1-2-3 block. Using the handwheel mode, the setup person cautiously touches the tool tip to the block. At this point the setup person places the mode switch back to automatic and reactivates the cycle. The G10 command automatically places the correct value into the correct tool offset (specified by #103) and then moves away in Z 0.500 inch (12.7 mm). Then #103 is stepped and the program is sent back to the test. If there are more tools to measure, the process is repeated for each tool. Eventually, after the last tool is measured, the control will come to a stop with the last tool positioned 0.500 inch away from the block.

While all of this may at first seem difficult, remember that all the setup person must do is touch tools to the block in Z. *All* other functions are automatic. This truly facilitates the setup person's task of measuring tool lengths on the CNC machining center and, in turn, reduces setup time. Though this technique is shown in Custom Macro B format, keep in mind that many versions of parametric programming allow the same results through similar commands.

HOW TOOL LIFE MANAGEMENT SYSTEMS AFFECT SETUP TIME

All cutting tools used by CNC machines will eventually wear out and some form of tool maintenance will eventually be necessary. Worn inserts will have to be changed or indexed and dull tools will have to be replaced. Depending on the kind of operation being performed and the material being machined, the tools used in a CNC program will commonly last differing lengths of time. In long production runs, it is likely that tool maintenance will be required before the production run is completed.

Tool-life management systems are used for two important reasons. Both have to do with tool maintenance. First, they can be used to prolong the length of time before tool maintenance (changing inserts, replacing dull tools, etc.) must be done. This is commonly required on bar-feed turning centers to allow the entire bar (or several bars) to run completely unattended. In this application, the cutting tool maintenance is simply being postponed until the bars have to be changed.

The second and more impressive reason for using tool-life management systems is to eliminate the on-line task of tool maintenance during the production run. We offer much more information on tool-life management during our presentation of *cycle time* reduction techniques in Chapter Nine.

All applications for tool-life management systems involve utilizing multiple identical (or at least very similar) cutting tools. The cutting tools that wear out fastest (like roughing tools) receive the highest consideration. If, for example, a rough milling cutter wears out three times faster than all other tools in the setup, the tool-life management system will allow three times as many rough milling cutters as other tools in the setup.

While tool-life management systems can greatly improve CNC machine utilization through cycle time improvements, this duplication of cutting tools can have an adverse effect on setup time. The more tools that must be assembled, measured, and loaded, the longer the setup time. Fortunately, tool-life management is primarily used for higher production quantities, when setup time is not so critical. And any additional setup time required with tool-life management can be easily made up in minimized tool maintenance during the production run. However, the more additional tools required, the more important it becomes that cutting tool setup procedures be done off line.

SPECIAL RECOMMENDATIONS FOR TROUBLESOME TOOL TYPES

As you study the video of your setup people assembling, measuring, loading, and entering tool offsets for cutting tools, you may notice that certain tool types present especially difficult problems. Here we offer a few suggestions for handling common cutting tool problems.

Taper reamers

Taper reamers are among the most difficult tools to work with in the CNC environment. First of all, they can be difficult to program, because the depth of the taper-reamed hole must be calculated from the taper angle and the small diameter of the taper reamer. Second, if the reamer is sharpened, the diameter of the reamer changes, meaning either the program or tool-length compensation offset for the tool must be changed. Additionally, when verifying the program, the setup person may have difficulty determining *how much* to change the offset in order to machine to the proper depth. The problems associated with taper reaming can be overcome by making a setup gage as shown in Figure 8-8.

Notice that the gage is made to a known length (1-inch [25.4mm] long in our example). It is machined to the same diameter in the reamer gage as the hole to be machined in the workpiece. For the tool-length compensation value, the tool length measurement is taken from the end of the reamer gage, not the end of the reamer. The programmed depth for the taper reamer will simply be the length of the reamer gage. Here is the canned cycle command, in the format for one popular CNC control, the taper reaming operation.

N050 G81 X3.0 Y3.0 R0.5 Z-1.0 F4.5

Notice that the Z word in the canned cycle is simply the length of the reamer gage and assumes the Z program zero point to be the surface being machined. Depending on how much of the reamer is protruding from the end of the reamer gage during measurement, the rapid plane (specified by the R word) can also be easily determined once the program is running. Note that if the reamer needs to be sharpened, the program will not have to be changed. As long as the reamer gage is used for measuring the length of the sharpened reamer, the operator or

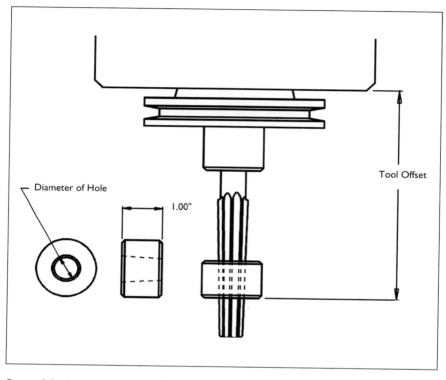

Figure 8-8. A taper reamer's tool-length compensation value can be easily measured and matched to the CNC program.

setup person can rest assured that the program will still machine the workpiece correctly. This technique facilitates the off-line task of measuring the reamer's length and also eliminates the on-line task of trial machining with the reamer to test for hole size during the program's verification.

Taper taps

Taper tapping also presents special problems for setup people during measurement and program verification. As with taper reaming, the depth the taper tap will go can be difficult to predict before machining and usually requires trial machining during program verification.

One way to eliminate these problems is to use a gage similar to the one in our discussion of taper reaming. The gage, made to a precise length, can be tapped to the same depth as the hole to be tapped in the workpiece. If the gage is used during the taper tap's length measurement, the tapped hole can be programmed to a depth equal to the length of the taper tap gage. With this technique, the setup person can rest assured that the tapped hole will be machined to the correct depth *before* the program is run.

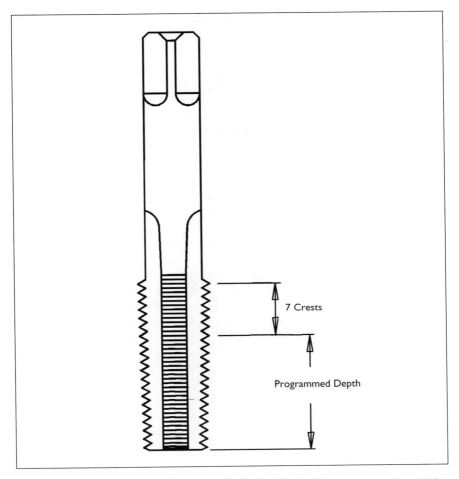

Figure 8-9. If you count back seven crests from the shank end of the taper tap, what is left is the depth of the tapped hole.

Figure 8-9 demonstrates another, slightly easier method used by many companies to determine taper tap depth. With most taper taps, you can simply count back seven crests from the shank end of the tap. The distance from the seventh crest to the end of the tap is the depth of the tap to program. With this method, the tool-length compensation value is the overall length of the tap (to the tip of the tap). Though you may have to test this method for the particular taps you use, this eliminates the need to make a special tool-length measuring gage.

Boring bars for machining center use

Setting a boring bar to machine to a precise diameter can be very difficult, especially for finish boring. Even if a tool presetting gage is used, the tool pressure induced by the machining operation will likely cause the boring bar to machine slightly smaller than its intended size.

Adjusting the diameter a boring bar will machine requires tedious and error prone on-line trial machining during the program's verification. In our discussions of program verification, we will show a helpful programming technique for facilitating the trial-boring operation.

PROGRAM ZERO ASSIGNMENT

One of the most important CNC programming features ever developed is the ability to define an origin point (which we call program zero) from which all coordinates within the program can be taken. While working in the absolute mode, the programmer can easily specify movements within the program from a logical position, usually from the point that dimensions on the blueprint are taken. In many cases, this lets the CNC programmer specify programmed coordinates right from the blueprint.

As important a CNC feature as floating zero point programming is, it can be a setup time waster if not efficiently used. Consider that in the early days of NC (before CNC), when floating zero point was not yet available, an NC programmer had to know where the workpiece would be located within the travels of a machine *before* the NC program could be written. For machining centers, this meant the workholding device had to be keyed to the table and the programmer had to know precisely how far it was from the machine's reference position to each of the table's key slots. The distance from each key to the vise or fixture location surfaces also had to be known. Only then, by using highly cumbersome techniques, could coordinates needed for the program be calculated. Since all calculations involved distances that were independent of the particular machine being programmed, coordinates going into the program made little sense, and programs could not be run on more than one machine tool.

While this was a terrible way of writing programs, and I would never recommend ignoring the floating zero point feature, think about how quickly the setup person could make the setup. Since there was no program zero assignment to be concerned with, the setup person could simply place the fixture to the table in the related key slots and clamp it down. Given that the program was based on the keyed position of the fixture, no on-line time was spent to specify the location of program zero. As difficult as it was to write programs without program zero, it truly minimized setup time.

With *every* convenience feature aimed at helping CNC programmers, there is usually a tradeoff of some kind if the feature is not adopted wisely. When it comes to floating zero point programming, for example, many companies have adopted the habit of having their setup people measure the program zero point's location for *every* setup made. With early CNC machines allowing floating zero point programming, this was truly the method of choice. CNC people were so thrilled with the ability to specify the location of the program zero point, setup time was quickly and willingly compromised to gain the convenience. In this section, we are asking you to reconsider the methods your setup people use to assign program zero, especially if you require your setup people to take measurements to determine the location of the program zero point.

Admittedly, CNC machine tool builders vary widely in the techniques they recommend for assigning program zero. What many end users do not know, however, is that most CNC controls allow you to assign program zero in several different ways. Though very few machine tool builders will describe every method of assigning program zero in their basic programming courses, you probably can choose a program zero assignment method that is best for your application. Newer controls are especially flexible in this regard.

As noted, how program zero is assigned can have a major impact on setup time. In many cases, you can dramatically improve on the generic methods commonly taught in a machine tool builder's standard courses. You may have to refer to your CNC control manufacturer's manual, or contact one of your machine tool builder's applications engineers in order to learn more about the specific alternatives you have when assigning program zero for your particular CNC controls. Many of the techniques we offer in this section assume you are willing to look into your alternatives. With the potential for eliminating the on-line task of assigning program zero altogether, it should be well worth your time to do so.

MACHINING CENTER ALTERNATIVES

Since the spindle position remains consistent from one tool to the next, machining centers require only one set of program zero setting values per program zero point. Deviations in tool lengths are handled with tool length compensation. This makes it relatively easy to assign program zero for machining centers (compared to turning centers, where each tool requires its own program zero setting values). In this section, we explore the alternatives you have for assigning program zero points for machining centers. Our goal is to completely eliminate all on-line measurements needed for program zero assignment.

Fixture offsets versus G92

The two most common methods used to assign program zero on machining centers involve using a G92 command within the program or some form of coordinate system shifting (commonly called *fixture offsets*). Generally speaking, the use of G92 within the program is the older and more cumbersome method. It requires that the distance from program zero to the *current* spindle position of the machine be specified within the CNC program. The use of G92 has several limitations, and we recommend against using it unless your machining centers are not equipped with fixture offsets. One especially troublesome problem that affects setup tasks has to do with how many program zero points can be easily assigned.

Since G92 specifies the current distance between the program zero point and the spindle, it is very difficult to assign more than one program zero point, as is required when more than one workpiece must be machined in the setup.

The use of fixture offsets overcomes this problem. Several program zero points can be easily assigned, one per fixture offset. For this and many other advantages over the use of G92, we strongly recommend using fixture offsets even when there is only one program zero point to assign. In fact, the only time we recommend using G92 to assign program zero is with older CNC machining centers that do not have the fixture offset feature.

Choosing a point of reference

Another advantage of fixture offsets over G92 (on most CNC controls) is the ability to change the point of reference used to designate the program zero points. Most machine tool builders recommend using the machine's reference position (commonly called *the zero return position, home position,* or *grid zero*) as the point of reference for program zero entries. With this method, the distances in X, Y, and Z between the program zero point and the spindle while it is resting in the machine's reference position are used as the program zero setting values. Figure 8-10 demonstrates this. (While the figure demonstrates the values needed to assign program zero in only the X and Y axes, the same principles apply for the Z axis).

Though many machine tool builders teach this method of assigning program zero in their basic courses, when using the machine's reference position as the point of reference for program zero assignment, the values being used to assign program zero will make very little sense. Additionally, these values will be difficult to calculate prior to making the actual setup, meaning setup people are almost forced into taking on-line program zero finding measurements during setup.

Most CNC controls allow the end user to change the point of reference for program zero assignment. Unfortunately, control manufacturers vary with regard to how this is done. One popular control manufacturer uses a special fixture offset, called the *common* offset, to specify the distance from the machine's reference position to the location you wish to use as the point of reference for all fixture offset entries. Figure 8-11 illustrates this method.

Notice how the point of reference for fixture offset entry has been changed to a key location right on the fixture itself. This makes it very easy to predict the values needed for the assignment of fixture offsets. The only numbers that will not make much sense will be those used to assign the point of reference.

Some control manufactures use parameters to move the reference point for program zero assignment, making it somewhat more difficult to change this point of reference. Since you may not want your setup people manually changing parameters during each setup, keep in mind that many controls allow parameters to be changed through programmed commands. Most that do use the same command that is used to enter tool offset values (commonly G10).

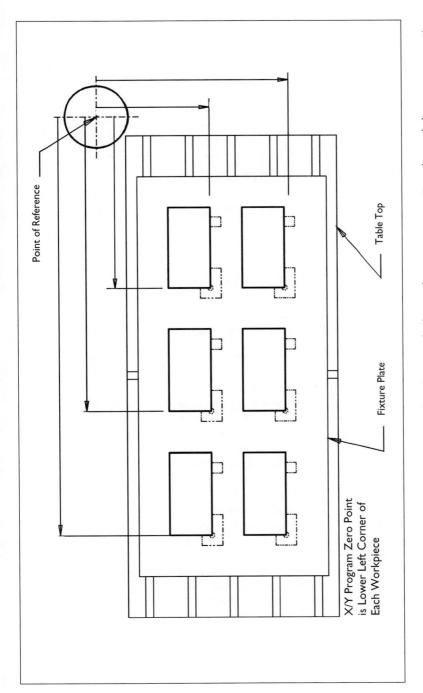

Point of Reference

Table Top

Fixture Plate

X/Y Program Zero Point
is Lower Left Corner of
Each Workpiece

Figure 8-10. When the machine's reference position is used as the point of reference for program zero setting, the needed program zero setting values will be difficult to determine and will involve numbers that make little sense.

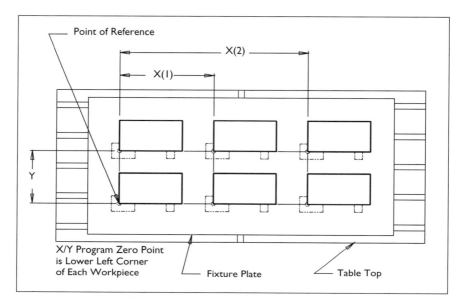

Figure 8-11. Moving the point of reference for program zero assignment for a vertical machining center makes it easy to predict the values needed for program zero assignment.

This ability to change the point of reference for program zero assignment is especially helpful for horizontal machining centers. Most horizontal machining centers utilize a rotary device within the table. This rotary device allows easy access to many surfaces of the workpiece being machined. To keep programmed coordinates logical and simple to calculate, many programmers assign a different program zero point for each side of the workpiece being machined. If the machine's reference position is used as the point of reference for assigning program zero, it will be very difficult to predict the values needed to assign program zero before the setup is made. Consider the techniques shown earlier for vertical machining centers that require the programmer to know several machine-related values in order to predict the programming zero point setting. These same cumbersome techniques are required by the use of an indexing device. Additionally, the resulting program zero setting values will make little sense.

If you change the program zero assignment point of reference to the center of table rotation in X and Z and to the top surface of the table in Y, fixture offset values will be much easier to determine. Since most fixture designers work from the center of rotation to dimension their fixture's location surfaces, program zero assigning values can be taken right from the fixture drawing, eliminating the need for on-line program zero point measuring. Figure 8-12 illustrates this.

Another time when changing the point of reference can be extremely helpful is when you are using subplates on vertical machining centers. The subplate can be designed to incorporate a central location position (usually the center of a dowel

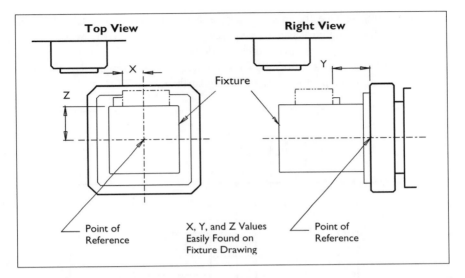

Figure 8-12. Moving the point of reference for horizontal machining centers to the center of rotation can make determining fixture offset assignment values much easier.

pin hole) from which all setups to be made on the subplate can be dimensioned. If you change the program zero assignment point of reference from the machine's reference position to the center of this dowel pin hole, all program zero assignment values can be taken right from the subplate drawing, making it very easy to predict fixture offset values.

You must be prepared for the possibility that changing the point of reference for program zero assignment may be somewhat more difficult than we have shown. As mentioned, it may be necessary to change parameters in order to move the point of reference. If you are having difficulty determining how to do this, contact your machine tool builder or control manufacturer for help. With the knowledge that almost all current model CNC machining center controls have this ability—and the determination not to take no for an answer—your efforts will be rewarded with the ability to eliminate the on-line task of program zero measurements.

Qualified workholding setups—key to eliminating program zero measurements

Having the ability to move the point of reference for program zero assignment is only part of what you need to completely eliminate on-line program zero measurements. Additionally, you must have the ability to locate the workholding setup on the machine table in a predictable and repeatable manner. You will only have this ability if your workholding setups are *qualified*.

By qualified, we mean that the locating surfaces of the workholding setup (used as bump-stops for the workpiece itself) must be fixed and predictable. This requires that the locating surfaces be in some way related to the machine's table slots or dowel pin holes. Though it usually requires little effort to qualify workholding tooling, it is amazing how many companies do not. Figure 8-13 shows a qualified vise with milled soft jaws. Notice how this vise can be easily replaced on the table with no change in the position of its locating surfaces.

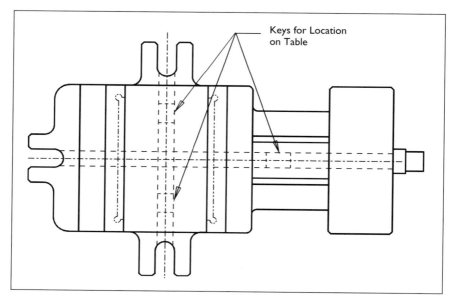

Figure 8-13. A qualified view setup. Note how this vise can be removed and replaced on the table with no change in the program zero point location.

Programming the program zero setting values. With any qualified workholding device, the values used to assign program zero will remain the same *every time* the workholding setup is made. Regardless of whether you calculate these values up front or whether you have your setup people take measurements to determine these values at the time of setup, remember that most controls provide a way of entering offsets (including fixture offsets) by programmed command, so that you can completely eliminate the on-line task of fixture offset entry. For fixture offset values you calculate up front, simply include a command in your program to assign the value of each fixture offset. For fixture offset values you have your setup person measure, modify the CNC program after setup to include the fixture offset setting command to eliminate future measurements.

As mentioned earlier, G10 is a common command used for this purpose. The following example command sets fixture offset 1 for one popular control.

N010 G90 G10 L2 P1 X-12.3305 Y-14.3235 Z-10.1245

More on program zero assignment numbers

As stated, when fixtures (and other qualified devices) are used as workholding devices, many times the fixture designer will specify right on the fixture drawing the distance from the center of keys or keyways (used for mounting to the machine table) to location surfaces on the fixture (used as bump-stops for the workpiece). If the CNC programmer has access to the fixture drawing, these location values will be readily available. This, combined with a knowledge of where table slots are located on the machine table (relative to the machine's reference position), makes it possible for the CNC programmer to predict the values of program zero assigning numbers. If G10 is used to enter fixture offset values, all on-line measurements previously needed to assign program zero can be eliminated.

Where to place the G10 commands. Since the fixture offset setting commands need be executed only one time, it is not necessary to include them in the body of the CNC program that actually machines the workpiece. Some programmers, for example, develop a special setup program (especially if tool-length and cutter-radius compensation offsets are also entered with G10 commands) that the setup person executes one time as part of making the setup. Here is an example program that sets several tool-length and cutter-radius compensation offsets as well as two fixture offsets.

```
O0001 (Program number)
G90 G10 L1 P1 R5.3664 (Set offset 1)
G10 L1 P2 R3.4532 (Set offset 2)
G10 L1 P3 R8.7653 (Set offset 3)
G10 L1 P4 R6.8655 (Set offset 4)
G10 L1 P5 R6.4556 (Set offset 5)
G10 L1 P6 R6.5455 (Set offset 6)
G10 L1 P7 R4.6655 (Set offset 7)
G10 L1 P8 R8.7655 (Set offset 8)
G10 L1 P9 R6.7786 (Set offset 9)
G10 L1 P10 R4.5677 (Set offset 10)
G10 L1 P31 R0.5 (Set cutter-radius compensation offset)
G10 L1 P32 R0.375 (Set cutter-radius compensation offset)
G10 L2 P1 X-12.2588 Y-14.1245 Z-12.1255 (Set fixture offset 1)
G10 L2 P2 X-10.1255 Y-12.2455 Z-11.1244 (Set fixture offset 2)
M30
```

Other programmers like to include the fixture offset setting commands in the CNC program that machines the workpiece; yet, to save program execution time, they do not allow the commands to be executed every time the program is run. Consider these commands:

```
O0001 (Program number)
N005 T01 M06 (Program begins)
N010 G90 S2300 M03 T02
```

.
. (Machine entire workpiece)
.

N450 M30 (End of CNC program to machine workpiece)
N999 G10 L2 P1 X-12.2588 Y-14.1245 Z-12.1255 (Set fixture offset 1)
G10 L2 P2 X-10.1255 Y-12.2455 Z-11.1244 (Set fixture offset 2)
M30

During setup, the setup person is told to scan into the program to line N999 and execute the program from there. This section of the program will set the fixture offset values and then return to the beginning of the program. The control will never see the N999 sequence again, since the first M30 in the program in line N450 will rewind the program to the beginning. This technique allows the CNC programmer to include the fixture offset setting commands within the CNC program, yet keeps them from being executed every time the program is run.

Facilitating the measurement of program zero

Unfortunately, not all machining center applications lend themselves to completely eliminating the program zero measurements. To this point we have assumed, for example, that setups being made today will have to be repeated at some future date. We have also assumed that you can justify the design and fabrication of qualified fixturing to hold workpieces. In some applications, and especially in contract shops, this is simply not feasible and the program zero point *must* be measured on line. Here we discuss two methods to facilitate the on-line task of measuring the program zero points during setup when it is impossible to use qualified workholding tooling.

Spindle probes for program zero measurement. One of the best applications for spindle probes is to help with setup-related functions that must be performed on line, including the measurement and entry of program zero assignment values. Regardless of how or where program zero is located, an automatic probing routine can be written to measure its position. In fact, most probe manufacturers supply their spindle probe with a standard set of *pickup routines* aimed at measuring the program zero point. For example, most supply standard routines for corner pickup, hole centerline pickup, slot center pickup, and round boss pickup. Additionally, many will develop special routines for your own requirements. Once the program zero point has been measured, most probing routines will automatically place the measured values into their corresponding fixture offsets.

As with tool-length measuring probes, keep in mind that, as good as spindle probes are at measuring program zero points, they are not as efficient as the techniques discussed earlier that completely eliminate the need to measure program zero points on line. You should reserve the use of spindle probes for times when it is impossible to eliminate the on-line measurement of program zero.

Helping the setup person measure program zero with a common edge finder. If the CNC machine does not have a spindle probe, and if program zero points must be measured on line, most setup people resort to conventional methods of measuring the program zero position. They commonly use the position display page to monitor the machine's current position (like using a digital readout on a manual milling machine). Many use a common edge finder to locate flat surfaces or a dial indicator to locate the center of round holes or bosses. If an edge finder is used, these techniques also involve doing some math based on the edge finder's radius to come up with the program zero setting values. For beginners, these techniques can be time-consuming and difficult to master, especially when using a dial indicator to locate the center of a round hole or boss. Once the program zero setting values are determined, the setup person must manually enter them into corresponding fixture offsets.

If your setup people are using these very cumbersome techniques to measure program zero, a great deal of setup time can be wasted, especially if more than one program zero point must be measured per setup. Additionally, there is always the potential for error when the program zero setting numbers are calculated and entered. Anything you can do to facilitate this procedure should be done.

As we discussed during our presentation of tool-length compensation offset measurement, if your machine has parametric programming capabilities, you have access to certain functions not commonly associated with normal CNC programming. Most versions of parametric programming give you access to machine position as well as to *fixture offsets*. With this knowledge, you can develop a series of manual pickup routines designed to help the setup person measure program zero points with a common edge finder. We recommend using a conductivity-style edge finder for this purpose since it is very easy to use. In fact, any routine a fully automatic spindle probe can handle can be accomplished with a comparable manual edge-finding parametric program. The only difference is that each touch of the edge finder must be done manually. You can, for instance, develop a center-finding routine that locates the center of a round hole or boss with a standard edge finder.

Here is a corner pickup routine written in Custom Macro version B. It assumes the radius of the edge finder is stored (once) in permanent common variable #500 and the length of the edge finder is stored in permanent common variable #501. It also assumes that the setup person will manually position the edge finder within 0.50 inch (13 mm) or so of the corner to be picked up in all axes. Since we are picking up the lower left hand corner in X and Y and the top surface of the workpiece in Z, the operator positions the edge finder minus of the corner in X, minus of the corner in Y, and plus of the corner in Z. Then this program is run.

```
O9051 (Corner pickup routine for lower left corner)
N1 G91 G01 Y0.75 Z-0.75 F30. (Move to first touch position)
N2 #3006 = 101 (TOUCH LEFT SIDE IN X)
N3 G90 G10 L2 P1 X[#5021 + #500] (Set fixture offset X)
N4 G91 G01 X-0.2 (Move away in X)
```

N5 Y-0.75 (Move down in *Y*)
N6 X0.55 (Move to second touch position)
N7 #3006 = 101 (TOUCH BOTTOM SURFACE IN *Y*)
N8 G90 G10 L2 P1 Y[#5022 + #500] Set fixture offset *Y*)
N9 G91 G01 Y-0.2 (Move away in *Y*)
N10 Z0.75 (Move up in *Z*)
N11 Y0.55 (Move to third touch position)
N12 #3006 = 101 (TOUCH TOP SURFACE IN *Z*)
N13 G90 G10 L2 P1 Z[#5023 - #501] (Set fixture offset *Z*)
N14 G91 G01 Z.5 (Move away in *Z*)
N15 X-0.75 Y-0.75 (Move away in *X* and *Y*)
N16 M30 (End of program)

When this program is run, the machine will first move plus in *Y* and minus in *Z* to a position that allows the setup person to easily touch the left side of the workpiece. In line N2, the machine will stop (just like M00) and the message "TOUCH LEFT SIDE IN *X*" will be displayed on the screen. At this point, the setup person will place the mode switch to the handwheel mode and manually move the edge finder over to the surface in *X*. If a conductivity-style edge finder is used, a light within the edge finder will come on to indicate that it is touching the surface. At this point, the setup person moves the mode switch back to automatic mode and presses cycle start. In line N3, the G10 command stores the distance from the machine's reference position to the surface just touched into the *X* value of fixture offset 1. Notice that the custom macro even does the calculation needed to take into consideration the radius of the edge finder.

In lines N4, N5, and N6, the edge finder is positioned to touch the *Y* surface. The machine stops again in line N7. The operator repeats the manual touching, this time in *Y*. When the cycle is reactivated, the *Y* value of fixture offset 1 is automatically set (from line N8).

In lines N9, N10, and N11, the edge finder is positioned above the *Z* surface to touch and the machine stops again in line N12. The operator places the machine in the handwheel mode and manually touches the edge finder to the *Z* surface. When the cycle is reactivated, line N13 stores the correct distance from the machine's reference position to the surface being touched in *Z* into the fixture offset *Z* value (even considering the length of the edge finder). Finally the machine moves the edge finder to a clearance position and stops the cycle.

Notice how easy this makes it for your setup people. Their only concern will be touching the edge finder to each surface. No more will they have to calculate and enter fixture offset values. This technique could save as much as 10 to 15 minutes per program zero measurement, depending on the experience level of your setup people.

Again, remember that you can now use edge finders as if they were spindle probes. One very important occasion when this can be useful is when you are finding the center of a round hole or boss. Most setup people would agree that a dial indicator is a cumbersome device. Here is another custom macro that finds the center of a hole in X and Y with a conductivity-style edge finder. The setup person simply loads the edge finder and manually positions it in the approximate vicinity of the hole center. Then this custom macro is run.

```
O9050 (Hole center pickup routine)
#101 = #5022 (Memorize start position in Y)
#3006 = 101 (TOUCH LEFT SIDE IN X)
#102 = #5021 (Memorize left side position)
#3006 = 101 (TOUCH RIGHT SIDE IN X)
IF [#5022 EQ #101] GOTO 1 (Test for a Y move)
#3000 = 101 (INVALID Y MOVE - TRY AGAIN)
N1 G91 G01 X-[[#5021 - #102]/2] F30. (Move to X center)
#101 = #5021 (Memorize current position in X)
#3006 = 101 (TOUCH LOWER SIDE IN Y)
#102 = #5022 (Memorize bottom Y position)
#3006 = 101 (TOUCH UPPER SIDE IN Y)
IF [#5021 EQ #101] GOTO 2 (Test for an X move)
#3000 = 101 (INVALID X MOVE - TRY AGAIN)
N2 G91 G01 Y-[[#5022 - #102]/2] F30. (Move to Y center)
G90 G10 L2 P1 X#5021 Y#5022 (Set fixture offset 1 X and Y to current
machine position)
M30 (End of program)
```

Again, this kind of pickup routine can be developed for *any* kind of program zero measurement. In fact, these routines can be developed for *any* application in which a true probing system can be used. The only difference between the edge finder and the probing system is that the probing system is totally automatic. The manual edge-finding routines require manual touching.

TURNING CENTER ALTERNATIVES

Though there are some similarities between program zero assignment for turning centers and machining centers, turning centers present their own special challenges to achieving the goal of eliminating on-line program zero measurement and entry tasks. Generally speaking, the challenges stem from the fact that each cutting tool style in the turning center setup will have its own program zero assigning values. The cutting edge of a turning tool, for example, will have a different position than the cutting edge of a boring bar. The cutting edge of a drill, or any center cutting tool for that matter, will have yet another position. Most turning center controls handle this problem by requiring the setup person to assign program zero for each cutting tool in the setup.

Geometry offsets versus G50

As with machining center program zero assignment, there are two ways by which program zero can be assigned on a turning center. The older and more difficult method involves assigning program zero within the program. One command commonly used for this purpose is G50 (though some controls use G92 to maintain consistency with machining center program zero assignment). Also, as with machining center program zero assignment, the use of offsets (commonly called *geometry offsets*) allows much greater potential for setup time reduction in program zero assignment. The only valid reason for assigning program zero in the program is if the control does not have geometry offsets, as may be the case with older controls. The balance of this discussion assumes you can assign program zero for each tool with geometry offsets.

Our primary goal is to completely eliminate any measurements needed to measure and assign program zero. When this is not feasible, our secondary goal is to facilitate the setup person's ability to quickly measure and enter program zero assignment values.

Turning center program zero assignment

Single-point turning tools such as rough turning tools, finish turning tools, grooving tools, threading tools, rough boring bars, and finish boring bars, are quickly affected by tool wear. Even if the program zero assigning numbers are perfectly set and the tool *begins* machining the workpiece correctly, tool wear will soon affect the size of the workpiece being machined by the tool. With the close tolerances held by today's turning centers, it does not take much tool wear to cause the surface machined by a tool to go out of tolerance. This, of course, is one reason why wear offsets are required—to let the operator allow for imperfections in workpiece size caused by tool wear.

Additionally, most techniques used to determine the program zero values for a given tool (whether on line or off line) are *static*. They do not take into consideration the dynamics involved with the machining operation they perform. The program zero assignment numbers calculated for a given turning tool may be the exact distances between the tip of the tool to the program zero point in X and Z, but they do not consider the tool pressure induced during the machining operation. If the tool deflects the workpiece even 0.0002 inch (5 µm), it may be enough to throw the workpiece out of tolerance. This is a second reason for wear offsets— to allow for the unpredictable dynamics of machining that cause program zero assigning values to be incorrect.

For these reasons, program zero assignment values and wear offsets are very closely related. In fact, one way to think about wear offsets is to visualize them actually changing the program zero point by the amount of tool wear or imperfection in the program zero assignment numbers. Wear offsets truly work in conjunction with program zero assignment values to give the control the accurate location of program zero, taking into account the dynamics of the machining operation.

As you consider reducing the setup time related to program zero assignment on turning centers, you must remember the impact of wear offsets on program zero assignment. No matter how well you determine the program zero values, there will probably be something extra to be done during setup to perfectly size each tool. These sizing techniques are commonly done during the program's verification. Later in this chapter we offer several suggestions for facilitating the setup person's ability to size workpieces. For now, we present ways to target the location of the program zero point for each tool as closely as possible, without taking on-line time to do so.

Qualified turning center tools

One of the keys to eliminating program zero measurements on turning centers is to use qualified tooling. Qualified tools are made to precise specifications published in the tooling manufacturer's catalog. A qualified finish turning tool, for example, may be precisely 6.0000 inches long and 1.2500 inches wide (152.4 mm and 31.75 mm). This makes the tool completely interchangeable with other qualified finish turning tools. As long as the turning center's turret (or other toolholding device) allows the turning tool to be bumped against solid stops in both directions (X and Z), the program zero assignment numbers will remain consistent for every qualified tool that is precisely 6.0000 inches long and 1.2500 inches wide.

Two factors contribute to the interchangeability of turning center tooling, even for qualified tools:

- Indexable insert tools incorporate clamping mechanisms to hold inserts in position. The method by which these clamps are tightened has a great deal to do with how accurately the insert will be positioned in the holder. An experienced CNC setup person or operator can replace inserts with excellent repeatability.
- If the radius of the insert is changed from one setup to the next, there will be a substantial difference in the position of the tool tip.

These are two more reasons why eliminating the measurement and entry of program zero setting values must be considered simply a targeting procedure. There will likely be more to do when sizing the first workpiece to be machined.

Some tools commonly used on turning centers are not perfectly qualified. Boring bars, for example, may have a qualified distance from the tip of the boring bar insert to the center of the boring bar, making it possible to predict the program zero assignment value in the X axis. However, boring bars, being straight-shank tools, can be moved easily along the Z axis in most tool-holder bushings. Most setup people will push a boring bar into its bushing until there is just enough length left to machine with the boring bar to the necessary depth. This technique keeps the boring bar as rigid as possible, but it makes predicting its program zero assignment numbers more difficult. The same is true for almost all other forms of internal cutting tools, including internal grooving tools, internal threading tools, drills, and reamers.

One way to handle this problem is to set the tip of the internal tool to a precise distance relative to the face of the internal tool holder. If the programmer specifies this distance on the setup sheet, and the setup person correctly sets the tool as specified, the Z-axis program zero assignment value can be determined up front. As long as the tool is assembled and set off line, setup time can be minimized. However, if the setup person must set internal tools to precise lengths on line, setup time can be wasted. It may be possible for setup persons to actually measure the Z-axis program zero assignment value faster than they can accurately set the position of the internal tool in its holder on line.

Quick-change tooling. During our presentation of cutting tool issues, we noted the impact of quick-change tooling on tool loading time. Since this feature incorporates a machining-center-like tool-clamping system, loading tools is as easy as pushing a button, placing the tool into the holder, and pushing the button again. Tool loading can be done in under five seconds per tool.

Like qualified tooling, quick-change tooling makes it possible to determine program zero assignment values up front. And since most internal tools are qualified in X and Z, both program zero assignment values can be predicted.

Even for quick-change tools that are not qualified, such as drills, it is very easy to assemble the tool off line and measure its gage length (just as machining center tools are measured). This gage length value can even be used as the program zero assignment value in Z.

Tool touch-off probes

Tool touch-off probes (Figure 8-14) are designed to measure program zero assignment numbers, and many machine tool builders are supplying their machines with them. For each tool to be measured, the tool is driven into the probe surfaces

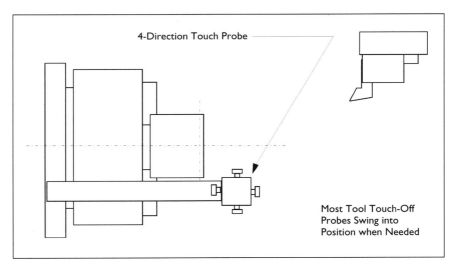

Figure 8-14. Depicted is a tool touch-off probe typical of those supplied with turning centers. Note the four directions of probing.

twice, once for an X-axis measurement and once for a Z-axis measurement. Using constant values internal to the probing system, the control automatically calculates the program zero assignment values and stores them in the related (geometry) offsets.

Tool touch-off probes on turning centers are comparable to tool-length measurement probes on machining centers. Though each tool's position can be accurately determined (more accurately than just about any other method), on-line time is being taken to measure tool positions. Since tool touch-off probes render highly accurate program zero value measurements, and since the time required to measure and enter program zero values is relatively short, many companies simply allow this on-line time to gain the benefits of the tool touch-off probe. If, however, your goal is to completely *eliminate* on-line program zero value measurement and entry time, tool touch-off probes will not suit your needs.

Eliminating on-line measurements to determine program zero assignment values

The balance of this discussion focuses on completely eliminating the on-line task of program zero measurements. We start by describing the most common method of determining program zero assignment values (techniques commonly taught by machine tool builders) and work toward the more efficient techniques.

By far, the most common methods used to determine program zero assignment values involve taking on-line measurements. Say, for example, the setup person wishes to determine the program zero assignment values for a finish turning tool. With one popular method, the setup person will follow this procedure:

1. Index the turret to the finish turning tool.
2. Load the first piece of raw material into the chuck.
3. Start the spindle and skim-cut the outside diameter of the workpiece, removing just enough material to measure the diameter just machined.
4. Without moving the machine in the X axis, set the X-axis position display to the diameter just machined.
5. Send the machine to its X-axis reference position.

The position display screen will follow along, and when the machine reaches its reference position in X, the display screen will show the distance from program zero to the tip of the tool at the machine's reference position, which is commonly the value used to assign program zero in X.

This entire procedure is then repeated for the tool in the Z axis. It must then be repeated for all other tools in the setup. Depending on the ability level of the setup person, it could take from 2 to 10 minutes per tool just to determine program zero assignment numbers. While some turning center control manufacturers have developed features to make these measurements easier to take, it remains a tedious, time-consuming, and error-prone method of determining program zero assignment numbers.

If you dedicate your turning center to running only one or two different workpieces, or if you never have to change tools from one setup to the next, this may be an acceptable way to determine program zero assignment values, since measurements will seldom be needed. If, on the other hand, your setup people are constantly removing and loading tools, this is a poor way to determine program zero assignment values.

If you use this method to determine program zero assignment numbers, you will want to minimize the number of times measurements must be taken. One way to accomplish this is to use the standard tool station techniques discussed earlier (during our discussion of cutting-tool issues). At least for tools that get used from one production run to the next, there will be no measurements to take, assuming your programmer is working from a logical program zero point. Let's say you are currently using this method to assign program zero, but would like to improve.

In the X axis, the program zero point is *always* the center of the workpiece. That is, the program zero point in X will remain the same for every program ever run. For this reason, the program zero assignment value in X will remain the same for a given tool from one production run to the next (as will the wear offset value), as long as the tool is not removed from the turret.

However, care must be taken when determining the program zero point for the Z axis. Many programmers specify the program zero point in Z as the right end of the finished workpiece. While this makes it very easy for the programmer to write programs, it makes it quite difficult for the setup person to assign program zero in Z from one setup to the next. Since the length of the workpiece usually changes from one production run to the next, it is possible that program zero assignment values in Z *for all tools* (including those used in the previous production run) will have to be determined again, resulting in wasteful duplication of effort.

To overcome this problem, most CNC control manufacturers incorporate a *work shift* feature. This feature lets the programmer continue placing the program zero point wherever desired (commonly the right end of the finished workpiece), yet lets the program zero assignment numbers be taken from a more consistent surface on the machine tool itself, say the face of the chuck. The work shift value is the distance from the consistent surface (face of the chuck) to the actual program zero point. The use of this feature keeps programming simple while minimizing the number of program zero measurements required during setup. With work shift, one value shifts the coordinate system for all tools from the face of the chuck (or any consistent Z surface) to the program zero point used by the programmer. While we have not (yet) eliminated measurements to this point, with standard tool station techniques, we can dramatically reduce the number of needed on-line measurements when tools are used from one production run to the next.

Setting offset values by programmed command. If qualified (or quick-change) tools are used, and if the Z-axis point of reference for program zero assignment is the face of the chuck, the program zero assignment values will remain the same for qualified tools placed back into the turret at some future date. For example, today your setup person must make a setup that includes a qualified outside-diameter threading tool. The program zero assignment values are measured (using the techniques described earlier). Program zero assignment values come out to 10.2074 inches (259.268 mm) in the X axis and 8.1857 inches (207.917 mm) in the Z axis (Z is to the face of the chuck).

After the setup is made, production is run. Eventually the production run is completed. For the next production run, the setup person must remove the threading tool to make room in the turret for another tool, but at some future date, the (qualified) threading tool will be needed in another production run. If no concern is given to retaining the program zero assignment values for future use, the setup person will have to repeat the program zero measurements when the threading tool is eventually needed.

Just like machining center controls, most turning center controls allow you to enter offset values through program commands. Again, the G10 command is the commonly used command for this purpose. In the previous scenario, if a geometry offset setting command is added to the CNC program to specify the value of the threading tool's geometry offset values, you can eliminate the setup person's future need to duplicate the threading tool's measurement. Keep in mind that as long as your programmer works from a fixed position in Z (the face of the chuck, for example), the program zero setting command for this threading tool can be added to every program in which the threading tool is used, eliminating all future measurements and geometry offset entries for the tool.

Here is the G10 offset setting format for one popular turning center control:

G10 L2 P3 X10.2074 Z8.1857
G10 L1 P3 X0 Z0

For this particular control, the L word specifies which kind of data we are setting (L2 specifies geometry offsets for this particular control). The P word specifies which geometry offset we are setting (offset 3). The X and Z values are the values going into the geometry offset. Since any value currently in the wear offset is related to a tool that is no longer in the turret, it is not a bad idea to get into the habit of clearing the wear offset of any tool having its geometry offset set (which is done in the second G10 command).

If this technique is repeated for all qualified tools, a great deal of program zero measuring time can be eliminated. Even for nonqualified tools like boring bars, it will usually be possible to eliminate at least one of the needed measurements (for the X-axis program zero assignment value).

Since offset setting commands need only be executed once, many programmers like to include them at the very end of the CNC program and even after the M30 command. The setup person must scan to them and execute them once during setup. Additionally, the programmer can document exactly which tools need to be measured in this section of the program. The following is an example that shows this method:

O0004 (Program number)
N005 T0101 M41 (Index turret, select spindle range)
N010 G96 S500 M03 (Start spindle)
N015 . . .
.
.
.
N475 M30 (End of machining program)
N999 (Setup related commands)
(To the setup person: You must manually measure the X values of stations 4, 6, and 8)
G10 L2 P1 X12.1255 Z10.3386 (Set geometry offset 1)
G10 L2 P2 X12.1253 Z10.3385 (Set geometry offset 2)
G10 L2 P3 X12.1265 Z10.3379 (Set geometry offset 3)
G10 L2 P4 X9.1857 (Set X of geometry offset 4)
G10 L2 P6 X9.5048 (Set X of geometry offset 6)
G10 L2 P8 X8.5886 (Set X of geometry offset 8)
M30 (End of program)

If you are using work shift, remember that most controls also allow the work shift value to be entered through a programmed command. This means if your workholding setup will be done the same way the next time, you can eliminate not only the setup person's need to measure and enter geometry values for all qualified tools, but also the work shift value. Here is a command that does so in the format for one popular control:

G10 P0 Z2.3763

For this particular control, the P0 specifies that the work shift value is being set. This command can, of course, be added to the offset setting commands at the end of the program.

Can you predict program zero assignment values? If you can calculate the values used for program zero assignment, you can use offset setting commands in the program to enter geometry offsets into the control. This will completely eliminate the need to measure and enter program zero assignment values. However, because of the differences in tooling styles mentioned earlier, it can be somewhat cumbersome to calculate program zero assignment values, and the values will make little sense.

This is because, as with many machining centers, the *point of reference* for program zero assignment is commonly the turning center's reference position (also called *zero return, grid zero,* and *home position*). Under normal circumstances, most turning centers require that the distance from each tool tip to the program zero point in X and Z be stored in the corresponding geometry offset. These values are not at all easy to predict.

If you can change the point of reference to a more logical location, your geometry offset values will make much more sense. While the technique we now show can help with any form of qualified tooling, it is especially helpful with quick-change tooling. Tooling manufacturers qualify their quick-change tools from a datum surface in both directions (X and Z). The dimension from the datum point to the tip of each tool is published in the tooling manufacturer's catalog. If the point of reference for assigning program zero can be moved from the centerline of the spindle in X and the face of the chuck in Z to the datum point of quick-change tooling, geometry offsets can be taken from the tooling manufacturer's catalog. This makes it possible to determine the values of geometry offsets up front; it also ensures that geometry offset values will make sense.

Most turning center controls that allow changes in the reference point for program zero assignment do so with parameters. One parameter stores the distance from the spindle centerline to the quick-change datum point in X. Another parameter is used to store the distance from the face of the chuck to the datum point in Z. (Work shift is still used to specify the distance from the face of the chuck to the program zero point in Z.) Once these parameters are correctly set, geometry offsets can be specified from the quick-change tooling datum point.

Special program zero assignment considerations based on machine type

The discussions given to this point have assumed you are using universal-style turning centers that perform all kinds of turning work. There are other program-zero-related points we need to make for more specific turning applications.

Bar feeding turning centers. Many turning centers are dedicated to bar work. Indeed, many are designed specifically for bar work. Though there are exceptions, cutting tools used on bar-feed turning centers tend to be quite consistent from one production run to the next. In fact, there are those companies that use the same cutting tools for *all* production runs, meaning cutting tools are seldom, if ever, removed from the machine.

Would you believe zero setup time? By incorporating the techniques we have shown thus far (as well as some special programming techniques), you can actually eliminate *all* setup time for bar-feed turning centers when changes in cutting tools, bar diameter, and bar material are not required from one production run to the next. With our technique, the first workpiece of the next production run will come out of the machine immediately after the last workpiece of the previous production run with absolutely no work stoppage.

As the machine goes from one production run to the next (since tools remain in the turret), their program zero setting and wear offset values will remain the same. If needed, the work shift value can be changed with a simple G10 command. Here is a simple *controlling program* that tells the control which programs are involved and how many workpieces to machine of each:

```
O0005 (Main program)
N005 M98 P1000 L10 (Run 10 parts from program O1000)
N010 M98 P1001 L15 (Run 15 parts from program O1001)
```

N015 M98 P1002 L20 (Run 20 parts from program O1002)
N020 M98 P1003 L8 (Run 8 parts from program O1003)
M30

With this example, four different production runs are being machined. First, 10 workpieces will be machined from program O1000, then 15 from program O1001, 20 from program O1002, and finally, 8 from program O1003. Any length differences that require work shift changes are handled with G10 commands in programs O1000 through O1003. With this method, you could string together *all* workpieces being machined from a bar of given diameter and material as long as no cutting tool changes are necessary.

Though our control program is using simple subprogramming techniques (specified by M98), keep in mind that much more elaborate techniques are possible with parametric programming. For example, our simple program assumes that all four production runs can be done from a single bar. With parametric programming techniques, you can allow for greater numbers of workpieces, letting the control keep track of which program is currently active should the machine have to be shut down for a new bar, tool maintenance, or the end of the shift. These techniques are especially helpful when you are trying to attain unattended operation, as would be the case when you allow the machine to run through the night.

Gang-style turning centers. Gang-style turning centers utilize a *tooling table* instead of a turret to hold cutting tools. The advantage of this kind of machine is that tool changing time (normally requiring a turret index) can be completely eliminated. Chip-to-chip time is simply the time it takes to rapid the next tool on the tooling table to its first machining position.

For machines that require program zero to be assigned in the program (commonly with G50 commands), this kind of machine is extremely difficult to work with. The distance from one tool to the next must be known in order to calculate the G50 commands. While geometry offsets have made handling program zero assignment for gang-style turning centers much easier to deal with, most users still measure geometry offset values on line, while the machine is down between production runs.

As long as production quantities justify it, tooling subplates can be used to minimize the time it takes to load, measure, and input geometry offsets for cutting tools. If you evaluate the different workpieces you machine on your gang-style turning center, you may find that only a few tooling subplates are required to handle all workpieces you machine. If a subplate is made for each set of tooling, cutting tool changeover time will be the length of time it takes to swap subplates. Once the program zero setting numbers have been determined, you can use offset setting commands within your programs to specify the geometry offset values needed by the current subplate for future production runs (this assumes the subplate is keyed to the tooling table).

Assigning program zero with index chucks. Commonly used for machining plumbing fittings, index chucks allow the workpiece to be indexed within the setup during the machining cycle. Similar to an indexer on a machining center, this technique allows one program to machine several surfaces of the workpiece in one operation.

Our suggestion for index chucks is that you place the program zero point in Z at the center of index instead of the face of the workpiece being machined. By doing so, you can eliminate the need for the work shift feature. Simply specify your program zero assigning Z value to the center of index.

PROGRAM DEVELOPMENT

It should almost go without saying that if setup time reduction is given a high priority, CNC programs should be prepared off line. There are any number of excellent CAM systems available to help with even the most complicated program preparation. For simple work, the programmer may elect to write manual programs. Once written, a simple and inexpensive personal computer-based CNC text editor can be used to actually create the CNC program. If this is used in conjunction with a distributive numerical control system, your setup people will not have to type the CNC program into the control while the machine is down between production runs. Regardless of *how* the CNC program is created, if it is done while production is being run, no on-line setup time will be wasted while the program is prepared.

Though preparing programs off line is the ideal way of handling the task of CNC programming, there are applications for CNC machine tools whose production quantities are so low that *nothing* can be done off line. There are also those companies that expect the CNC operator to do all tasks related to processing, tooling, programming, setting up, verifying program, and running production. In these companies, CNC program development is being done on line.

As with any on-line setup-related task, if you cannot eliminate the task of programming or move it off line, you must do your best to facilitate the setup person's ability to perform the programming task on line. In an effort to help with preparing the CNC program on line, most CNC control manufacturers have developed *conversational* CNC controls.

Conversational controls (also referred to as *shop-floor-programmed controls*) are intended to make it as easy as possible for the setup person to develop the CNC program. Though conversational controls vary greatly from one control manufacturer to the next, most achieve this goal and dramatically simplify the programming task. Most will automatically calculate cutting conditions (feeds and speeds), allow easy input of complex geometry, allow quick and easy definition of machining operations, and even display a tool path for program verification purposes that lets the setup person see exactly what each tool will do. Once the conversational program is finished, many conversational controls will use it to actually create a true CNC program, identical to one prepared manually or by a CAM system. It is from this CNC program that production is actually run. Because conversational programming functions are so similar to a CAM system's, you can think of a conversational control as just like a normal CNC control with a built-in CAM system specifically designed for the control.

Regardless of how simple conversational controls are to work with, or how quickly they allow CNC programs to be prepared, they should be utilized only in those CNC environments where CNC programs cannot be prepared while production is running. In this sense, you can think of conversational controls as convenience features that make CNC machines easier to work with, but they do require a setup- or cycle-time-related penalty to be paid for their use (consider our earlier discussions of tool-length compensation or using a floating program zero point). Many companies misapply conversational controls, programming on line during setups simply because programming in this manner is so convenient, even though their applications are much better suited to off-line programming.

Some conversational controls allow programming to be done while production is running. This means, theoretically, that an operator, setup person, or programmer can be preparing a program while machining workpieces. However, most controls that allow this are somewhat cumbersome to work with. Since there is only one keyboard and one display screen, both must be simultaneously used for programming and operation tasks. The CNC operator will commonly need to adjust offsets, edit programs, and in general, manipulate many functions of the control while a setup person or programmer is trying to input a program. This conflict makes it difficult to perform all programming and operation functions in an efficient manner. In many cases, production must be halted while the program is completed.

PROGRAM LOADING AND SAVING

One common bottleneck in the setup procedure is getting programs in and out of the CNC control. Though we discuss distributive numerical control systems in much greater detail in Chapter Eleven, how programs are transferred to and from the CNC control plays an important role in determining setup time. Traditionally, CNC controls have been rather cumbersome in this regard. Until quite recently, the program storage capacity of CNC controls was extremely limited. In many, only a few programs could be placed in memory before memory space was exhausted. CNC users were forced to seek other means of program storage. All alternatives require programs to be transferred in and out of the control to and from some other device. This form of program transfer is known as distributive numerical control (one form of DNC).

The most common form of distributive numerical control requires RS-232C serial communications. Though RS-232C is a standard for communications protocol, anyone who has ever had to connect a CNC control to a distributive numerical control device knows that the actual connections are far from standardized.

Additionally, RS-232C has several limitations including maximum cable length and maximum transmission rate. For these reasons, CNC control manufacturers are moving away from RS-232C for transferring CNC programs.

More and more control manufacturers are trending toward personal computer-based CNC controls. With the ample memory available within today's personal computers (including RAM, hard drive, and floppy diskettes), CNC program storage within the control becomes a nonissue for most CNC users. In many cases, *all programs* the machine will ever run can be stored within the CNC control. For companies that do still need to transfer programs to and from a host computer (possibly for getting new programs into the control and for backup purposes), distributive numerical control is much easier to handle. Standard personal computer networking systems can be used. Now, not only can the computer communicate with the serving distributive numerical control computer, it can communicate with any computer in the network. This opens the door to many other applications for the distributive numerical control system (discussed in Chapter Eleven).

In the not-too-distant future, standard networking will be available with *all* new CNC controls. However, there are still many thousands of CNC machine tools in use that require RS-232C for program transfer. It will be several years before these machines are replaced and the problems related to program transfers for these machines disappear. For this reason, we offer several suggestions related to how you can facilitate the on-line task of program transfer with RS-232C communications-based distributive numerical control as efficiently as possible.

As with all on-line setup tasks, the best way to deal with the task of program loading is to eliminate it completely. Depending on your particular application and type of control, this goal may actually be achievable. If, for example, you dedicate a given CNC machine tool to machining a limited number of different workpieces, it is possible that all programs to be run by the machine will fit into the control's memory. For borderline cases, adding memory, though expensive, will completely eliminate this on-line task and should be easy to justify.

Since the program storage capacity of traditional CNC controls is quite limited, most companies are forced to perform program transfers every time a new setup must be made. As with all on-line tasks, if you cannot eliminate the task of program loading, your second choice is to move the task off line. Many CNC controls allow one program to be loaded as another program is running. Most control manufacturers call this feature *background edit*. If your control has this feature you can use it to minimize the time the machine is down between production runs. The time it actually takes to perform the loading task will not be critical, since the machine is running production while the program is being loaded. As long as your control has enough memory capacity to allow the programs for both jobs, the program for the next production run will be ready and waiting.

Unfortunately, if your control does not have background edit, and if you must load programs from one production run to the next, the task of loading programs must be done on line. In this case, you will want to do anything you can to facilitate the setup person's ability to load programs as efficiently as possible. This will involve some form of distributive numerical control system. DNC systems vary dramatically with regard to how easy they are to operate and how long they take for program transfers.

Manual DNC systems require the setup person to manually set both the CNC control and the distributive numerical control device to get ready for program transfer. There may be some communications protocol that must be correctly adjusted before program transfers are possible, especially for *portable* DNC systems. Since the time it takes to perform program transfers with manual systems can be quite long, manual DNC systems are best applied when your CNC controls have the background edit feature, allowing program transfers while production is running. Other time-consuming problems related to manual DNC systems include finding the portable DNC device, waiting for someone else to finish using it, walking back and forth to a stationary DNC system, and nature-related delays that can occur while programs should be transferred (talking to friends, getting coffee, bathroom breaks, etc.). If you are currently using a manual-style DNC system, especially a portable type, and if you study how long it truly takes to upload and download CNC programs, you may be unpleasantly surprised. What you may think is taking only a few minutes is probably taking a great deal longer.

Automatic distributive numerical control systems allow the entire task of program transfer to take place from the CNC control—no searching for the portable device, no walking back and forth to the stationary device, no waiting for someone else to finish with the device, no cryptic communications protocol. With these benefits, current-model automatic DNC systems allow CNC programs to be completely transferred in less than 30 seconds. The more program transfers your company makes, the easier it should be to justify an automatic DNC system.

Keep in mind that loading programs *into* the CNC control is only half of the program transfer problem. During program verification, it is likely that changes will be made to the CNC program that make the original program obsolete. At the very least, cutting-condition changes for the purpose of program optimizing will probably be done. In order to keep a corrected version of the CNC program, the program must be sent back to the DNC device. All time-related points apply equally to transferring programs back to the DNC system, meaning your justification of the automatic distributive numerical control system may be quite easy.

PROGRAM VERIFICATION

One on-line task that can be extremely time-consuming is program verification. We define on-line program verification as *everything that happens from the time the program is ready to run to the time the first good workpiece is machined.* The time it takes to perform this task will vary with the machine type, the complexity of the program, the method by which the CNC program is prepared, and the skill level of the people involved.

Note that by our rather broad definition, the CNC program is not only being checked for correctness, the first workpiece is actually being run. As any experienced CNC programmer would agree, even a perfectly prepared CNC program (maybe one that has run many times) will behave poorly if mistakes are made during setup. An incorrect tool-length offset or an improperly measured program zero point, for example, can cause a disaster if cautious program verification techniques are not followed. Even when the program appears to be running properly, the very tight tolerances CNC machines are expected to hold mandate that certain precautions be taken to ensure that the first workpiece comes out on size.

You must exercise caution as you attempt to reduce the time it takes to verify programs. Many procedures specified by machine tool builders are designed for safety and should not be ignored. Though we give many time-saving suggestions in this section, at no time should you attempt to reduce program verification time by skipping or shortcutting your machine tool builder's recommendations.

We break our discussion of program verification into three categories:

1. The verification procedure for new programs.
2. Verification procedures for programs that have been previously run.
3. Techniques to facilitate the running of the first good workpiece.

Throughout this discussion, we assume the CNC program to be workable. While it may not be perfect, at least the process used by the program to machine the workpiece is correct and only relatively minor modifications to the program are needed to perfect the program. As experienced CNC people, we know that mistakes made in processing, workholding selection, and cutting-tool selection can make it impossible to run good workpieces. Mistakes of this nature mean starting all over.

VERIFYING NEW CNC PROGRAMS

New CNC programs are the most difficult to verify. Since they have never run good workpieces, everything about the new program must be considered as suspect and checked with extreme caution. Generally speaking, manually written CNC programs are the most prone to error, especially when prepared by entry-level CNC programmers. Since manual programmers are human beings, and likely to make basic mistakes, every command within the manually written CNC program must be cautiously checked. Additionally, the mistakes made by beginning manual programmers can be *extremely* basic in nature. A beginning programmer may forget important things. Examples of mistakes of omission include forgetting to include decimal points in critical CNC words, forgetting to turn the spindle on, forgetting to drill a hole before tapping it, forgetting to program a cutting command after rapid, and forgetting to turn the coolant on. Any one of these mistakes of omission could result in disaster.

On the other hand, CAM-prepared CNC programs tend to be less dangerous to run, since once the system is properly tailored to a given CNC machine, the CAM system will not make the kind of basic mistakes a manual programmer is prone to. Additionally, most CAM systems provide some form of tool-path checking, giving the CNC programmer a chance to see the motions the program will generate. However, the setup person must still be quite careful when verifying new CAM-generated CNC programs. The programmer can still miss mistakes in the cutter path during the program's preparation, and there could be mistakes in the program that will not show up on any tool-path display (cutting conditions and setup mistakes, for example).

Moving tool-path verification off line

A CNC machine tool makes a terrible tool-path verification device. Many CNC machines do very little to show the setup person what the motions of the program will look like until the CNC program is actually run and the machine starts moving. With intricate motions, it can be impossible to tell what the machine is doing during the program's execution.

For this reason, more and more CNC controls are being equipped with graphic capabilities that allow setup people to see the tool path a CNC program will generate before the axes start moving. However, many that do this cannot show the tool path for one program while the machine is running another. For these machines, tool-path verification must be done on line. While this is a very nice feature, keep in mind that if tool-path verification is done on line, so must any *corrections* to the CNC program be made on line if mistakes are found. Depending on the severity of the mistake, a great deal of setup time can be wasted while the programmer corrects the mistake.

You can easily move the task of tool-path verification (as well as the associated CNC program corrections) off line. There are any number of excellent personal computer-based software programs designed to help with tool-path verification. In essence, the personal computer-based program verification software will simulate the tool-path display of your CNC controls. Most are reasonably priced and can show the CNC programmer exactly what each tool in the program will do. Most will also check for basic syntax mistakes in the CNC program, generating CNC control-like alarms if mistakes are found. Some can even *animate* the machining process, showing the programmer what the workpiece will look like when the program is completed. While there could still be problems during on-line program verification of tool offsets, program zero setting, and cutting conditions, and the setup person must still exercise caution when running the program, at least the setup person can rest assured that each tool's basic motions are correct.

Procedures related to verifying new programs

Machine tool builders make specific recommendations with regard to how new CNC programs should be verified for their specific machines. In all cases, we bow to their recommendations. Here we make some time-saving suggestions about the procedures to verify new programs.

Machine-lock dry run. If CNC programs are written manually, and if no off-line tool-path verification software is used, it is possible that some basic syntax mistakes may exist in the CNC program. Many machine tool builders recommend that the setup person check the program for syntax mistakes by letting the control scan the program. Many control manufacturers call the feature that gives this ability *machine lock*.

In the machine lock mode, the control will be allowed to run through the program, but no actual axis motions will be made. Many machines that allow this function will execute the CNC program in exactly the same manner as when allowing the axes to move, meaning certain functions like dry run rate and rapid rate affect how quickly the machine-lock dry run will be done. Since there is no danger of crashes or interference during the machine-lock dry run, the setup person must set dry-run and rapid rates to their maximum settings before performing the machine-lock dry run. This will allow the CNC control to execute the program as quickly as possible.

Free-flowing dry run. Once the setup person has confirmed that the basic syntax of the program is correct (either with off-line software for tool-path verification or with a machine-lock dry run), many machine tool builders recommend a free-flowing dry run. With this program verification procedure, the machine will respond to the CNC program, but the setup person can completely control the motion rate of every movement. Usually a multiposition switch works as a kind of rheostat to control the rate of all motions. As a tool nears an obstruction, the setup person can slow the motion rate. As the tool moves away from obstructions, the motion rate can be increased.

When verifying new programs, and especially manually prepared programs that have not been checked with an off-line tool-path verification system, the setup person will have to be *very* careful with all movements the program generates. Every motion command in the program could contain a motion mistake.

Normal air-cutting run. Most CNC controls make it impossible for the setup person to tell the difference between rapid motions and cutting motions during a free-flowing dry run. If the CNC program is prepared manually, the programmer may forget to specify a cutting command (like G01, G02, or G03) after having the tool rapid up to the workpiece. In this case, the control will maintain the (modal) rapid command and crash the tool into the workpiece.

Many machine tool builders recommend one more verification procedure to check for this problem. The setup people run the program one more time, just as if a workpiece were being machined, but with no workpiece in the setup. During this normal air-cutting run, the setup person can check to be sure cutting commands are included in the commands as they are needed.

This on-line procedure can be eliminated if off-line tool-path verification software is used. Most tool-path verifiers make it easy for the programmer to tell the difference between rapid and cutting commands. Most depict rapid motions with dotted lines and cutting motions with solid lines.

Actually running the first workpiece. At this point, the setup person is ready to machine a workpiece. If all previous recommendations are followed, there should be no catastrophic problems left in the program, and the program can almost be treated as if it has been run before. While there may still be some subtle motion mistakes that would cause machining problems, and the setup person should be more careful than when verifying a previously run program, the setup person can assume at this point that the program has the ability to make a good workpiece. A little later, we will describe several techniques that can be used to efficiently machine the first workpiece to size.

VERIFYING PREVIOUSLY RUN CNC PROGRAMS

Many manufacturing people who do not work directly with CNC do not understand why programs that have been run before must still be verified. These people (especially uninformed managers) tend to pressure CNC setup people to rush the program verification procedure. To avoid misunderstanding, everyone in the CNC environment must understand how setup-related mistakes can cause dangerous situations for the people running the CNC equipment.

Remember that the more manual intervention there is with setup tasks, the more potential there will be for setup-related mistakes and the harder it will be to verify previously run CNC programs. Tool-length compensation offsets on a machining center, for example, control the Z position to which each tool is sent. While a CNC program may correctly tell the control to send a tool to a Z-axis clearance position, if the tool's length compensation value is incorrectly measured or entered, it is possible that the tool will be crashed into the workpiece or workholding setup. The potential for this kind of mistake is dramatically reduced if tool-length compensation offset values are determined *off line* with the kind of tool-length measuring gage that automatically creates the offset setting CNC program. As compared to manual tool-offset measurement and entry, the setup person can be much more confident that each tool's first Z-axis approach to the workpiece will be correct.

Other examples of potential mistakes during setup that make it necessary to verify previously run CNC programs include those made with the assignment of program zero, the placement of workholding devices, the assembly and placement of tools into the machine's magazine or turret, and even something as basic as calling up the correct version of the CNC program itself. In almost all cases, if you can automate or eliminate the task, you can minimize the amount of time and work it will take to verify previously run CNC programs.

Procedures related to previously run CNC programs

Though some companies go to great lengths to eliminate the need to verify previously run CNC programs (by minimizing or eliminating manual intervention during setup), the vast majority of companies still have their setup people repeat at least some of the program verification procedures even for previously run CNC programs. While it is *much* easier to verify programs run previously, it still takes time. We urge you to look into steps that may be taken in your particular CNC environment to minimize the work required to perform the procedures we now discuss.

Free-flowing dry run. Many setup people feel the need to repeat the free-flowing dry run even when verifying previously run programs. While we do not want to encourage techniques that are not entirely comfortable to your setup people, if a program has been run successfully in the past, it will run correctly again as long as no program changes have been made since its last running. While the setup person will have to exercise extreme caution while actually running the first workpiece, as long as single-block, dry-run, and distance-to-go functions are used on each tool's first approach to the workpiece, any setup-related mistake can be safely and easily found, eliminating the need for a free-flowing dry run.

Running the first good workpiece. Regardless of whether they are verifying a new or previously run CNC program, setup people must take certain precautions if they expect the first workpiece to come out perfectly to size. From a safety standpoint, it is necessary for setup people to be extremely cautious with each tool's first approach to the workpiece. Since the machine will be in the rapid mode during this movement, and since the approach distance is usually very small, there will be little the setup person can do if the program and/or tool offset values are incorrect.

For this reason, we recommend that the setup person get into the habit of using single-block and dry-run on *every* tool's first approach to the workpiece for the first workpiece being run. We also recommend using a display screen feature available on almost all CNC controls called *distance to go*. The distance-to-go feature dynamically shows the setup person how much farther the machine will move in the current CNC command and makes it obvious if a command will cause the tool to bump into the workpiece. With these features, the setup person can easily control each tool's approach and safely stop the machine before a mishap if mistakes are found.

Making the first workpiece correctly

While this may sound like a odd statement, several factors affect how important it is to make the first part come out to size. If, for example, the cost of raw material is very low and production quantities are high enough to allow extra workpieces, the setup person may not be overly concerned if the first few workpieces are scrapped. For small workpieces run from bar on bar-feed turning centers, for example, the raw stock cost may be less than 20 cents per workpiece. People in the CNC environment may feel that getting the first workpiece to come out perfectly will cost more than the workpiece value. For this reason, some CNC setup people will be unconcerned with making the first few workpieces correctly, as long as they learn enough from quickly machining the workpieces to eventually make the necessary adjustment (usually in the form of offset settings). People who do this will contend that the overall cost and time related to completing the setup can be reduced by this method.

While a good case can be made for minimal concern about scrapping a few inexpensive workpieces, we strongly feel that every setup person should at least *have the ability* to make the first workpiece come out perfectly to size. As the cost of the raw material grows, so does the importance related to making good workpieces. Additionally, the time may come when your setup person has no additional raw material on which to practice. In this case, the setup person must make the first workpiece a good one. All of this does not even consider the potential for dangerous situations that can arise if the machining operations performed by each tool in the program are not cautiously checked.

Understanding the need for trial machining

CNC equipment is used to hold extremely close tolerances. The closer the tolerance, the more difficult it is to ensure that the machine will hold the tolerance without special considerations. It is not unusual, for example, for a CNC turning center to hold an overall diameter tolerance of less than 0.0004 inch (0.01 mm). While it is not unrealistic to expect the CNC turning center to hold this tolerance in production, it *is* unrealistic to expect this tolerance to be held on the very first workpiece if trial machining is not performed. The only exception would be if the tool used to machine the close tolerance diameter was used in the previous production run and was machining properly at the completion of the job.

When faced with holding close tolerances, there will *always* be a way for the setup person to guarantee that the first workpiece will come out on size for *every tool* by using trial-machining techniques. Traditionally, trial machining involves considering the offset values used to machine the close tolerance dimension *before* machining takes place. There is *always* a way to adjust the tool's offset in such a way that excess material is left on critical surfaces being machined, ensuring that the workpiece will not be scrapped the first time the tool is run. After the preliminary offset adjustment, the tool will then be run and the cycle stopped when the tool is finished (an optional stop M01 is commonly used for this purpose). The critical dimensions will then be measured. Based on the measurements, the setup person can determine how much excess stock is left on the critical surfaces. This will be the amount of offset required to machine the surfaces perfectly to size. After the offset change the tool will be rerun. As long as tool pressure does not affect the machining operation, the tool will machine the critical surfaces perfectly to size. These important principles can be applied to both machining centers and turning centers any time a critical surface must be machined.

Trial machining is a helpful technique used by many setup people to ensure that each tool machines the workpiece in the correct manner. However, there are setup-time-related considerations that make this very common and helpful technique rather wasteful. And there are many things a CNC programmer can do within CNC programs to speed the setup person through the trial-machining process for critical machining operations.

Facilitating trial machining

If a given tool's machining time is relatively short, rerunning the entire tool is the best alternative. Say, for example, a turning center's grooving tool takes only 15 seconds to perform its necking operation. The setup person can use common trial-machining techniques and quickly rerun the entire grooving tool to make it cut to size in about 15 seconds.

However, as the machining time for a given tool grows, so does the length of time needed for trial machining. Consider a rough turning tool used on a turning center to machine a large shaft. Say this tool machines for more than 30 minutes. In this case, using traditional trial-machining techniques to ensure that the rough turning tool will leave the correct amount of finishing stock will be rather wasteful. And rerunning the tool will take another 30 minutes. While it may be possible to speed the process somewhat with dry run, the setup person will have to be quite careful with its use.

Additionally, consider tools with which dry run cannot be used to speed up the tool's (air-cutting) movements. With lengthy threading operations, for example, the feed rate must remain perfectly synchronized with the spindle speed. If machining a coarse multiple-start thread, for example, the time needed to completely remachine the thread after trial machining could be very lengthy.

Another turning center sizing problem has to do with finishing critical surfaces. If common trial machining techniques are used when finishing, the setup person will first finish the workpiece with an offset that causes a small amount of excess stock to be left on each surface. This small amount of stock is *not* the intended depth of cut for the finishing tool. After trial machining, the setup person knows how much to offset the tool for the *next* workpiece. However, if this offset is used for the current workpiece, the reduced tool pressure may cause the tool to remove too much material. This can be a highly troublesome problem when machining somewhat flimsy workpieces having extremely close tolerances.

A CNC programmer can do many things to facilitate the setup person's ability to quickly size for each tool, and completely eliminate the need to rerun the tool. Most of the techniques we show involve the use of optional block skip (also called *block delete*) to allow the setup person to execute a series of trial machining commands during setup that are ignored during the production run. While we show several specific examples of how this can be done, remember that similar techniques can be used whenever rerunning the entire tool is wasteful.

Facilitating sizing for lengthy rough turning and boring operations. Figure 8-15 is a drawing of a large shaft requiring about 15 minutes of rough turning time. If common trial machining techniques are used to ensure that the rough turning tool leaves the proper amount of finishing stock, the entire rough turning operation must be repeated. Fifteen minutes of program verification time will be

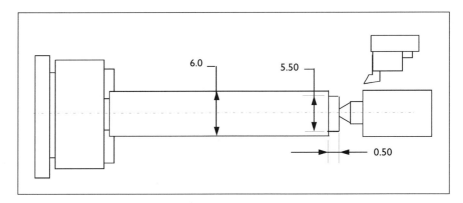

Figure 8-15. By setting the rough-turning tool's offset before the first workpiece is rough turned, substantial program verification time will be saved.

wasted. By using our recommended method, the setup person will be able to set the rough turning tool's offset *before* the first workpiece is rough turned. Since the amount of time needed to actually set the offset will remain essentially the same with our given method, the amount of program verification time that will be saved will be almost 15 minutes.

As you can see from Figure 8-15, the programmer is going to program a small rough turning pass under the influence of optional block skip. To ensure that tool pressure will remain consistent, this roughing pass should be at the same depth of cut used for the normal rough turning operation. This rough turning pass only needs to go far enough into the workpiece to allow a measurement to be taken.

Here is a portion of the program showing the trial-machining operation.

O0003 (Program number)
N005 T0101 M41(Select rough turning tool, offset, and spindle range)
N010 G96 S400 M03 (Start spindle clockwise [cw] at 400 sfm)
N015 G00 X6.0 Z0.1 (Rapid up to the workpiece)
/N020 X5.5 (Begin trial-machining operation)
/N025 G01 Z-0.3 F0.020 (Trial machine)
/N030 X6.0 (Feed up face)
/N035 G00 X8.0 Z3. (Rapid to convenient measuring position)
/N040 M00 (Stop for measurement, diameter should be 5.50 inches [139.7 mm])
/N045 T0101 M03 (Reinstate offset, restart spindle)
/N050 G00 X6. Z0.1 (Rapid back to starting point)
N055 G71 P060 Q160 D2500 U0.040 W0.005 F0.020 (Rough turn)
N060 . . .
.
.
.

In line N020, we begin the trial-turning operation. In lines N025, N030, and N035, the tool makes the trial-turning pass and rapids to a convenient measuring position. At this position the setup person can easily measure the workpiece. In line N040, the machine stops because of the M00. We strongly recommend that a message be included in the program at this point telling the setup person what the diameter (and if necessary, the Z face position) of the workpiece should currently be. The setup person measures and adjusts the offset accordingly. Line N045 reinstates the offset, based on the setup person's offset change. In line N050, the tool rapids back to its starting point. From line N055, the program continues in its normal manner. After setting the offset, the operator turns on the optional block skip function.

The optional block skip switch can be turned off whenever the setup person wishes to use the trial-machining sequence; therefore, the CNC operator will also have this sequence available should it be needed when changing (or indexing) the rough turning tool's insert during the production run.

Any lengthy roughing operation can be handled in much the same manner. For rough boring on a turning center, the only difference will be that the programmer may have to move the boring bar farther away from the workpiece to allow a measurement to be taken.

Eliminating tool pressure problems when finish turning or boring. The turning center programmer can also facilitate the setup person's ability to size for the finishing tool *before* the finishing operation takes place. By incorporating this technique, any tool-pressure-related problems caused by using more conventional trial-machining processes can be eliminated.

While we use the same long shaft shown in Figure 8-15 for this example program, keep in mind that this technique can also be used when the rough turning operation is quite short. If the goal is simply to perfectly size the finishing tool, however, you must still use a trial roughing operation to confirm that the roughing tool leaves the proper amount of stock for finishing.

With this technique, we simply include another set of commands under the influence of optional block skip for finish turning right after the trial rough turning commands (in the rough turning tool portion of the program). These commands will first remachine with the rough turning tool to ensure that the rough turning tool has left the correct amount of finishing stock. Then the program will index the turret to the finishing tool and continue machining on our practice surface. It is important to program the same depth of cut during the trial-finishing operation as will be used for the actual finishing operation. After machining, the machine will move to the convenient measuring position and stop again. At this point the setup person measures the surfaces and adjusts offsets accordingly. After this technique is used, the setup person can rest assured that the finishing tool will machine perfectly to size, even on the very first workpiece. Here is the example program.

O0003 (Program number)
N005 T0101 M41(Select rough turning tool, offset, and spindle range)
N010 G96 S400 M03 (Start spindle [clockwise] at 400 sfm)
N015 G00 X6.0 Z.1 (Rapid up to the workpiece)
/N020 X5.5 (Begin trial-machining operation)
/N025 G01 Z-0.3 F0.020 (Trial machine)
/N030 X6.0 (Feed up face)
/N035 G00 X8.0 Z3. (Rapid to convenient measuring position)
/N040 M00 (Stop for measurement, diameter should be 5.50 inches [139.7
mm])
/N045 T0101 M03 (Reinstate offset, restart spindle)
/N055 G00 X5.5 Z.1 (Rapid back to rough-turned diameter)
/N060 G01 Z-0.3 F0.020 (Ensure correct diameter)
/N065 X6.0 (Feed up face)
/N070 G00 X8.0 Z6.0 (Rapid to tool change position)
/N075 T0202 M42 (Index to finish turning tool, select range)
/N080 G96 S700 M03 (Select finish turning speed)
/N085 G00 X5.42 Z.1 (Rapid to trial diameter, 0.040 inch [1.016 mm] cut
depth)
/N090 G01 Z-0.3 F0.008 (Trial machine)
/N095 X6.0 (Feed up face)
/N100 G00 X8. Z5. (Rapid to tool change position)
/N105 M00 (Diameter should be 5.4200 inch [137.67 mm])
/N110 T0101 M41 (Reselect rough turning tool)
/N115 G96 S400 M03 (Reselect roughing speed)
/N120 G00 X6. Z0.1 (Rapid back to starting point)
N125 G71 P130 Q230 D2500 U0.040 W0.005 F0.020 (Rough turn)
N130 . . .
.
.
.

With this new technique, the setup person must confirm that the rough turning
tool will not cut undersize with its first pass, meaning a positive offset must be
placed into offset 1 to force some excess stock to be left. Additionally, the proper
speed and feed rate for actual finish turning must be used in trial finish turning. In
line N105 the setup person measures the diameter and adjusts the finishing offset
accordingly. The program then indexes back to the rough turning tool and begins
the actual rough turning operation.

After the trial-machining operation has been completed and the setup person is
sure that roughing and finishing will be done correctly, the optional block skip
switch can be turned on to skip the trial-machining operation in production. Whenever inserts are changed, these same techniques can be used again.

You may be questioning the wisdom of including the actual trial-machining commands in the program that machines the workpiece. Admittedly, if these techniques are used often, the CNC programmer may be cluttering the program with a great number of commands that are seldom used. Keep in mind that the trial-machining commands can be easily stored in a separate subprogram or parametric program, and invoked with one simple command from the main program. Here is how trial machining can be done in the format for one popular control. This particular control happens to use an M98 to call a subprogram. Within the M98 command, a P word specifies which subprogram is being called.

O0003 (Main program)
N005 T0101 M41(Select rough turning tool, offset, and spindle range)
N010 G96 S400 M03 (Start spindle clockwise at 400 sfm)
N015 G00 X6.0 Z0.1 (Rapid up to the workpiece)
/N020 M98 P1000 (Call trial-machining subprogram)
N025 G71 P130 Q230 D2500 U0.040 W0.005 F0.020 (Rough turn)
N030 . . .

.
.
.

O1000 (Subprogram)
N001 X5.5 (Begin trial-machining operation)
N002 G01 Z-0.3 F0.020 (Trial machine)
N003 X6.0 (Feed up face)
N004 G00 X8.0 Z3. (Rapid to convenient measuring position)
N005 M00 (Stop for measurement, diameter should be 5.50 inch [13.97 mm])
N006 T0101 M03 (Reinstate offset, restart spindle)
N007 G00 X5.5 Z0.1 (Rapid back to rough-turned diameter)
N008 G01 Z-0.3 F0.020 (Ensure correct diameter)
N009 X6.0 (Feed up face)
N010 G00 X8.0 Z6.0 (Rapid to tool-change position)
N075 T0202 M42 (Index to finish turning tool, select range)
N011 G96 S700 M03 (Select finish turning speed)
N012 G00 X5.42 Z0.1 (Rapid to trial diameter, 0.040 inch [1.016 mm] cut depth)
N013 G01 Z-0.3 F0.008 (Trial machine)
N014 X6.0 (Feed up face)
N015 G00 X8. Z5. (Rapid to tool-change position)
N016 M00 (Diameter should be 5.4200 inches [137.67 mm])
N017 T0101 M41 (Reselect rough turning tool)
N018 G96 S400 M03 (Reselect roughing speed)
N019 G00 X6. Z.1 (Rapid back to starting point)
N020 M99 (End of subprogram)

If similar trial-machining operations are to be done on different workpieces using this subprogram, several commands of the subprogram must be changed in order to go from one production run to the next. While this may not be extremely difficult to do, if your control allows parametric programming, you can simplify this task. You can have one program in your control (the parametric program) that does the trial machining for all workpieces. From each main program that machines a different workpiece, you can call the parametric program, specifying *variables* to tell the parametric program the diameter to be trial machined, along with other things that change from one production run to the next, like the tool stations involved, speeds and feeds, and other parameters.

Facilitating the sizing for lengthy threading operations. Many threads take very little time to machine. A very fine pitch, single-start, short thread on a small diameter, for example, may not require more than about 10 or 20 seconds to machine. In this case, use conventional trial-machining techniques to size the thread. However, the longer the thread takes to machine, the more time it will take to use conventional trial-machining techniques. Coarser threads, for example, require more passes (and more time) to machine. Multiple-start threads require even more passes, meaning even more time. A lengthy 4-start Acme thread on a large diameter, for instance, may take *20 to 30 minutes* to machine. If conventional trial-machining techniques are used, the entire thread must be run again, consuming from 20 to 30 minutes of wasted program verification time. By the way, since threading requires the spindle speed and feed rate to be perfectly synchronized, there will be no way to reduce this remachining time with dry run.

The same optional block skip techniques just shown can be used to minimize trial-machining time for lengthy threading operations. However, keep in mind that some CNC controls make it easy to specify how threads are to be machined within their standard canned cycles. If this is the case, it may be quite easy for the setup person to simply modify the threading command to minimize the number of threading passes needed to finish the thread after offset adjustment. Once the first thread has been machined to size, of course, the threading command must be changed back to its original state.

Unfortunately, modifying the CNC program to minimize the number of threading passes requires the setup person to thoroughly understand the threading command. Mistakes can result in disaster for the threading tool. For this reason, and since not all companies use canned standard threading cycles, it may be necessary to size threads by using optional block skip techniques if you wish to minimize program verification time.

Threading poses additional challenges to the development of trial-machining techniques. Since the depth of cut is reduced with each successive threading pass, the setup person must be rather careful when reducing the offset value during trial threading. If, for example, after trial threading during the program stop (M00), the setup person finds that the thread is still 0.008 inch (0.203 mm) oversize, it will not be feasible to expect the threading tool to completely machine the 0.008

inch in a single pass. The setup person will have to cause the machine to take several more threading passes. One method of handling the multiple additional passes problem follows. To keep the example simple, we demonstrate it on a single-start thread, but similar techniques can be used for any kind of thread with any number of thread starts.

O0004 (Program number)
N005 (Turning operation begins)
N010 . . .
.
.
.
N085 T0505 M41 (Threading begins)
N090 G97 S500 M03 (Start spindle cw at 500 rpm)
N095 G00 X1.2 Z0.2 (Rapid up to thread)
N100 G76 X0.838 Z-1.0 D0200 K.081 F.125 (Machine 1 inch -8 thread)
/N105 G00 X5. Z3. (Move to convenient measuring position)
/N110 M00 (Stop for measurement)
/N115 T0505 M03 (Reinstate offset for next pass, restart spindle)
/N120 G00 X0.838 Z0.200 (Rapid back to current pass position)
/N125 G32 Z-1.0 (Make another pass)
/N130 G00 X1.2 (Rapid up)
/N135 Z0.2 (Rapid back to start position)
/N140 GOTO 105 (Go back to line N105)
N145 G00 X8. Z7. (Rapid to tool change position)
N150 M01 (Optional stop at tool end)
N155 T0606 M42 (Program continues)
.
.
.

In line N140, we are assuming that your control allows *unconditional branching* (the GOTO statement). This command allows trial-threading passes to be done an unlimited number of times. In line N110, for example, if the setup person finds the thread to be 0.008 inch (0.203 mm) oversize, the offset can be reduced by a smaller amount, say 0.002 inch (0.051 mm). When the cycle is activated (with optional block skip still off), another threading pass will be made and the machine will stop again. The setup person can continue to reduce the offset in this manner to make trial-threading passes. Eventually, when the thread is to size, the setup person will turn *on* the optional block skip switch, and the control will stop executing trial-threading passes. The machine will return to its tool change position and continue with the next machining operation.

If your control allows parametric programming, this procedure can be made more automatic. Instead of repeatedly changing the threading offset, the setup person may simply set a variable telling the control how much more stock is to be removed. The parametric program can calculate the number of additional passes needed and make them automatically.

Facilitating finish boring on machining centers. Most machining center setup people would agree that adjusting a finish boring bar to machine perfectly can be one of the most challenging tasks related to making machining center setups. Regardless of how well the boring bar is adjusted prior to machining, tool pressure will usually affect the diameter the boring bar machines. For this reason, the boring bar is usually intentionally set undersize. During program verification, the setup person allows the boring bar to enter the hole just far enough to allow a measurement to be taken. Then the cycle is stopped, the hole diameter is checked, and the boring bar is adjusted for another try. Depending on the quality of the boring bar and the skill of the setup person, it can take several tries before the boring bar is properly set.

Anything a programmer can do to facilitate the setup person's ability to size for the finish boring bar will effectively reduce program verification time. If you perform boring operations on a regular basis, we recommend keeping a trial boring subprogram permanently stored in the control's memory. The subprogram we show can be used for any hole diameter in any location from any program. It can be called repeatedly, giving the setup person as many tries as are needed.

O0005 (Main program)
.
.
.

N355 T09 M06 (Place boring bar in spindle)
N360 G54 G90 S800 M03 (Startup for boring bar)
N365 G00 X5. Y3.5 (Rapid to hole location)
N370 G43 H01 Z0.1 (Instate tool length compensation, rapid just above work surface)
/N375 M98 P1000 (If optional block skip is on, make first trial-boring pass)
/N380 M98 P1000 (Make second trial-boring pass)
/N385 M98 P1000 (Make third trial-boring pass)
/N390 M98 P1000 (Make fourth trial-boring pass)
N395 G86 R0.1 Z-1.5 F3. (Bore hole to depth)
N400 . . .
.
.
.

This program allows up to four trial-boring passes. By simply repeating the subprogram call command (M98 in our case), your programmer can easily add as many trial-boring passes as deemed necessary. If your control allows unconditional branching, your programmer can provide an unlimited number of trial-boring passes. Consider these commands:

/N390 M98 P1000 (Make fourth trial-boring pass)
/N395 GOTO 390
N400 G86 R0.1 Z-1.5 F3 (Bore hole to depth)

With this technique, the control will continue to make trial-boring passes until the optional block skip switch is turned on. Here is the subprogram that actually does the trial boring. Notice that it is written in the incremental mode to allow it to function for any hole location.

O1000 (Program number)
N1 G91 G86 R0 Z-0.3 F3 (Bore to shallow depth)
N2 G80 (Cancel cycle)
N3 G00 Z3. (Rapid up three inches)
N4 G00 X4.0 Y 4.0 (Move to convenient measuring position)
N5 M00 (Program stop to take measurement and adjust bar)
N6 X-4.0 Y-4.0 M03 (Move back over hole, restart spindle)
N7 Z-3. (Move back to just above hole)
N8 G90 (Reselect absolute mode)
N9 M99 (End of subprogram)

Notice in line N1, the hole is being bored just deep enough to allow the measurement to be taken. The boring bar then moves to a convenient measuring position and stops. The setup person measures the hole size and adjusts the boring bar accordingly. The cycle is reactivated and the control returns to the hole location.

Other potential applications for trial machining. The specific techniques we have shown for trial machining should illustrate how easily the programmer can facilitate program verification. The key is to let the programmer know up front that some program verification problem exists. *Your programmers will consider making changes to facilitate program verification only if they watch setup people making setups.* Truly, any time a setup person must take more than about a minute to size a tool is a good time to consider using program verification facilitating techniques. Though it may take a little ingenuity and determination on the programmer's part, there will always be something the programmer can do to help the setup person.

Other examples of times when program verification facilitating can be done include:

- Lengthy rough milling operations. The programmer can program a short milling pass under the influence of optional block skip. The surfaces milled by this pass can be used as a point of reference for a measurement. Based on the measurement, the setup person will adjust a tool-length compensation and possibly a cutter-radius compensation offset during the program stop. When the cycle is reactivated, the rough milling cutter will correctly machine the first workpiece to its programmed size, allowing the perfect amount of finishing stock.

- Slot-milling operations. Many times if the tool-length compensation off-set for a slot-milling cutter is not perfectly set, a slot will not be machined in its proper location. With trial-machining techniques, a small portion of the slot can be machined to a very shallow depth (just deep enough for the measurement). During an M00 program stop, the setup person can measure and adjust the slotting cutter's tool-length compensation offset accordingly. The slot will then be machined in its proper position and to its correct depth.

CHANGING MACHINING ORDER

There may be times during a program's verification that severe mistakes are discovered in the process. For example, say a program is prepared for a turning center that follows this process:

1. Rough turn
2. Finish turn
3. Drill 2-inch hole
4. Rough bore
5. Finish bore

Experienced machinists would agree that you should *rough everything before you finish anything.* In this process, the finish turning is being done before the drilling and rough boring. It is quite likely that during these operations, the workpiece will be deformed slightly, and the turned diameter will probably not be concentric to the bored diameter.

While this may be a relatively obvious processing mistake, there are times when your CNC machine will be down waiting for the program to be changed to reflect a correction in machining order. In these cases, the programmer must usually go back to the office and call up the program within a CNC text editor. Using cut-and-paste techniques, the programmer will modify the original program to reflect the process change and download the corrected program to the CNC machine. Depending on the programmer's availability, the programmer's aptitude, whether the computer is available, and how quickly programs can be transferred, this wasteful on-line task could take from 15 to 45 minutes.

Keep in mind that newer CNC controls have very powerful text editors that allow cut-and-paste techniques to be done right on the CNC control. When available, these editors should be used to minimize the CNC machine's downtime. Most are simple enough to use that the setup person can make the required program changes without needing the programmer's help.

For those machines that do not have cut-and-paste features, remember that many controls allow unconditional branching (usually specified with a GOTO command). The setup person can quickly and easily change the program to match the desired machining order. Here is an example to stress how easy it is. First we show the *incorrect* program based on the previous process.

O0008 (Program with incorrect process)
N005 T0101 M41 **(Rough turning tool)**
N010 G96 S400 M03 (Start spindle cw at 400 sfm)
N015 G00 X3.040 Z0.1 (Rapid to rough turn diameter)
N020 G01 Z-1.995 F0.017 (Rough turn)
N025 X3.25 (Feed up face)
N030 G00 X6.0 Z5.0 (Rapid to tool change position)
N035 M01 (Optional stop)
N040 T0202 M42 **(Finish turning tool)**
N045 G96 S600 M03 (Start spindle cw at 600 sfm)
N050 G00 X3. Z0.1 (Rapid to diameter to be turned)
N055 G01 Z-2. F0.006 (Finish turn)
N060 X3.25 (Feed up face)
N065 G00 X6.0 Z5.0 (Rapid to tool change position)
N070 M01 (Optional stop)
N075 T0303 M41 **(2-inch [51-mm] drill)**
N080 G97 S300 M03 (Start spindle cw at 300 rpm)
N085 G00 X0 Z0.1 (Rapid to position)
N090 G01 Z-2.6 F0.009 (Feed to depth)
N095 G00 Z0.1 (Rapid out of hole)
N100 G00 X6.0 Z5.0 (Rapid to tool change position)
N105 M01 (Optional stop)
N110 T0404 M41 **(1.5-inch [38-mm] rough boring bar)**
N115 G96 S400 M03 (Start spindle cw at 400 sfm)
N120 G00 X2.085 Z0.1 (Rapid to rough bore diameter)
N125 G01 Z-1.995 F0.010 (Rough bore)
N130 X2.0 (Feed down face)
N135 G00 Z0.1 (Rapid out of hole)
N140 X6.0 Z5.0 (Rapid to tool change position)
N145 M01 (Optional stop)
N150 T0505 M42 **(1.5-inch finish boring bar)**
N155 G96 S600 M03 (Start spindle cw at 600 sfm)
N160 G00 X1.125 Z0.1 (Rapid to position)
N165 G01 Z-2.0 F0.006 (Finish bore)
N170 X2.0 (Feed down face)
N175 G00 Z0.1 (Rapid out of hole)
N180 G00 X6.0 Z5.0 (Rapid to tool change position)
N185 M30 (End of program)

Here is the corrected version of the program that uses the desired process.

O0008 (Program with incorrect process)
N005 T0101 M41 **(Rough turning tool)**
N010 G96 S400 M03 (Start spindle cw at 400 sfm)
N015 G00 X3.040 Z0.1 (Rapid to rough turn diameter)

N020 G01 Z-1.995 F0.017 (Rough turn)
N025 X3.25 (Feed up face)
N030 G00 X6.0 Z5.0 (Rapid to tool change position)
N035 M01 (Optional stop)
N038 GOTO 075 (Jump to line N075)
N040 T0202 M42 **(Finish turning tool)**
N045 G96 S600 M03 (Start spindle cw at 600 sfm)
N050 G00 X3. Z0.1 (Rapid to diameter to be turned)
N055 G01 Z-2.0 F0.006 (Finish turn)
N060 X3.25 (Feed up face)
N065 G00 X6.0 Z5.0 (Rapid to tool change position)
N070 M01 (Optional stop)
N073 GOTO 185 (Jump to end of program)
N075 T0303 M41 **(2-inch [51-mm] drill)**
N080 G97 S300 M03 (Start spindle cw at 300 rpm)
N085 G00 X0 Z0.1 (Rapid to position)
N090 G01 Z-2.6 F0.009 (Feed to depth)
N095 G00 Z0.1 (Rapid out of hole)
N100 G00 X6.0 Z5 (Rapid to tool change position)
N105 M01 (Optional stop)
N110 T0404 M41 **(1.5-inch [38-mm] rough boring bar)**
N115 G96 S400 M03 (Start spindle cw at 400 sfm)
N120 G00 X2.085 Z0.1 (Rapid to rough bore diameter)
N125 G01 Z-1.995 F0.010 (Rough bore)
N130 X2.0 (Feed down face)
N135 G00 Z0.1 (Rapid out of hole)
N140 X6.0 Z5.0 (Rapid to tool change position)
N145 M01 (Optional stop)
N150 T0505 M42 **(1.5-inch finish boring bar)**
N155 G96 S600 M03 (Start spindle cw at 600 sfm)
N160 G00 X1.125 Z0.1 (Rapid to position)
N165 G01 Z-2.0 F0.006 (Finish bore)
N170 X2.0 (Feed down face)
N175 G00 Z0.1 (Rapid out of hole)
N180 G00 X6.0 Z5.0 (Rapid to tool change position)
N183 GOTO 040 (Go back to line N040 to finish turn)
N185 M30 (End of program)

With three simple commands (lines N038, N073, and N183), we have changed the machining order to match our desired process. Admittedly, this program is rather difficult to follow since it is not executed in sequential order. For this reason, we still recommend that the programmer eventually changes it in the conventional manner. But since the machine is running production, this can be done at the programmer's leisure.

PROGRAM OPTIMIZING

If you are running a small quantity of workpieces with short cycle times, just about any CNC program that makes acceptable workpieces will suffice. However, as production quantities grow, it becomes more and more important to reduce cycle time. The higher the production quantities, the more important it is that you optimize your CNC programs. We devote an entire chapter of this text to cycle-time reduction techniques. For now, we simply address the setup-task-related functions of program optimizing.

As may be fairly obvious, any time taken between production runs for the purpose of optimizing the way a CNC program runs adds to setup time. While this time may be easily made up during the production run, anything that can be done to facilitate the setup person's ability to optimize will effectively reduce setup time.

Optimizing CNC programs can be broken into three steps (as stated, the degree to which your setup people optimize must be based on production quantities):

1. The setup person should optimize the program's format to remove commands that waste program execution time. These commands are usually easy to spot when watching the program run. For example, the machine may be making wasteful movements; M codes may not be programmed efficiently; or, in general, noticeable pauses exist at points during the program's execution. Wasteful program formatting tends to be repeated, meaning the CNC programmer must be made aware of habitual techniques that waste program execution time.

2. Cutting conditions must be optimized. By monitoring the machining of several workpieces, the setup person may be able to spot operations that can be accomplished faster by simple speed and feed-rate changes. Or it may be possible to improve cycle time by using a different style tool. However, the setup person must be on the lookout for tradeoffs when changing cutting conditions. It is likely that more aggressive cutting conditions will adversely affect tool life.

3. For ultrahigh production quantities, the *process* must be optimized. Just as a CNC programmer may overlook wasteful motions while developing a CNC program, so the process engineer may miss potential time-saving improvements to the process during its development. After seeing the process in action, it may be possible to spot areas that can be improved. For example, it may be possible to hold more workpieces on the table of a machining center during machining, to minimize the number of tool changes necessary per workpiece. Optimizing in this area often means backtracking to a certain extent. It may be necessary, for instance, to modify fixtures, purchase different cutting tools, and in general, spend more money to improve the method by which workpieces are produced.

PROGRAMMING WITH VARIABLES

The CNC programmer can facilitate the setup person's ability to optimize cutting conditions. More and more CNC controls allow parametric programming techniques. One of the most important features of parametric programming is the use of variables. Variables allow the CNC programmer to specify values in such a manner that they can be easily changed. With one version of parametric programming, called Custom Macro version B, variables are specified with a pound sign (#). For example, #100 is a common variable. The value of #100 can be set one time at the beginning of the program and referenced easily during the program's execution. Consider these commands:

N005 #100 = 500

.

.

.

N150 S#100 M03

When the first command is executed, the value of #100 is set to 500. In line N150, when the spindle command is read, the control will read the command as if the S word is set to 500. This has fantastic implications, especially for program optimizing purposes.

Programming a variable rapid approach distance

Many programmers are taught to rapid tools to a clearance position of 0.100 inch (2.54 mm) from the surface to be machined. While this is a very safe position, especially for beginners, it tends to be a bit excessive if the surface being approached is qualified (more on this in Chapter Nine). If the setup person wishes to change the rapid approach distance to, say, 0.050 inch (1.27 mm) in order to reduce air-cutting time, there could be *many* commands in the program that will have to be changed. Also, without intensive study of the program, the setup person will not know which approaches can be *safely* changed. Here is a portion of a program that uses Custom Macro version B, one of the most popular versions of parametric programming, to program the rapid approach distance as a variable:

O0001 (Program number)
#100 = 0.100 (Rapid approach distance)
N005 T01 M06 (Place center drill in spindle)
N010 G54 G90 S500 M03 T02 (Select coordinate system, absolute mode, start spindle clockwise at 500 rpm)
N015 G00 X4.5 Y2.25 (Rapid to first X/Y position)
N020 G43 H01 **Z#100** (Instate tool-length compensation, rapid to approach position)
N025 G81 **R#100** Z0-.25 F5.0 (Center drill first hole)
N030 X5 (Center drill second hole)
N035 X5.5 (Center drill third hole)

N040 G80 (Cancel cycle)

N045 G91 G28 Z0 M19 (Return to tool change position, orient spindle)

N050 M01 (Optional stop)

N055 T02 M06 (Place 1-inch [25.4-mm] milling cutter in spindle)

N060 G54 G90 S600 M03 T03 (Select coordinate system, absolute mode, start spindle cw at 600 rpm, get tool 3 ready)

N065 G00 X7.50 **Y-[0.5 +#100]** (Rapid to first *X/Y* position)

N070 G43 H01 Z-0.55 (Instate tool-length compensation, move to mill surface in *Z*)

N075 G01 **Y[5.0 + #100]** F4.5 (Mill right side of workpiece)

N080 G00 **Z#100** (Rapid up in *Z*)

N085 G91 G28 Z0 M19 (Rapid to tool change position, orient spindle)

.

.

.

Notice that in lines N020 and N025, the *Z* approach and the rapid plane are specified with #100. Since #100 is set to a value of 0.100 at the very beginning of the program, the control will read these words as Z0.100 and R0.100. For these two commands, the programmer is simply sending the tool to a position in *Z* of 0.100. Additionally, there will be many times when the rapid approach distance must be added to a position used within the program. This is shown in lines N065 and N075. For controls with parametric programming capabilities, arithmetic calculations can be done within CNC program commands. For the Y word in line N065, for instance, first the control will add 0.5 plus 0.1. The result (0.6) will be taken as a minus Y departure.

As long as the programmer consistently uses this technique throughout the CNC program for all approach positions for qualified surfaces, the setup person can easily modify this rapid approach distance during program optimization.

Programming variables for speeds and feeds

The same technique can be used for speeds and feeds within the program. Though many setup people are quite comfortable with scanning through the program and editing the related feed rates and spindle speeds, this technique makes it much easier and less error-prone, especially if the feed-rate word must be changed in several positions for each tool. These techniques are especially helpful when proving a new *process*. When programming new tools to machine new materials, the programmer may be unsure of what cutting conditions will work best. Consider this program:

O0006 (Program number)

#100 = 0.100 (Rapid approach distance)

#101 = 5.5 (Feed rate for slotting cutter)

#102 = 750 (Spindle speed for slotting cutter)

#103 = 6.0 (Feed rate for center drill)

#104 = 1000 (Spindle speed for center drill)
N005 T01 M06 (Place slotting cutter in spindle)
N010 G54 G90 **S#102** M03 T02 (Note variable speed word)
N015 G00 X1. **Y-[0.5 + #100]** (Rapid to approach position)
N020 G43 H01 Z-0.25 (Instate tool-length compensation, rapid to first Z position)
N025 G01 **Y[3.5 + #100] F#101** (Mill first slot)
N030 **Z#100** F50. (Fast feed up to clearance position in Z)
N035 G00 X3. **Y-[0.5 + #100]** (Rapid to approach position)
N040 Z-0.25 (Rapid to surface level)
N045 G01 **Y[3.5 + #100] F#101** (Mill second slot)
N050 **Z#100** F50. (Fast feed up to clearance position in Z)
N055 G00 X5.0 **Y-[0.5 + #100]** (Rapid to approach position)
N060 Z-0.25 (Rapid to surface level)
N065 G01 **Y[3.5 + #100] F#101** (Mill third slot)
N070 **Z#100** F50 (Fast feed up to clearance position in Z)
.
.
.

Notice that the message provided in each variable setting command makes it easy for the setup person to know exactly what each variable means. While this technique makes the program more difficult for the CNC programmer to prepare (especially if a CAM system is used), the setup person can easily modify the speed or feed rate for any tool in the program.

HOW CERTAIN CANNED CYCLES ALLOW EASY OPTIMIZING

Almost all CNC controls allow the use of some very special programming features aimed at making (manual) programming much simpler. These features (commonly called *canned cycles*) dramatically reduce the number of commands the manual programmer must input into the program. For turning centers, for example, most control manufacturers provide a series of roughing cycles aimed at almost eliminating the entire rough turning, facing, or boring commands. Here is an example of one popular control manufacturer's rough turning command.

N050 G71 P055 Q150 U0.04 W0.005 D.125 F0.012

For this particular command, the control will scan between lines N055 and N150 (specified with P and Q) to see the finish pass definition. It will leave 0.040 inch (1.016 mm) on all diameters (U), and 0.005 inch (0.127 mm) on all faces (W), machining with a 0.012 ipm (0.305 mm/min) feed rate for the entire rough turning cycle.

The D word in this example command specifies *the depth of cut per rough turning pass,* which must usually be changed when optimizing the rough turning operation. While there is quite a controversy brewing on whether it is wise to use the control manufacturer's canned cycles (as opposed to using a CAM system), nothing beats the ease of optimizing the roughing depth of cut than this simple canned cycle command. With it, changing only one word in the CNC program will modify the depth per pass of the entire rough turning cycle.

With most CAM-generated CNC programs, the programmer must go back to the CAM system and have it generate a new CNC program based on the new depth of cut per pass, wasting precious program optimizing time. The only exception to this is if the CAM system can generate rough turning commands taking advantage of the CNC control's canned cycle for roughing (as some can).

The same principles apply to turning centers for threading operations. Most turning center control manufacturers offer excellent internal 1-line threading commands. These commands make it very easy to optimize the threading cycle by changing feed angle, number of passes, depth per pass, etc.

Similarly, more and more machining center control manufacturers are offering special milling cycles for thread milling, pocket milling, and face milling. These cycles make optimizing relatively easy, allowing milling considerations like percentage of cutter overlap and depth per pass to be easily changed.

Many companies use these canned cycles solely for optimizing purposes. When they have a new process to prove (based on tooling, workpiece material, and other optimizing considerations), they use the control's canned cycles. Once they determine how machining must be done, they use these criteria for future programming with their CAM systems.

FIRST WORKPIECE INSPECTION

Even after a program is cautiously verified, most companies require that the first workpiece be thoroughly checked. Companies vary dramatically with regard to who actually does this checking as well as where and when it is done.

The first point we offer is related to how many workpieces are machined per cycle. If only one workpiece is to be machined per cycle, no special consideration need be given to the actual running of the first workpiece. However, if multiple workpieces are being run (as is often the case on machining centers), it would be foolish to machine all of them only to find that they do not pass inspection.

For multiple workpiece cycles, since the setup person must eventually get the first workpiece inspected, many programmers facilitate the setup person's ability to quickly run only one of the workpieces being machined per cycle. One simple way to do this is to use the optional block skip function to skip all commands except those needed to machine one workpiece. While this may sound somewhat difficult to program, if subprogramming techniques are used for the purpose of repeating the redundant commands needed when machining multiple identical workpieces, it is really quite easy. During the program's verification, the setup person simply turns on the optional block skip function. After the first workpiece has passed inspection, optional block skip is turned off to run production. Depending on the cycle time needed to machine all workpieces, this facilitating technique can dramatically reduce program verification time.

WHO DOES THE INSPECTION?

Smaller companies that utilize their CNC people to perform many tasks may actually have the setup person perform the first workpiece inspection. In this case, the same basic suggestions made for getting ready to make setups off line apply to first workpiece inspection. Before the setup person even *begins* to make the setup, all cleaning supplies, deburring tools, and gaging tools needed to perform the first workpiece inspection must be prepared and readily available.

In larger companies, first workpiece inspection usually falls under the responsibility of the quality control department. A quality control inspector will perform the workpiece inspection and report the result. Since this kind of company usually maintains an inspection room stocked with all gage tooling required to check every workpiece the company machines, the inspectors should always have everything they need to check the first workpiece.

It is imperative that the inspector place a high priority on first workpiece inspections. A very expensive CNC machine tool is down, waiting for the inspector's findings. For this reason, most companies ask their inspectors to stop whatever they are doing when a first workpiece inspection is requested.

Since first workpiece inspection is always an on-line task, you should look for ways to enable the setup person to deal with any problems found with the first workpiece. If anything is wrong, modifications must be made, probably to the CNC program or related tool offsets. Depending on the skill level of the setup person and the severity of the problem, it may be wise to have the CNC programmer, process engineer, and tool designer on call to help with needed corrections.

SUMMARY OF ALL SETUP TIME REDUCTION TECHNIQUES

To help you quickly scan the suggestions offered in this chapter, we include the following summary organized by type: eliminating on-line tasks, moving on-line tasks off line, and facilitating off-line tasks. The techniques are listed within each category in the same order as they appear in the chapter.

ELIMINATE ON-LINE TASKS

1. Eliminate setup documentation interpretation time by preparing clear and concise setup documentation.

2. Eliminate wasted time by ensuring that all documents needed for setup are available *before* a setup is started.

3. Eliminate as much of the setup person's personal time as possible while a machine is down for setup.

4. Eliminate cutting tool loading by utilizing standard tool stations for often-used cutting tools.

5. Eliminate many tasks related to tearing down and making setups by organizing the order in which you machine workpieces. Run all similar workpieces consecutively.

6. Eliminate the on-line task of adjusting for taper induced by the tailstock on turning centers by purchasing turning centers with tailstocks that maintain alignment throughout their entire travel.

7. Eliminate the on-line task (or at least move off line the task) of bar feeder changeover on turning centers by purchasing multiple-bar-diameter bar feeders.

8. Eliminate the on-line task of cutting tool removal, loading, measurement, and offset entry by minimizing the number of different cutting tools used by a given CNC machine tool.

9. Eliminate the on-line task of measuring tool-length compensation offset values for tools used from one production run to the next by making the tool's length the offset value.

10. Eliminate the on-line task of manually entering tool-length and cutter-radius compensation by programming tool offset entering commands.

11. Eliminate the on-line task of entering tool-nose radius compensation offset values (for the tool radius and type) by having your CAM system generate movements based on the tool-nose radius size you intend to use.

12. Eliminate the on-line task of trial machining with taper reamers by using a setup gage during the reamer's length measurement.

13. Eliminate the on-line task of trial machining with taper taps by easily finding the depth the taper tap should go. Simply count back seven crests from the shank end of the tap. What is left is the depth the tap should go.

14. Eliminate the on-line task of making program zero measurements by utilizing qualified workholding setups.

15. Eliminate the on-line task of entering fixture offset values by programming them within your CNC program.

16. Give your company the potential to eliminate the on-line task of measuring program zero assignment values on turning centers by using geometry offsets instead of G50 commands in the program to assign program zero.

17. Eliminate the on-line task of measuring the Z-axis program zero assignment values on turning centers by using the work shift feature. This allows geometry offsets (in X and Z) to remain the same from one production run to the next.

18. Eliminate the on-line task on turning centers of measuring the program zero assignment numbers for qualified tooling more than one time. Use G10 commands to specify the geometry offset values for the future use of qualified tools.

19. Completely eliminate on-line program zero assignment measurements for quick-change tooling on turning centers by shifting the point of reference for geometry offset entry to the gage position of the quick-change tooling. Geometry offsets in X and Z then become the gage position of each tool and can be easily entered with offset setting commands.

20. Eliminate setup time altogether for bar-feed turning centers when machining different workpieces from the same bar material and diameter (assuming the cutting tools remain the same). By using our recommendations related to program zero assignment, you can make two consecutive different workpieces with absolutely no setup time between them.

21. Eliminate program zero assignment measurements on gang-style turning centers by utilizing tooling subplates.

22. Eliminate program zero assignment measurements for turning centers with index chucks by making the program zero point the center of index in the Z axis.

23. Eliminate the on-line task of program loading for CNC machines dedicated to running a limited number of workpieces by loading all needed CNC programs into the control's memory. If necessary add memory to the CNC control.

24. Minimize the on-line time it takes to perform machine-lock dry runs by making sure the dry-run rate and rapid rate are set to their maximums.

25. Minimize the on-line time it takes to perform free-flowing dry runs by utilizing off-line tool-path verification software.

26. Eliminate the on-line task of performing a normal air-cutting run by utilizing off-line tool-path verification software that can display the difference between rapid motions and cutting motions.

27. Eliminate the on-line task of performing free-flowing dry runs for previously verified CNC programs. Any mistakes in setup that cause problems can be safely found during the running of the first workpiece.

MOVE ON-LINE TASKS OFF LINE

1. Move the task of locating and organizing tools (hand tools, cutting tools, gages, and fixtures) off line by preparing a *tool cart* that includes all tools needed during the setup.

2. Effectively move tasks normally classified as on line to an off-line status by staggering the working hours of your setup people. Perform as many setups as possible during nonproduction times (breaks, lunch, or off shift).

3. Incorporate subplates and pallet changers on machining centers to move the workholding portion of setups from on line to off line.

4. Assemble, measure, and develop offset setting programs for cutting tools off line to minimize on-line tasks related to cutting tools.

5. Measure tool-length compensation values off line to free the setup person from having to do so when the machine is down during production runs.

6. Move the task of program loading off line by utilizing background edit. This feature allows you to load one program into the CNC control's memory while running another.

7. Move the task of tool-path verification for new CNC programs off line by utilizing computer-based program verification software.

FACILITATE ON-LINE TASKS

1. Minimize losses when two or more machines are down by having clear policies about the *pecking order* in which machines get set up first, second, third, etc.

2. Facilitate the task of assembling cutting tools by providing adequate cutting-tool documentation.

3. Facilitate the task of mounting top tooling to turning center chucks by utilizing quick-change chucks.

4. Facilitate the task of boring soft jaws by programming the jaw boring operation. Further facilitate this process by using parametric programming techniques to allow the setup person to simply specify needed jaw boring variables (starting diameter, ending diameter, etc.)

5. Facilitate the on-line task of measuring tool lengths by incorporating a tool-length measurement probe.

6. If no tool-length measuring probe is available, use a parametric program to facilitate the on-line task of measuring tool lengths and entering tool-length compensation values.

7. Facilitate the on-line task of measuring program zero by using probing systems to take the measurement and enter the fixture offset values.

8. If you have no probing system, use parametric programming techniques to facilitate the setup person's ability to measure program zero points with a standard edge finder.

9. Facilitate the setup person's ability to take program zero assignment measurements on turning centers by utilizing a tool touch-off probe.

10. Facilitate the on-line task of preparing CNC programs by utilizing conversational controls. (This is only for companies that have such low production quantities that CNC programs cannot be prepared off line.)

11. Facilitate the on-line task of program loading by utilizing an automatic distributive numerical control system. Doing so can reduce the time it takes to transfer programs to under 30 seconds.

12. Facilitate the turning center setup person's ability to size for lengthy rough turning and boring by using optional block skip to machine just a small portion of the roughing operation. Based on this trial machining, the offset value for the roughing tool can be determined.

13. Facilitate the turning center setup person's ability to size for critical finishing operations when tool pressure presents problems by trial *rough and finish* machining before the roughing operation begins.

14. Facilitate the setup person's ability to size for lengthy threading operations by using optional block skip; this will give additional finish threading passes.

15. Facilitate the setup person's ability to size for critical boring operations on machining centers by using optional block skip to allow trial boring passes to be made for the purpose of adjusting the boring bar.

16. Facilitate the setup person's ability to change rapid approach distance by programming the rapid approach distance as a variable.

17. Facilitate the setup person's ability to modify feed rates and spindle speeds for optimizing purposes by using variables for speeds and feeds instead of hard and fixed constant numbers.

18. Facilitate the setup person's ability to modify other cutting conditions (like depth of cut in rough turning, rough boring, rough facing, and threading) by using the control manufacturer's powerful canned cycles.

19. Facilitate the setup person's ability to run just one of several workpieces machined by a cycle for the purpose of first workpiece inspection, by using optional block skip to ignore all commands except those needed to machine one workpiece.

Chapter Nine

Cycle Time Reduction Techniques

D uring our introduction to setup time reduction techniques, we discussed how workpiece quantities dictate the effort that goes into the process. As you know, the higher the production quantities, the more elaborate the process should be. The better the process, the shorter the cycle time. We cannot stress this enough. *The higher the production quantities, the greater the emphasis that must be placed on minimizing cycle time.* How short a given CNC cycle can be is based primarily on the quality of processing that goes into the operation. While we freely admit that processing is the most important facet of the CNC operation, as we said in the introduction to setup time reduction principles, it is not within the scope of this text to address processing-related issues. We limit our discussion of cycle time reduction techniques to those that can be applied *after* a good process has been developed.

Videotaping is a technique introduced in our discussion of setup time reduction principles that applies equally well to cycle time reduction. While most CNC cycles do not require the kind of manual intervention associated with making setups, you can learn a great deal about how your operators work and come up with many ideas for improvement by studying your CNC operators' current methods. To get a true understanding of what happens during each cycle, be sure to videotape several activations of the CNC cycle. You will need to see the operator do more than simply load and unload a few workpieces. You must videotape enough to see what happens during all tool maintenance (tool replacement, insert indexes, offset setting, etc.) in order to find any bottlenecks in the flow of production.

CYCLE TIME REDUCTION PRINCIPLES

The principles of cycle time reduction are quite similar to those for setup time reduction. Armed with a firm understanding of setup time reduction principles, you should find these concepts quite easy to understand. Nevertheless, you should be able to easily follow all presentations in this chapter even if you have not read the setup time reduction presentations in the previous two chapters. However, we did eliminate redundant presentations on topics already discussed during setup time reduction.

CYCLE TIME DEFINED

We offer two definitions of cycle time. First, many CNC people consider cycle time as *the interval that passes from a given event in one cycle to the same event of the next cycle.* For manually activated cycles, the activation of the CNC program is commonly the event used to measure cycle time. In this case, cycle time is considered as the time that passes from one pressing of Cycle Start to the next. For completely automatic cycles, like those for CNC turning centers equipped with automatic workpiece-loading devices, the event used to gage cycle time might be the closing of the chuck jaws to clamp the workpiece.

Measuring cycle time by this elementary definition is very easy. Anyone with a stopwatch can go out to the machine and simply time the cycle.

While this definition is the one commonly used for cycle time, remember that there are other tasks that add to the time it takes to complete a production run that are not included in this simplistic definition. These tasks are not necessarily performed during every cycle. If the quantity of workpieces is large enough, for example, tool maintenance must be done during the production run. This is commonly done on line, while the machine sits idle. Other examples of on-line intermittent tasks include the loading of bars into bar-feed turning centers and checking workpieces on a sampling basis.

These are at least relatively productive tasks, done to keep production flowing. However, there may be other, far less productive things going on in your CNC environment that add to production time, yet are not truly part of your CNC cycle. If, for example, a CNC machine tool breaks down during a production run and requires corrective maintenance, the time it takes to complete the production run increases. If an operator must halt production to take a phone call, production time increases. If the machine is shut down during breaks, lunch, or any other personal time, production time increases. Truly, *any time* the machine is down for *any reason* during a production run increases the time it takes to complete the production run and must be considered in the definition of cycle time.

For this reason, we offer a much more realistic definition of cycle time. *Cycle time is the overall length of time required to complete a production run divided by the number of cycles needed to complete the production run.* Unfortunately, by this definition, cycle time is much more difficult to measure. While certain intermittent tasks like tool maintenance can be factored in to help calculate cycle time, certain things, like machine failures, make it impossible to perfectly predict cycle time from one production run to the next. With this broader definition, anything that adds to the time it takes to complete a production run is considered in the definition of cycle time—and is fair game for your cycle time reduction program.

As with setup time reduction principles, there are two types of tasks related to cycle time reduction. *On-line* tasks are those tasks actually performed on the CNC machine during the cycle. In fact, the sum of on-line tasks *is the cycle time. Off-line* tasks are those tasks performed outside the machine tool but *internal* to the machining cycle. In order to further define cycle time, we divide cycle time tasks into four categories.

1. A CNC machine tool is only truly productive while chips are being cut. We call tasks that contribute to actually machining chips *productive on-line tasks*. The actual machining operations themselves make up the bulk of productive on-line tasks and are commonly fully automatic tasks, completely controlled by the CNC program.

2. *Nonproductive on-line tasks* occur within the machine tool during the cycle but do not actually produce chips. These can be fully automatic tasks or manual tasks. Examples of fully automatic nonproductive on-line tasks include tool changing (by an automatic tool changer on a machining center or turret index on a turning center), rapid approach and retract motions, and air-cutting motions while tools approach surfaces to be machined. Examples of manual nonproductive on-line tasks include manual workpiece loading and unloading, tool maintenance, and any manual intervention that occurs during a program stop (blowing chips from holes before tapping, breaking clamps loose for finishing, changing chucking pressure on turning centers before finishing, etc.).

3. *Nonproductive off-line tasks* are usually manual tasks that the CNC operator performs to maintain the machining cycle while the machine is running workpieces. Examples of nonproductive off-line tasks include offset changes to hold size on turning centers for the purpose of dealing with tool wear, workpiece inspection, and SPC data recording.

4. *Productive off-line tasks* are tasks the CNC operator performs during the CNC cycle that actually further the completion of machining operations on workpieces. Simple tasks like deburring, cleaning, and polishing, as well as more complex secondary operations, fall into this category.

Just as with setup time, remember that cycle time is the sum total of on-line tasks. Also, as with setup time reduction, the goal in cycle time reduction will be to eliminate on-line tasks, move on-line tasks off line, or facilitate on-line tasks.

THE FOUR WAYS TO REDUCE CYCLE TIME

While we offer many cycle time reducing techniques in this chapter, our broader intention is for you to be able to develop your own cycle time reduction techniques. Regardless of how many techniques we show, it is likely that your company's special needs will require that you develop many more. As long as you understand the four ways we show to reduce cycle time, you should be able to modify our given techniques, or come up with your own, to tackle the challenges that await you in your own CNC environment.

Improve the efficiency of productive on-line tasks

The first general technique we offer is *to improve the efficiency of productive on-line tasks*. Generally speaking, this means optimizing the CNC program's execution. In Chapter Eight, we offered three basic steps to program optimizing.

- First, eliminate inefficiencies caused by the program's basic format.
- Second, optimize the cutting conditions.
- Third, optimize the machining process.

How much optimizing you do during a given production run must be based on the production quantities. For small production quantities, it may be difficult to justify *any* optimizing. For example, if you are only running a few workpieces with a short cycle time, it is unlikely that anything can be done to reduce cycle time that will shorten the overall production run. In this case, making good workpieces may be the *only* criterion for the production run. Maybe the best you can hope for is that your people learn enough during this production run to improve cutting conditions for *future* times when the same workpiece (or material) must be machined.

On the other hand, as production quantities grow, optimizing can have a major impact on the overall production time needed to complete the job. Say, for instance, you must machine 10,000 workpieces with a 2-minute cycle. If you come up with an improvement that saves just one second per cycle, your production run will be shortened by over 2.5 hours (10,000 seconds divided by 60 is 166.6 minutes, or 2.7 hours). In this case, more than two hours of optimizing time could be justified before production is run.

Most CNC programmers would agree that *any* new CNC program can be improved. However, as human beings, we tend to leave well enough alone. If a CNC program is machining workpieces in an acceptable manner, it can be difficult to change *anything*. You know the saying, "If it ain't broke, don't fix it!" However, given the great potential for savings, CNC people must avoid the natural tendency to leave well enough alone.

Minimize nonproductive on-line tasks

Our second general cycle time reduction technique is to *minimize nonproductive on-line tasks*. Remember that these can be manual tasks or automatic tasks. Since anything that happens on line during the cycle is fair game, this category often offers the greatest potential for improving cycle time.

One example for reducing automatic nonproductive time is related to tool changing on machining centers. Regardless of how fast your automatic tool changer changes tools, tool changing is nonproductive on-line time. If multiple workpieces are run during the machining cycle, the tool changing time can be averaged over several workpieces. Another nonproductive on-line task that offers a great potential for savings is workpiece loading. Anything that can be done to reduce workpiece loading time (whether manual or automatic) will effectively reduce cycle time.

Move nonproductive on-line tasks off line

The third general technique is *to move nonproductive on-line tasks off line*. While this assumes the cycle time is long enough that certain tasks can be performed off line, anything that reduces on-line time reduces overall cycle time.

Tool maintenance is one prime candidate for this kind of savings. If the machine is down while a CNC operator performs basic tool maintenance like indexing inserts for carbide insert tooling, a great deal of production time can be wasted. While this task does not occur in every cycle, when tool maintenance *is* performed, it will add to the length of time required to complete the production run and must be considered as part of cycle time. As you will see later in this chapter, there are several things that can be done to help an operator perform tool maintenance tasks off line.

Move productive on-line tasks off line

Another general technique that can save a great deal of production time is to *move productive on-line tasks off line*. This category is most important with CNC cycles having extremely long cycle times. As long as CNC operators have sufficient time during the CNC machining cycle, they can perform machining operations *internal* to the CNC operation. This not only keeps the operators busy, but dramatically reduces overall cycle time.

One common way companies keep their CNC operators busy is to have them perform secondary machining operations on the workpieces being machined by the CNC machine tool. In many CNC environments, for example, the machining center operator performs tapping operations off line. Since tapping tends to be a troublesome operation, and since it demands no great skill or accuracy (it can be done on a simple drill press and requires almost no setup time during changeovers), even an unskilled CNC operator can perform this secondary operation with a minimum of training. When this technique is used, the overall cycle time can be reduced by the time it would take the CNC machine to perform the tapping operations.

While this is an excellent technique that can dramatically reduce cycle time, keep in mind that the priority must be placed on *reducing cycle time*, not keeping the operator busy. The secondary operations must be well planned in order to ensure that the operator has time to perform secondary operations *during* the machining cycle and avoid conflicts with the operator's other CNC-related tasks.

If the goal is simply to keep the operator busy, it may be wiser to seek an alternative method. Instead of forcing the operators to rush through the secondary operations in order to keep up with the CNC machine tool, have them perform secondary operations on less urgent workpieces. Or have them perform tasks that require less skill, like cleaning, polishing, and deburring.

THE 1-SECOND RULE

To stress the importance of reducing cycle time and to offer a simple method of calculating the potential savings related to incorporating a cycle time reducing technique, we suggest this simple rule of thumb. We will be using it throughout this chapter to show how seemingly minor changes can result in dramatic production time savings. *For every second you can remove from cycle time, you save 16.6 minutes per 1000 cycles* (1000 seconds ÷ 60). While at first glance this may not sound like much, you will be surprised at how fast the savings can add up. For approximating purposes, you can round the 16.6 minutes down to a quarter-hour (15 minutes). By applying the 1-second rule, if you can save only four seconds per cycle, you will save over one hour of production time per thousand cycles.

Other time-related formulas

There are several other formulas that are helpful when you are calculating a machining operation's influence on cycle time. We will be using them throughout this chapter. While they assume you are working in the inch system, similar formulas can be developed for use with the metric system.

Time in minutes = length of motion in inches ÷ the ipm motion rate
One second = 0.01666 minutes
Inches per minute feed rate = inches per revolution (ipr) feed rate × rpm
Rpm = 3.82 × sfm ÷ machining diameter

Though it may be somewhat obvious at this point, keep in mind that feed rate is directly proportional to spindle speed (in rpm). If you double spindle speed and maintain inches per revolution feed rate, feed rate will double. For example, if you wish to machine at 0.010 ipr, at 1000 rpm, your inches per minute feed rate will be 10 ipm (0.010 × 1000). If you double the spindle speed to 2000 rpm and maintain 0.010 ipr, feed rate will increase to 20 ipm.

Always remember that cycle time is inversely proportional to feed rate. If the feed rate for a given operation is increased, the cycle time required for the operation will be reduced in equal proportion.

Knowing that feed rate is inversely proportional to cycle time is very helpful when you are approximating the effect a change in cutting conditions will have on cycle time. Consider a milling operation that requires a motion distance of 10 inches (254 mm). At 10 ipm (254 mm/min), this operation will take precisely one minute to complete (10-inch distance ÷ 10 ipm). If you double the feed rate to 20 ipm (508 mm), the required time will be cut in half to 30 seconds (10 inches of motion ÷ 20 ipm).

How fast can your machines rapid?

The rapid rates of today's CNC machines, especially smaller ones, are amazingly high. It is not uncommon, for instance, for a small machining center to have rapid rates of 1200 ipm (3050 cm/min) or more. With these very fast rapid rates, manufacturing people (and especially CNC programmers) tend to ignore the impact of rapid movements on cycle time. Remember that rapid movements do not occur instantaneously. And during many rapid movements, the machine will never reach its true rapid rate. All CNC machine tools utilize an acceleration and deceleration function during rapid to protect the machine's servo drive systems. Though this protection varies with machine size, it is unlikely, for example, that any CNC machine will ever reach its full rapid rate during a small 0.500-inch (13-mm) motion.

Regardless of how fast your machines can rapid, any motions that occur add to cycle time. Since rapid motions tend to get overlooked when cycle time reduction techniques are implemented, we employ the time calculation formulas to demonstrate the impact of rapid motions on cycle time.

To help you determine the impact of rapid movements on your own CNC machines, calculate how far the machine needs to rapid before one second is added to your cycle time. This distance will differ from one machine to another, depending on the machine's rapid rate. Say, for example, you have a small machining center with a rapid rate of 1200 ipm (3050 cm/sec). For this machine, one second will be added to cycle time for every 20 inches (510 mm) of rapid motion (20 ÷ 1200 = 0.0167 minutes, which is just slightly more than one second).

It is not unusual for a 10-tool machining center program to require as much as 150 inches of rapid motion, even on a relatively small machine tool. At 1200 ipm, that's about 7.5 seconds of rapid motion per cycle (150 ÷ 1200 = 0.125 minutes or 7.5 seconds). In a production run of 1000 workpieces, it equates to just over *two hours* of production time (7500 seconds), just for rapid movements. Remember, this calculation does not take into account the acceleration and deceleration of the machine. In reality, even more time will be required. Also, this calculation is for one of the fastest rapid rates currently available. With older equipment having slower rapid rates, the impact of rapid motions on cycle time will be even greater. The total rapid time for a machine that can only rapid at 400 ipm (1000 cm/min) in the previous example will be well over six hours.

We urge you to use this technique to calculate the impact your rapid motions have on your own CNC machine tools. The examples should easily emphasize the need to minimize rapid motions whenever possible within your CNC programs.

REDUCING WORKPIECE LOADING, UNLOADING TIME

As stated, the time it takes to load and unload workpieces fits into the category of nonproductive on-line cycle time. With almost all CNC equipment, regardless of how workpieces are loaded (manually or automatically), the machine must be sitting idle during this task. *Any* technique that reduces on-line workpiece loading time will effectively reduce cycle time. And as with any facet of cycle time reduction, the higher the production quantities, the more important it is to reduce loading and unloading time.

More and more CNC machine tools are being equipped with automatic loading and unloading devices that free the CNC operator from the load/unload task. However, since these devices usually have fixed load/unload times (not much can be done to improve their efficiency), and since most CNC machines still require at least some operator intervention, we limit our discussions to operator-assisted workpiece loading and unloading.

As with our discussion of setup time reduction principles, anything that can be done to facilitate nonproductive on-line tasks will have a positive impact on workpiece loading time. Steps should be taken to make it as easy as possible for the CNC operator to efficiently complete the workpiece loading process. Also as with our discussion of setup time reduction principles, one excellent aid to brainstorming for workpiece loading and unloading ideas is a videotape of the workpiece loading process.

Certain enhancements to workpiece loading/unloading should be easy to spot. For example, workpieces run on CNC machine tools can be very heavy. This, combined with the fact that CNC operators will become fatigued during a day's work (and prone to injury), makes it essential that every step be taken to help the operator move heavy workpieces safely and efficiently. Overhead and boom cranes should be used for heavy workpieces. Even with these devices to assist them, workpiece loading/unloading can still be difficult. Is your operator struggling with the cumbersome straps used with most cranes? Perhaps a magnetic attachment will make the task easier to complete. Is the headstock of the machine (or any other machine component) interfering with the operator's ability to use the crane? Perhaps the program can be changed to position the machine's axes so that interference problems are eliminated. Any number of possible improvements can be made to facilitate the operator's ability to perform the load/unload task. In this section, we offer a few suggestions.

TURNING CENTER SUGGESTIONS

The difficulties associated with turning center workpiece loading vary greatly with the machine's application. Generally speaking, because of the nature of their workholding devices (usually an automatic chuck of some kind), turning centers tend to be easier than machining centers to load and unload. Commonly a foot pedal is used to activate the open/close action of the chuck jaws. Additionally, the chuck is usually mounted to the machine in a horizontal attitude, meaning chips formed during the machining cycle will not impede the workpiece loading and unloading process.

While none of the suggestions we offer related to workpiece loading are revolutionary, they highlight many of the most common problems CNC turning center users face when it comes to loading and unloading workpieces. As you view the videotape of your CNC turning center operators loading workpieces, it is likely that you will spot similar problems.

Bar-feed turning centers

Bar-feed turning centers almost eliminate the need for manual workpiece loading. The machine will be in automatic cycle for the entire length of (at least) one 12-foot- (3.66-m-) long bar. The workpiece loading will be automatic, and equal to the time it takes to feed the bar for the next workpiece (generally well under 10 seconds). Workpiece unloading will also be automatic, and equal to the time it takes to part the workpiece from the bar with a cutoff tool so that it falls into the (automatic) part catcher.

Once a bar is exhausted, however, a new bar must be loaded. If the bar feeder has an automatic bar loading device, even this nonproductive on-line task will be done in automatic fashion. Time for this function with automatic multiple bar feeders is usually less than 20 seconds.

If your company does not utilize automatic bar loaders for your bar-feed turning centers, you will need to make the task of manual bar loading as easy and efficient as possible. As you view the videotape of your operators loading bars, watch for anything that can be improved. Is the raw material stored in proximity to the bar feeder? Is the area around the bar feeder clear of obstructions that can impede bar loading? Are all tools required to load the next bar readily available? If the bar requires a chamfering operation prior to loading (as many bar feeders do), is it being done *off line*, during the CNC cycle of machining previous bars?

Chuck-style turning centers

Manually loaded chuck-style turning centers require much more operator intervention. Depending on the size, shape, and weight of the workpiece, unloading and loading can be as simple as removing the completed workpiece with one hand and loading the unmachined workpiece with the other; for very large, heavy workpieces, it can involve using some form of crane.

For extremely large, heavy workpieces, the vertical-style turning center can dramatically simplify the workpiece loading and unloading process. With this kind of turning center, the spindle is positioned in a vertical attitude, making it very easy to place workpieces in the chuck. Additionally, the workpiece's weight actually adds to the capability of the workholding device.

Shaft-style turning centers

For turning centers used for shaft applications, the workpiece must be loaded into the chuck, and a tailstock center of some kind must also be engaged for support. If the shaft is very heavy, it can be quite difficult for the operator to adequately support the workpiece while the tailstock is engaged. For this reason, many companies utilize a temporary work support device to help with large-shaft unloading and loading.

This simple device resembles a cart on rollers and has two work support positions (one for the previously machined workpiece and one for the next workpiece to be machined). It is adjustable for different-diameter shafts, allowing the centerline of the workpieces to be positioned at the same height as the spindle centerline. To use the device, the operator loads the next heavy shaft onto the work support cart (possibly with an overhead crane) off line while the previous workpiece is being machined. At the completion of the machining cycle, the work support cart is rolled into the work area. With the empty work support station of the cart under the finished workpiece, the tailstock center is disengaged, the chuck jaws are opened, and the workpiece is removed from the chuck and placed on the cart. The cart is then repositioned to allow easy loading of the next workpiece. Following this, the jaws are closed, the tailstock center is engaged, and the work support cart is removed.

For 4-axis (single-spindle) turning centers used in shaft applications, many companies use the lower turret as the work support device. A special work support device (resembling a vee block) is mounted to one of the lower turret stations. For workpiece loading, a program is run that indexes the turret to the work support station. The turret then rapids to a position at which the work support device is positioned directly under the spindle centerline. The program is then stopped (with M00). At this point, the operator can easily remove the previous workpiece, load the next one into the chuck, and engage the tailstock. Once workpiece loading is completed, the cycle is reactivated. The lower turret will move away for the beginning of the machining cycle.

MACHINING CENTER SUGGESTIONS

While turning center workpiece loading and unloading is usually straightforward and simple, machining center workpiece loading can be much more challenging. Some setups, like vise setups, allow rather quick and easy loading/unloading, but the bulk of machining center setups require much more. Fortunately, unlike most turning center applications, many of the tasks related to machining center workpiece loading can be moved from the status of on-line tasks to off-line tasks. For those tasks that can be moved off line and done during the machining cycle, you effectively reduce cycle time. This, of course, assumes that the program's execution time is long enough to permit off-line workpiece loading tasks.

Pallet changing devices

Once only found on horizontal machining centers, pallet changing devices are becoming popular even on vertical machining centers. The workpiece load/unload advantages of pallet changers remain the same regardless of the machine type. A pallet changing device allows most of the tasks related to workpiece loading and unloading to be moved off line. As long as machining time (program running time) is long enough to allow the workpiece to be completely loaded and unloaded, the workpiece load/unload time will be consistent and equal to the amount of time it takes the pallet changer to change pallets. The bulk of workpiece loading time will be internal to machining time.

Keep in mind that pallet changers (especially manually activated pallet changers) can be purchased and attached to any CNC machining center at any time.

While the machine tool builder is normally regarded as the primary supplier of pallet changers (especially automatic pallet changers), any number of aftermarket suppliers can provide excellent pallet changers for a reasonable price. When you understand what they can do for workpiece load/unload time, you should readily agree that their purchase price can be easily justified for almost any CNC machining center.

As an example of how pallet changers can help, let's assume you have a very simple vise setup on the table that has a 4-minute program execution time (that is, the CNC program runs for four minutes). Say the CNC operator is currently loading workpieces on line, meaning the machine is down during unloading and loading. The operator must (manually) unclamp the vise, remove the finished workpiece, brush away the chips from the parallels of the vise and clean location surfaces, load the next workpiece into the vise, clamp the vise, and give the workpiece a good whack with a lead hammer to ensure that it is flush with the parallels.

Depending on the skill (and motivation) of your CNC operators, the time required for this on-line task may vary dramatically. Regardless of how long it takes, the machine is down waiting for the next workpiece to be loaded. Knowing this, the operator may even try to hurry the process, causing a mistake during the workpiece loading procedure.

If the average workpiece load and unload for this example setup takes one minute, in a run of 1000 workpieces, more than 16 hours are required just for workpiece loading (1000 minutes ÷ 60 = 16.6 hours). That's more than two 8-hour shifts just to load and unload workpieces.

Most manual pallet changers can be easily activated in well under 30 seconds. And the manufacturers of some automatic pallet changers are boasting pallet change times under 10 seconds. Even with a manual pallet changer that takes 30 seconds to activate, in the previous example program loading and unloading time will be cut in half, saving over eight hours of production time. At a shop rate of just $50 per hour, an $8000 manual pallet changer will pay for itself in just over 20 shifts.

Another benefit of pallet changers, and especially automatic pallet changers, is that workpiece loading and unloading time will remain consistent. When an operator must load and unload workpieces completely on line, the manual intervention involved will cause inconsistent cycle times. While one operator may take two minutes to do it, another may take only 30 seconds. Additionally, operator fatigue may cause inconsistent loading times even during one operator's shift. While fresh, the operator may be able to load and unload workpieces quickly. As the day wears on, workpiece load and unload time may increase. This inconsistency in cycle time makes it very difficult for production control people to schedule jobs moving through the shop. With pallet changers, the operator will have ample time to actually unload and load workpieces while the machine is in cycle. There will be no pressure to load workpieces quickly. When the machining cycle is finished, the workpiece loading time will be consistent, equal to the time it takes to activate the pallet changer.

Remember that machining time must be greater than workpiece loading time in order to reap the full workpiece loading and unloading benefits of pallet changers. In our example, the machining time is four minutes. Off-line workpiece loading time averages one minute. That leaves three minutes for the operator to do other things.

There may be times with short machining cycles when it takes longer to load workpieces than it takes to machine them. If, in the previous example, machining time is only 50 seconds, the operator will not have completed the setup (off line) by the time machining is completed. The operator will have 10 more seconds of workpiece loading to finish (if workpiece loading time is one minute) as well as the 30 seconds to activate the (30-second) pallet changer. In this case, 20 seconds is still saved over changing workpieces completely on line (one minute on line versus 40 seconds with the pallet changer). Though pallet changers may be harder to justify in this situation, they still offer an excellent alternative to loading workpieces completely on line.

To this point, we have been primarily discussing manual pallet changers. Automatic pallet changers, though they must usually be purchased directly from the machine tool builder, boast amazingly short pallet changing times. Many automatic pallet changers can be activated in less than 15 seconds and there are even some that can change pallets in under five seconds. With a 5-second pallet changer being used in our previous example, what would take one minute of manual loading on-line time would now take only five seconds of on-line time, saving 55 seconds per cycle. In a thousand workpieces, that amounts to a production savings of over 15 hours. Said another way, if the shop rate for this machine is $100 per hour, more than $1500 is saved during this production run of a thousand workpieces.

Other workpiece loading and unloading devices

If workpiece loading is done on line, it is imperative that you do everything possible to facilitate the workpiece loading process. However, even if your operators are loading workpieces off line, it is still important that you make it as easy as possible. While the CNC machine's shop rate is certainly more than the typical CNC operator's hourly wage, do not underestimate the operator's free time during the machining cycle.

As discussed earlier in this chapter, many companies keep their CNC operators busy during the execution of CNC programs. Many even have their operators running two or more CNC machines simultaneously. For these reasons, the workpiece loading and unloading functions should be made as easy as possible, even if loading is done off line, in order to gain more of the operator's time to perform other tasks.

Automatic hand tools. One inexpensive way to make the CNC operators' tasks easier is to provide them with automatic hand tools whenever possible. Most setups, for example, and especially fixture setups, require that at least some clamping bolts be tightened. It is faster and easier to tighten bolts consistently with pneumatic torque wrenches than with standard box-end wrenches. Additionally, you can ensure much more consistent torque with automatic wrenches than with manual hand tools, since one operator may be stronger than another.

Automatic clamping. One of the reasons most turning centers are so easy to load and unload is that most incorporate automatic clamping. A hydraulic (or pneumatic) chuck can be easily clamped and unclamped with a foot pedal. The operator simply places the workpiece into the chuck jaws and presses the foot pedal, and the workpiece is clamped.

Though somewhat more difficult to employ on machining centers, since workholding tooling usually changes with each setup, the same principles do apply. As production quantities grow, it will be possible to justify the purchase of hydraulic clamping devices for fixtures. This will allow the operator to engage clamps by throwing a switch or pressing a button instead of manually tightening several clamps.

REDUCING CNC PROGRAM EXECUTION TIME

The time it takes the CNC program to run is an extremely important part of the overall cycle time. In fact, we have already discussed cases when the CNC program execution time *is* the cycle time (or very close to it). With bar-feed turning centers, for example, we have shown that the program's execution time is the cycle time unless tool maintenance is required on line during the production run. Because the program's execution time usually has the largest single impact on cycle time, we devote a lengthy discussion in this text to how the execution time for the CNC program can be reduced.

EFFICIENT PROGRAM FORMATTING

A well-formatted CNC program will execute faster than a poorly formatted one. There are many functions of CNC programming that cause subtle, yet substantial, differences in execution time, based on how they are programmed. Given the number of commands in the typical CNC program, it is essential that your programmers develop good program formatting habits if program execution time is to be minimized.

There is no excuse for your programmers *not* to incorporate efficient program formatting techniques. They cost nothing. They require little, if any, additional work on the programmer's part. And, even for companies that run rather small production quantities, the amount of time they save over poor formatting techniques during the course of a CNC machine's life is substantial.

Unfortunately, most machine tool builders do not stress the most efficient programming techniques in their programming courses, for two reasons:

1. The most efficient method and the safest method of handling a given programming problem often conflict. When faced with safety conflicts, machine tool builders (rightfully) teach beginners the safe way to handle program formatting. However, as CNC users gain experience and confidence, they may be willing (if not forced) to compromise on certain safety issues in order to minimize program execution time (just as a child will eventually compromise safety for speed and remove the training wheels from a bicycle).

2. Most beginning programmers find it easier to learn programming functions if only one major function is happening per CNC command. First they turn the spindle on. Then they rapid the tool to its first approach position. Then they turn the coolant on. And so on. Step by step. While this makes it very easy to teach and visualize what is going on in a program, it is not a very efficient manner of programming. For CNC controls that wait for the completion of commanded functions in one command *before* going on to the next, including two or more functions within one command will effectively reduce program execution time.

Even computer-aided manufacturing systems vary dramatically with regard to how efficiently they can format CNC programs. Though CAM systems are getting better and better at formatting CNC programs in an efficient manner, given the variety of ways by which different CNC controls require commands to be given, it is unlikely that any CAM system will ever be able to format CNC programs in the most efficient manner possible for every CNC machine tool. You may find after reading this section that there are things about the way your CAM system generates CNC programs that must be changed if program execution time is to be kept as short as possible.

The effects of special programming features

There is a saying that applies to any form of computer device, including the computers within CNC controls. *Any time you make a computer think, it takes time.* While computers cannot really think, they do make calculations similar to those made when we think. The computers within CNC controls are capable of making calculations very quickly, and as time goes on, they will only continue to improve in this regard. But even with current-model CNC controls, you may still notice substantial changes in the control's execution time when certain high-level features are used.

Here is a test you can perform on one of your own CNC machining centers that illustrates how time can be wasted as your CNC *thinks* about high-level commands (a similar test can be easily developed for turning centers). First write a program that makes 1000 consecutive 0.002-inch (0.051-mm) incremental movements along the X axis (2 inches [51 mm] of total movement). If you have a CNC text editor that allows copy-and-paste functions, this program should be rather easy to create. Set the programmed feed rate to 2.0 ipm (51 mm/min). Here is a portion of this test program in the format for one popular control. Once you are finished, download this test program to the CNC control.

O0001 (Program number)
N001 G91 G01 X0.002 F2.0 (Move 0.002 inch in X)
N002 X0.002 (0.002 inch more in X)
N003 X0.002 (0.002 inch more in X)
N004 X0.002 (0.002 inch more in X)
.
.
.
N1000 X0.002 (0.002 inch more in X)
N1001 M00 (Program stop for timing)
N1002 G00 X-2.0 (Go back to start point)
N1003 M30 (End of program)

Once this program is in the control, manually move the X axis so that there is at least two inches of movement possible in the plus X-axis direction. Then run and time the program. At two ipm, this program should take precisely one minute from the time you press Cycle Start to the time the program stop is reached (2 inches of motion at 2 ipm). If it does, increase the feed rate to 4 ipm (102 mm/min). Run and time the test program again. Now it should take exactly 30 seconds to run. Continue doubling the feed rate and timing the execution of the program. Each time you double feed rate, the time needed to run the program should be cut in half.

At some point, you will notice that the movements will not keep up with the programmed feed rate. For example, you may have specified the feed rate as 16 ipm (406 mm/min). It should take precisely 7.5 seconds to smoothly execute the program, but you notice choppy movements totaling more than 7.5 seconds. You have reached the saturation point of this control's ability to calculate its motions at 0.002-inch (0.051-mm) departures. The control is simply unable to keep up with the calculations necessary to maintain the desired feed rate with these tiny axis departures. This saturation point will vary from one control to another. With specially designed high-speed machining controls, it may even be over 200 ipm (5080 mm/min). With current CNC controls for general-purpose machining, it will likely occur at a feed rate of under 20 ipm (508 mm/min).

When you reach the control's saturation point (16 ipm in our example), make the control *think* some more. Execute a similar program commanding 0.002-inch departures, but use subprogramming techniques instead of one main program. Here is another sample test program in the format for one popular control manufacturer.

O0002 (Main program)
N001 M98 P0003 L1000 (Execute subprogram 00003 1000 times)
N002 M00 (Stop for timing)
N003 G00 X-2.0 (Move back to start point)
N004 M30

O0003 (Subprogram)
N001 G91 G01 X0.002 F16.0 (Move 0.002 inch in *X*)
N002 M99

This program is executing the subprogram 1000 times, making the control use a great deal more logic than it must when simply executing a normal CNC program. When you execute this program, what happens to execution time? If your control behaves like most, you will notice a substantial increase in the time it takes to execute.

Now instate cutter radius compensation in the main program prior to executing the subprogram 1000 times. Or turn on axis rotation prior to executing the subprogram, or scaling, or any other high-level programming feature (even tool length compensation). You may be very (negatively) surprised to see how poorly your CNC control responds when these high-level features are used.

While most general-purpose machining programs will not push your CNC control to its calculating limits as our tests do, these tests should nicely illustrate our point. *Any time you make the control think, it takes time.*

Subprogramming. Subprogramming techniques are notorious for adding to program execution time (as our test program illustrated). Generally speaking, if you are machining ultrahigh production quantities and cycle time is of the utmost importance, it is best to not use subprograms at all, especially if your CNC program makes extensive use of subprograms. (As the processing speed of CNC controls continues to increase, this statement may eventually become untrue.)

If you do wish to use subprograms, look into how your particular control functions with regard to subprograms. Most controls are affected by *how many* programs reside in the control's memory. The larger the number of programs, the longer it may take to find the subprogram being called. Also, those controls that incorporate program numbers will commonly look for subprograms in numerical order, meaning the control will find program 1 faster than program 1000 (assuming there are many programs in the control's memory).

Canned cycles. Almost all CNC controls come with a set of special programming features called *canned cycles*. The actual cycles themselves will vary from one control manufacturer to another. One popular machining center control, for example, comes with a complete set of hole machining canned cycles for drilling, tapping, reaming, boring, and so on. It also has a series of special milling canned cycles for circle milling, pocket milling, and face milling. One popular turning center control comes with a set of canned cycles for rough turning, rough boring, rough facing, finish turning, finish boring, finish facing, threading, and grooving. While these canned cycles dramatically shorten program length and make it *much* easier for the manual CNC programmer to prepare programs, if no consideration is given to how they affect the program's execution time, they can be cycle-time-wasting commands.

While transparent to most CNC users, almost all canned cycles are affected by the CNC control's set of *parameters*. For the chip-breaking drilling cycle on one popular machining center control, for example, a parameter controls how much the drill will back up after each peck to allow the chip to break. About 0.002 to 0.005 inch (0.051 to 0.127 mm) is sufficient to break the chip. However, if this parameter is improperly set, say to 0.100 inch (2.54 mm), the drill will back up an excessive amount between pecks, wasting program execution time.

Similar parameters affect how turning center canned cycles behave. With one popular control's rough turning, rough boring, and rough facing canned cycles, a parameter controls how far the roughing tool will retract from the cut before the tool starts its rapid motion back for the next roughing pass. Since the retract motion is done at the current feed rate, if it is excessive, a great deal of program execution time will be wasted.

Note that we are not discouraging the use of all special features of programming. We simply want to ensure that you consider the cycle-time-related impact they can have when used. If you confirm that they are being used in the most efficient manner possible, by all means use them. However, if cycle time is of the utmost importance, and if you discover that certain special features appear to be taking too much time to execute (you see noticeable pauses while the program executes), you can always go back to straightforward motion commands (with G00, G01, G02, and G03) to eliminate thinking time from the machining cycle.

Parametric programming. Parametric programming gives the CNC user the ability to use computer programming commands within the CNC program. Variables, arithmetic statements, and logic statements are among the many additional features parametric programming allows. Think of parametric programming as having the computer language BASIC included within your CNC control.

While parametric programming is commonly utilized for family-of-parts applications, it also provides the ability to machine complex geometric shapes, to drive optional devices (like probes and postprocess gages), and to develop special utility programs (like part counters and tool-life managers). It is a very powerful programming language that allows your programmers to do things not commonly associated with standard CNC programming.

Parametric programming is an excellent programming tool, and there are many applications in which it can truly streamline the long-term CNC programming process. When it comes to program execution time, however, parametric programming can be a real cycle-time-wasting function. If minimizing cycle time is of the utmost importance, generally speaking, parametric programming techniques should not be used.

Almost every parametric programming command makes the control think. When a variable is set to a value, it takes time. When an arithmetic command is executed, it takes time. When a logic statement is evaluated, it takes time. As the processing speed of CNC controls continues to improve, the day may come when parametric programming functions will have no impact on program execution time, but for now, parametric programs will take longer to execute than their standard G-code-level counterparts.

How M codes relate to program execution time

As you know, M codes, called *miscellaneous functions*, have two purposes. First, they allow the control manufacturer to provide special programming functions (M00—program stop, M01—optional stop, M30—end of program, M98—sub-program call, M99—subprogram return, etc.). M codes used in this manner are under the complete control of the control manufacturer.

Second, M codes allow machine tool builders (as well as accessory device manufacturers) to interface certain mechanical devices to the CNC machine tool. Once interfaced, they allow the automatic activation of these devices through programming commands. Examples of standard M codes used in this manner include M03—spindle forward start, M04—spindle reverse start, M05—spindle stop, M06—tool change, M08—flood coolant on, and M09—coolant off. While these M codes remain consistent from one machine tool builder to the next, machine tool builders vary widely with the values used to assign other miscellaneous functions. One turning center manufacturer may, for example, use M23 to specify low spindle range while another uses M41.

M codes used to activate machine devices have a dramatic impact on program execution time. Unfortunately, machine tool builders also differ with regard to what actually happens when a given M code is commanded. In extreme cases, we have even seen identical machines made by *the same machine tool builder* vary in this regard when a particular M code is given. For this reason, the CNC programmer may have to perform tests on the CNC machines your company owns in order to determine the most efficient manner by which to program the commonly used M codes. In extreme cases, the programmer may even have to contact the machine tool builder to change the function of time-wasting M codes.

How M codes work. Depending on the machine tool builder or accessory device manufacturer (or even the end user who adds special M codes), M codes can be fully interfaced or partially interfaced. Fully interfaced means that a *confirmation signal* is used to tell the control that the task commanded by the M code has been completed and it is all right to continue executing the CNC program.

For example, assume you have a *fully* interfaced M code-activated indexer mounted on the table of a vertical machining center. M13 is the command used to activate the indexer. When M13 is executed, the control will not continue to the next command in the program *until* the confirmation signal is received from the indexer. Since the rotation of the indexer will take time, and since there would be potential for danger if the control were to continue executing the program, this confirmation signal is very important. Once the indexer completes its rotation, it sends the confirmation signal, telling the control it has done so. At this point, the control will move on to the next command in the program.

A partially interfaced M code is one that does not utilize the confirmation signal. The control simply activates the device and (immediately) moves on to the next command in the CNC program. *The only time the machine tool builder, the accessory device manufacturer, or the end user should utilize a partially inter-*

faced M code is when there is absolutely no potential for danger due to the control continuing to the next command in the CNC program. Unfortunately, this rule is often broken. Many accessory devices are partially interfaced, even though they place the user in potentially dangerous situations. The installer of the M code simply tells the user to program a dwell command (usually G04) in the command after the M code, to cause the control to pause for a period of time long enough to allow the device to complete its function. While *many* devices are programmed in this manner, if something interferes with the device during its activation, the control will continue with the program at the completion of the dwell command even though the device is still trying to complete its function. Depending on what is coming up in the program, this can cause terrible problems.

The cycle-time-related implication of fully interfaced M codes is that the control will not continue to the next command until the function of the M code is completed. Consider these commands.

.
.
.

N055 G96 S500 M03 (Start spindle clockwise [cw] at 500 sfm [15 m/min])
N060 G00 X3. Z.1 (Move to position in *X* and *Z*)

.
.
.

N125 G00 X6.0 Z6.0 (Move to program ending position)
N130 M05 (Stop spindle)
N135 M30 (End of program)

In line N055 of this turning center program, the spindle is told to start at 500 sfm. As long as the M03 is a fully interfaced M code (as it is on most machines), the control will not proceed to line N060 until the spindle is running at 500 sfm. In similar fashion, in line N130, the spindle will have to come to a complete stop before the program end command will be read and executed in line N135. From a program execution time standpoint, this is a very inefficient manner of programming. Since there is no danger in having the spindle start during the tool's approach (in line N060), it is wiser to combine the motion command with the spindle start. Likewise at the end of the program, since there is no danger in having the spindle begin to stop during the tool's return to the tool change position, it is wiser to combine the rapid motion with the spindle stop. Here are the modified commands.

.
.
.

N055 G00 G96 X3. Z.1 S500 M03 (Move to position and start spindle cw at 500 sfm)

.

.
.

N125 G00 X6.0 Z6. M05 (Move to program ending position and stop spindle)
N130 M30 (End of program)

Most CNC turning center controls will execute these commands faster than those in the previous example. In line N055, most controls will allow the spindle to begin accelerating as the machine moves into position (though there are exceptions). This will cause whichever is shorter, the spindle startup or the motion, to be internal to the other. Given the fast rapid rates on current-model turning centers, it is likely that the rapid motion will be internal to the spindle startup.

The same will be true in line N125. With most turning centers, the spindle will begin to stop during the motion. Again, with most turning centers, this will mean the motion command will be internal to the spindle stop.

When are M codes activated? Machine tool builders also differ with regard to when certain M codes will *begin* to execute, even when they are included within commands that contain other functions. Either the M code will be activated at the very beginning of the command or at the very end of the command. Determining what happens with each M code from the machine tool builder's documentation can be difficult. It may actually require that tests be done on each CNC machine tool with each commonly used M code.

In our previous example, as long as both M03 and M05 are executed at the *beginning* of the command, combining them with motion commands will save program execution time. Depending on the distance of motion, the rapid rate, and how long it takes the spindle to respond, either the motion will be internal to the spindle start/stop or the spindle start/stop will be internal to the motion. Either way, program execution time is reduced.

This principle applies to *all* completely interfaced M codes. As long as it is safe to do so, include M codes within motion commands. Either the motion will be internal to the M code function or the M code function will be internal to the motion.

Unfortunately, you must be prepared for those machine tool builders that interface M codes in such a way that the M code does not begin to take place until the completion of the CNC command. In this case, the machine tool will still respond in the same (slow) manner regardless of whether M codes are included within motion command or not. If you come across this kind of machine, first check the machine's list of M codes. You may find that there are actually two more M codes that control whether other M codes take place at the beginning or end of a CNC command.

If there are no such M codes, you will have to contact your machine tool builder for help. As long as the builders understand your reasoning and agree that your desired change to the M codes in question poses no potentially dangerous problems, most will be willing to change the way the M codes are interfaced to the

CNC control. However, at least one popular control manufacturer actually requires its users to sign a waiver stipulating that the control manufacturer will not be held responsible for any mishaps before it will modify the function of M codes. Most control manufacturers are quite reasonable in this regard, and, for newer equipment, modifying the time when M codes begin executing involves a simple change to the machine's programmable controller.

Reducing rapid approach distance

Most programmers are taught to rapid the cutting tool, when approaching qualified surfaces, to within 0.100 inch (or 2.54 mm when programming in the metric mode) of the surface. For unqualified surfaces (like cast surfaces), most programmers increase this distance to 0.25 inch (6.4 mm). While these distances provide a very safe approach (especially good for beginning programmers), you must consider the impact your rapid approach distance has on cycle time. The machine is usually moving at a very slow feed rate (as opposed to rapid) during the time it takes the tool to move from the approach position until it begins cutting. While we *do not* recommend changing your approach distance to unqualified surfaces, when approaching finished surfaces, we consider the 0.100-inch approach distance to be excessive.

Most current-model CNC controls boast an accuracy and repeatability capabilities of well under 0.0005 inch (13 μm). When the control is told to position the tool to a given location, it will do so with amazing precision. As long as the approach surface is qualified (machined by the CNC machine itself), the control will be able to stop the tool a smaller distance away from the surface, say 0.050 inch (1.3 mm), just as easily as it can stop the tool 0.100 inch from the surface. This means you can dramatically reduce your approach distance for qualified surfaces to something just slightly larger than imperfections or inconsistencies (possibly caused by tool wear) in the qualified surface.

During my presentation to experienced programmers in live seminars, reducing rapid approach distance has met with mixed reaction. There seems to be a certain comfort zone related to the 0.100-inch approach distance that is difficult to overcome. However, I contend that if something is wrong with the CNC program, the tool offset, or the program zero setting numbers that would cause the tool to crash into a workpiece surface (even with the 0.100-inch approach distance), the CNC operator would not be able to stop the machine in time to save a crash. At a rapid rate of 400 ipm (1000 cm/sec) (considered a slow rapid rate by today's standards), it only takes the machine 0.015 second to move 0.100 inch.

Additionally, during the program's verification, the setup person will have ample control (with dry run, rapid override, and single block) of each tool's approach. If there are problems that would cause a tool to crash into a surface, the setup person will be able to find them just as safely and easily with a 0.050-inch approach (or even smaller) as with a 0.100-inch approach.

Another point that students make during the live seminars when we discuss reducing rapid approach is that doing so really doesn't have much of an impact on cycle time. Let's just see about that.

Say you have a block of steel to machine on a vertical machining center. After face milling (qualifying the top surface), you need to machine 50 holes. The holes are to be spot-drilled (chamfering the hole to a diameter bigger than the tapped hole size), drilled, and tapped. Using conventional methods, the spot drill and drill are programmed to approach within 0.100 inch of the surface and the tap to within 0.25 inch (6.35 mm) (using a tension/compression holder for the tap). Say the holes are 1/4-inch–20 tapped holes machined in tool steel. You determine that the spot drill should feed at 5.0 ipm (127 mm/min), the 13/64-inch drill at 8.0 ipm (203 mm/min), and the 1/4-inch–20 tap at 25 ipm (635 mm/min).

For the spot-drilling operation, the total distance the tool will be feeding through air while it approaches the work surface is 5 inches (50 holes × 0.100-inch approach). At 5 ipm, the tool will be air cutting for exactly one minute. If we reduce the rapid approach distance to 0.050 inch, we can immediately remove 30 seconds from our machining cycle. In a production run of 1000 workpieces, this will amount to a savings of more than eight hours.

After a hole is spot-drilled (especially when done to a diameter larger than the drill), a great deal of approach clearance is machined into the workpiece. In our case, say the (90°) spot drill is machining the hole to a diameter of 0.3125 inch (7.9 mm) (a 1/32-inch chamfer on the 0.250-inch hole). In this case, our 13/64-inch drill will not even begin cutting until it is well below the surface being drilled. The actual clearance in this case will be even more than the lead of the drill (the lead of a 13/64-inch 118-point angle drill is 0.060 inch [1.52 mm]). Figure 9-1 illustrates this. The programmer could rapid the tool to a position *below* the surface being machined by 0.025 inch (0.635 mm) and still maintain 0.090 inch (2.29 mm) of clearance before the drill starts machining.

Since the rapid plane (usually specified with the letter address R in the canned cycle command) is now below the surface being drilled, the programmer must confirm that the drill will come out of the hole to a position above the work surface before moving to the next hole. Most controls allow this through standard canned cycle commands (using something called the initial plane specification). However, even if one or more of your controls does not allow this obstruction clearing feature, the time that can be saved by incorporating this technique is well worth the additional work your programmer must do to program the clearance motions in more longhand fashion.

In our example, if we rapid to 0.025 inch (0.635 mm) below the surface being drilled, we will save 0.125 inch (3.175 mm) of cutting motion per hole (as opposed to making the rapid approach position 0.100 inch [2.54 mm] above the work surface). For 50 holes, that's 6.25 inches (159 mm) of saved motion, and at 8 ipm (203 mm/min), 46 seconds of saved program execution time per workpiece. In our production run of 1000 workpieces, we just saved another 12.7 hours of production time.

Keep in mind that similar techniques are possible if you are center drilling (not spot drilling). However, since most programmers do not center drill to a diameter larger than the drill size (most machinists would agree that doing so defeats the purpose of center drilling), it will not be possible to rapid the drill below the surface being drilled. However, it is still possible to rapid the drill *to* the surface being machined and still maintain ample clearance.

When it comes to our tapping operation, since we are using an older tension/compression-style tapping holder, we must ensure that the tool comes out of the hole far enough to guarantee that the tap truly comes out of the hole. (If your machine has the feature *rigid tapping* [also called *synchronous tapping*], you will be able to rapid the tap much closer to the work surface. Doing so eliminates the need to incorporate the technique we now show.) Remember the reason *why* we must program the tension/compression-tap style holder with the larger approach distance. Again, it has to do with what happens *after* the hole is tapped. As the tap machines the hole, it is possible that the tap will extend in the holder, making the tap's overall length longer than it was before the tapping operation started. As the tap feeds back out of the hole, it must come out to a position in *Z* that ensures that the tap will snap back to its neutral position in the tap holder. Though most standard tapping canned cycles do not allow this, the programmer could program the *approach* distance much closer (say 0.100 inch [2.54 mm]). On the tool's *retract* from the hole, the programmer will use the 0.25 inch (6.35 mm) clearance distance. Using this technique on most controls will require very cumbersome and tedious longhand programming (not simple canned cycles). Is it worth it? If this technique is used for our example, we will save 0.15 inch (3.81 mm) of motion per hole. That's 7.5 inches (191 mm) of motion per workpiece, or 18 seconds at 25 ipm. In a run of 1000 workpieces, that's 5 more hours of saved production time.

The sum total of reducing rapid approach distance for our example is 1 minute, 34 seconds per workpiece, or more than 26 hours of production time.

Our example used hole machining operations to stress how easy it is to reduce program execution time by reducing rapid approach distance. However, these principles apply with any machining operation that requires a tool to be moved to an approach position relative to a qualified surface.

Programming efficient motions

This rather obvious suggestion has to do with confirming that all motions the program causes the machine to make are the most efficient possible. While it should almost go without saying, I am always amazed at the number of programs I see that break this basic rule. Many times programmers ignore this rule simply because they feel that, since today's CNC machines have extremely fast rapid rates, even inefficient rapid motion commands will not waste much time. Additionally, some special programming features that make programming much simpler cause rather wasteful motions.

For example, many programmers use subprogramming techniques in conjunction with hole-machining operations on machining centers. A hole may have to be center drilled, drilled, and tapped. The more holes there are to machine, the more repeated commands must be in the CNC program. While subprogramming can dramatically reduce the number of redundant commands, if the programmer is not careful with this feature, it is likely that wasteful motions will be commanded. Say the holes to be machined are along a straight line. The subprogram is written to machine the holes in one direction along the line. After the center drill has machined all the holes, using subprogramming usually requires the machine to rapid back to the start of the pattern, wasting machine movement.

Here is a list of rapid motion distances that will take precisely one second to complete. Notice the slower your machine's rapid rate, the more important it should be to minimize wasteful rapid motions within your program.

Rapid rate, ipm (cm/min):		One second's motion distance, inch (cm)	
100	(254)	1.666	(4.232)
200	(508)	3.332	(8.463)
300	(762)	4.998	(12.695)
400	(1016)	6.664	(16.927)
500	(1270)	8.330	(21.158)
600	(1524)	9.996	(25.390)
700	(1778)	11.662	(29.622)
800	(2032)	13.328	(33.853)
900	(2286)	14.994	(38.085)
1000	(2540)	16.660	(42.316)
1100	(2794)	18.300	(46.482)
1200	(3048)	20.000	(50.800)
1300	(3302)	21.700	(55.118)
1400	(3556)	23.300	(59.182)
1500	(3810)	25.000	(63.500)

Keep in mind that this chart does not consider acceleration and deceleration time. During short movements, it is likely that the machine will never reach its actual rapid rate, making it even more important to program motions efficiently.

Again we urge you to consider the effects of rapid motions on cycle time. During optimizing, the programmer and setup person should be on the lookout for wasteful motions. Here we suggest a few easy-to-incorporate techniques.

Approach and retract in all axes. Machining center programmers are commonly taught to make a 2-step approach with each tool. First they command the X/Y position, then they move in Z. While this kind of approach is easy to visualize and tends to make programming easier, it is a rather wasteful way to approach. Since the shortest distance between two points is a straight line, moving in all axes simultaneously will be faster than making the 2-step approach.

Admittedly, when a machining center tool is approaching to within a 0.100-inch (2.54-mm) position (or less) relative to the work surface, it can be very scary to watch, especially when the tool is still moving in X and Y. For this reason, we

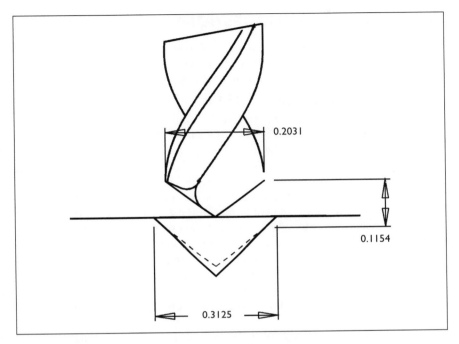

Figure 9-1. How spot drill makes ample clearance for drill.

recommend that you maintain a safe clearance position above the workpiece when making 3-axis approaches. Many programmers will rapid in all axes to within 2 inches (51 mm) of the work surface in the Z axis. This makes for an efficient approach, yet still keeps the movement from worrying the operator. In similar fashion, if there is a need to retract to the tool changing position in more than one axis, be sure to do so in all axes simultaneously.

Start and end each tool as close to the workpiece as possible. While many machining centers require that the machine be sent to a hard-and-fixed tool changing position to make tool changes, most turning centers allow turret indexing to occur at any position along the machine's travels. With current programming methods, it is possible to program tool-changing positions with extreme ease and efficiency. Most turning centers use some form of offset to assign the tool's program zero position. For these machines, it is very easy to retract the tool just far enough away from the workpiece to allow a safe index. At this position, the tool change can take place, minimizing the machine's travel distance.

Even with older machine tools that require program zero to be assigned in the program, it is possible to move the program's starting position (and tool-changing position) closer to the workpiece. Though this technique requires more caution from the operator with older machines, it can dramatically reduce program execution time.

TURNING CENTER SUGGESTIONS

We now discuss techniques that can be used to reduce program execution time specifically for turning centers. Since most require nothing but changes in programming, they are very simple and cost nothing to implement.

Moving warm-up time off line

As any machine tool warms up, there will be minor growth in critical components. As the spindle continues to run, for example, the area around the spindle will grow slightly because of the heat buildup caused by friction in the spindle bearings. In similar fashion, as the X and Z axes continue to move, the area around the axis drive motors, ball screws, and other axis components will also grow because of heat created by the friction between moving parts. While the points we make in this section can be applied to machining centers, the growth of turning center components tends to have a greater impact on the ability to hold size.

Though manufacturers do their best to design turning centers to compensate for thermal deviations during warm-up, most turning center users would admit they experience at least minor changes in workpieces (especially in diameters) during the machine's first hour of production after power-up. If they are trying to hold very close tolerances, this thermal deviation during warm-up can wreak havoc with workpiece consistency.

For this reason, many CNC turning center users accept an extremely wasteful machine warm-up period whenever a cold machine is started. They turn the machine on, start the spindle, and let it run for an hour or so prior to starting their production run. In a 3-shift environment, when the spindle of the machine is rarely allowed to completely cool down (possibly only during weekends and holidays), this warm-up procedure may not be terribly wasteful. However, if only one or two shifts are running, the machine will be turned off at the end of each day, meaning the warm-up will be required every day.

If the warm-up is done while the machine should be machining workpieces, you must consider this time as on-line nonproductive time and factor it into the overall calculation of cycle time. There are (at least) three ways to perform machine warm-up prior to the beginning of the production shift that effectively move machine warm-up off line.

One is to have someone come in at least an hour before the start of the first shift to start all CNC machines with thermal deviation problems. This person must be able to start each machine and activate a warm-up program, meaning the person must possess at least some operation skills. For this reason, one of the CNC operators or setup people may be the best choice. Here is an example of a start-up program in the format of one popular CNC control manufacturer:

O8000 (Program for warm-up)
N005 G28 U0 W0 (Go to machine's reference position)
N010 G97 S1000 M03 (Start spindle at 1000 rpm)

N015 G00 W-8.0 (Move 8 inches [20.32 cm] in the minus Z direction)
N020 G00 U-5.0 (Move 2.5 inches [6.35 cm] in the minus X direction)
N025 U5.0 (Move back to start point in X)
N030 W8.0 (Move back to start point in Z)
N035 M01 (Stop if optional stop is on)
N040 M99 (Return to beginning and run again)

When activated, this program will continue to run until the optional stop switch is turned on. The machine will first go to its reference position, the spindle will start, and the axes will move back and forth until the program is stopped by the optional stop switch. Of course, the spindle speed and the amount of travel distance can be modified according to the size of each machine in your shop.

As stated, this technique requires someone with CNC skills to come in and activate the warm-up program for each CNC machine tool. If you are willing to leave your CNC machines on during the off shift, you can eliminate this need. The CNC operator for the *last* shift will run a special program that will cause the machine to *dwell* for the off shift. The length of time for the dwell may vary, depending on when the last shift operator runs the program. One hour before the start of the first shift, the dwell period will end and the warm-up program will run. Here is an example that utilizes subprogramming to handle the dwell period in the format for a popular CNC control.

O8000 (Dwell and warm-up program)
N005 M98 P8001 L6 (Dwell for 6 hours)
N010 G28 U0 W0 (Move to machine's reference position)
N015 G97 S1000 M03 (Start spindle at 1000 rpm)
N020 G00 W-8.0 (Move 8 inches minus in Z)
N025 G00 U-5.0 (Move 2.5 inches minus in X)
N030 U5.0 (Move back to start point in X)
N035 W8.0 (Move back to start point in Z)
N040 M01 (Stop if optional stop is on)
N045 M99 P020 (Go back to line N020 and continue executing)

This program uses a subprogram (O8001) that dwells for one hour. The subprogram below assumes your control allows a 3600-second dwell (one hour). If it does not, you can string together several dwell commands.

O8001 (Dwell program)
N1 G04 X3600 (Dwell for 3600 seconds [one hour])
N2 M99 (End of subprogram)

In line N005 of program O8000, the operator sets the L word to the number of hours the machine must dwell prior to starting the warm-up procedure before the beginning of the first shift. For example, if the second shift ends at 11:00 p.m. and the first shift starts at 7:00 a.m., the second shift operator will set the L word to 7. At 6:00 a.m. the next day, the warm-up portion of the program will run. Because of the unconditional branching command in line N045 of program O8000, the control will continue performing the motions in X and Z until the first shift operator comes in and turns on the optional stop switch.

If you do not like the idea of leaving your CNC machines turned on during off shifts, a third way to incorporate this technique is to utilize an outside timer that will automatically turn the machine on and start the warm-up program. Interfacing this kind of device will probably require the help of your machine tool builder (or an outside consultant). However, once the device is connected, you can effectively move the on-line task of machine warm-up off line without requiring anyone to come in early or leaving the machine on during the off shift.

Programming spindle range changes efficiently

While more and more (especially smaller) turning centers are coming with high-torque spindle drive motors that have only one power range, there are still many that have two or more spindle ranges. M codes are commonly used to specify spindle range changes. For a machine with two spindle ranges, for example, M41 is commonly used to specify the low spindle range and M42 for the high spindle range.

Beginning programmers are commonly taught to perform roughing operations in the low spindle range and finishing operations in the high range. While this may be a relatively good rule of thumb, a better understanding of each turning center's power curve and maximum rpm in each spindle range is required to ensure efficient programming of spindle ranges.

Spindle range changes take time. While there are exceptions, many machines require the spindle to come to a complete stop before the spindle range change can take place. Then the spindle must be restarted to its required rpm. Depending on the speeds involved and the spindle's deceleration and acceleration properties, a spindle range change can take from 2 to 15 seconds to complete.

By understanding the spindle's horsepower and torque characteristics, the CNC programmer may be able to eliminate the need to make range changes for certain workpieces. Say for example, you have a 30-horsepower turning center with two spindle ranges. The low range runs from 30 to 1500 rpm. The high range runs from 30 to 4000 rpm. By studying the power curves for this machine's spindle (published by the machine tool builder), you determine that the spindle reaches full power at 300 rpm in the low range and at 2200 rpm in the high range.

When you are machining workpieces under 1 inch (25.4 mm) diameter with speeds of 600 sfm (180 m/min) or more, there will be absolutely no reason to machine in the low range. The speed in rpm for a 1-inch diameter workpiece at 600 sfm is 2292 rpm, which is greater than the rpm at which the spindle reaches full power in the high spindle range. Depending on the material being machined and the required horsepower for each operation, it may be possible to machine even larger diameters at slower spindle speeds in the high range without taxing the spindle drive motor.

Another instance in which CNC programmers tend to ignore the impact of spindle range changes on program execution time is in rough turning from a very large diameter to a very small diameter. Again, they tend to place the machine in the low range for the entire rough turning operation. Depending on the machine's spindle characteristics and the diameters involved, it may be wiser to use the low range only until the maximum rpm in the low range is reached. At this point, all CNC turning centers will *peak out*, not allowing the spindle to run any faster. By switching to the high range for the balance of the machining operation, it may be possible to actually reduce the overall length of time required for the rough turning operation (since the spindle speed will still continue to increase with each rough turning pass).

The effects of constant surface speed on cycle time

Constant surface speed is an excellent turning center feature that constantly and automatically adjusts spindle speed in rpm to the diameter being machined. It provides several benefits, including easier programming, consistent finishes on machined surfaces, and longer tool life. However, constant surface speed can be a real time waster if no consideration is given to how it is programmed. Consider the following program. It machines a piece of 1.5-inch (38.1-mm) diameter tubing with a 0.75-inch (19.05-mm) diameter hole.

O0001 (Program number)
(Note that the current diameter of the roughing tool is 8 inches [203.2 mm])
N005 T0101 (Index to rough face and turn tool)
N010 G96 S600 M03 (Start spindle cw at 600 sfm [180 m/min])
N015 G00 **X1.7** Z.005 (Rapid to approach position)
N020 G01 X0.55 F0.010 (Rough face)
N025 G00 X1.415 Z0.1 (Rapid back to rough-turn diameter)
N030 G01 Z-0.995 (Rough turn)
N035 X1.55 (Feed up face)
N040 G00 **X8.0** Z4.0 (Rapid back to tool change position)
N045 M01 (Optional stop)
N050 T0202 (Index to rough-boring bar)
N055 G96 S600 M03 (Start spindle cw at 600 sfm)
N060 G00 **X0.960** Z0.1 (Rapid to rough-bore position)
N065 G01 Z-0.995 F0.009 (Rough bore)
N070 X0.65 (Feed down face)
N075 G00 Z0.1 (Rapid out of hole)
N080 **X8.0** Z4.0 (Rapid back to tool change position)
N085 M01 (Optional stop)
N090 T0303 (Index to finish boring bar)
N095 G96 S700 M03 (Start spindle cw at 700 sfm [210 m/min])
N100 G00 **X1.125** Z0.1 (Rapid to starting position)
N105 G01 Z0 F0.005 (Feed flush to face)

N110 X1.0 Z-0.0625 (Form chamfer)
N115 Z1.0 (Finish bore)
N120 X0.75 (Feed down face)
N125 G00 Z0.1 (Rapid out of hole)
N130 **X8.0** Z4.0 (Rapid back to tool change position)
N135 M01 (Optional stop)
N140 T0404 (Index to finish turning tool)
N145 G96 S700 M03 (Start spindle cw at 700 sfm)
N150 G00 **X1.25** Z0.1 (Rapid to starting position)
N155 G01 Z0 F0.005 (Feed flush to face)
N160 X1.375 Z-0.0625 (Form chamfer)
N165 Z-1.0 (Finish turn)
N170 X1.5 (Feed up face)
N175 G00 **X8.0** Z4.0 (Rapid back to tool changing position)
N180 M30 (End of program)

When it comes to the use of constant surface speed, this program contains several time-wasting commands. Before reading further, can you tell why? We have shown the time-wasting commands in bold type. Though this format of programming is *very* wasteful, it *is* the format commonly taught by machine tool builders, since it is quite easy to break up the program in step-by-step fashion. It is also the format commonly used by computer-aided manufacturing systems to create CNC programs.

Remember that constant surface speed forces the machine's spindle to change rpm with diameter changes. Your turning center spindle cannot instantaneously respond to changes in rpm. The larger the rpm changes, the more time required for your spindle to respond. We strongly recommend that you have your people perform some tests to determine just how long it takes your turning centers to respond to rpm changes. Have them measure how long it takes to accelerate from zero to 1000 rpm. Then measure how long it takes to decelerate from 1000 rpm back down to zero. Repeat the test from zero to 2000, zero to 3000, and so on up to your machine's maximum rpm capability. You may be (unpleasantly) surprised at how long it takes your spindle to respond. A typical 30-horsepower turning center with a 10-inch (254-mm) 3-jaw chuck, for example, will take more than 10 seconds to accelerate from zero to 4000 rpm. It will take an equal amount of time to stop from 4000 rpm.

Notice that our inefficient program uses a tool changing position in X of 8 inches (203.2 mm) diameter. In line N010, the spindle is started at 600 sfm (183 m/min). At an 8-inch diameter, that equates to 225 rpm (3.82×600 sfm $\div$ 8). During the movement in line N015, the tool is sent to a 1.7-inch (43.18-mm) diameter. The spindle will accelerate during this motion to 1348 rpm ($3.82 \times 600 \div 1.7$). However, it is quite unlikely that it will be able to accelerate all the way up to 1348 rpm *during* the rapid motion. If the machine has a rapid rate of 800 ipm (2032 cm/min), as many current model turning centers do, the motion will take only 0.3 second. The machine will be sitting idle for the balance of the time it takes the spindle to accelerate to 1348 rpm.

Since the spindle is not running at the beginning of this program, about the best you can do to improve this first spindle acceleration is include it with the motion (combine lines N010 and N015). At least this will make the motion time internal to the spindle acceleration time.

In line N040, notice that the tool is returning to the 8-inch tool change diameter. Since the machine is still running at 600 sfm in the constant surface speed mode, the spindle will slow to 225 rpm. Again, it is likely that the spindle slowdown cannot be completed during the motion, meaning more program execution time will be wasted. This wasted time will vary with the response time of your spindle. For a 30-horsepower machine with a 10-inch (254-mm) chuck, it would not be unusual to experience at least a 2- to 3-second pause in line N040 while the spindle slows down.

In lines N055 and N060, the same problem will occur again. The spindle will need to accelerate from 225 rpm to 2387 rpm during a 4- to 5-inch (102- to 127-mm) motion. The machine will sit idle at the end of the motion while the spindle accelerates. This same problem will occur in every tool's approach to and retraction from the workpiece. Note that the finishing tools run at 700 sfm (210 m/min), requiring a greater change in rpm. The greater the spindle speed change, the more program execution time will be wasted.

Though they have nothing to do with program execution time reduction, there are two other detrimental results caused by having the spindle repeatedly accelerate and decelerate throughout the execution of the CNC program. First, any change in spindle rpm requires electricity. Even when decelerating, the machine uses electricity to apply an electronic brake to reduce the time it takes the spindle to slow. If you can eliminate the accelerations and decelerations, you also reduce the waste of electricity. Second, these repeated spindle speed changes cause undue wear and tear on the machine. Eliminate them and you will lengthen the period between spindle-related maintenance procedures.

There are two ways to eliminate the repeated spindle speed changes caused by this program. One is to keep the tool-changing position in X very close to the diameter of the workpiece. If no diameter changes are required, no rpm changes will be necessary. However, for this program, the tool-changing position in X would have to be at about 1.5-inch (38.1-mm) diameter. This would require that the tool be retracted in only the Z axis. Depending on the size of the machine, whether a tailstock is in use on the machine (causing interference), and how long internal tools project from the turret face, this may not always be feasible or, in some cases, even possible.

Our recommended way to handle the problem is to temporarily switch to the rpm mode during each tool's retraction to the tool-change position. During this motion, you can even specify that the rpm be changed to the rpm required for the *next* tool. Here is a modified version of the previous program that uses our recommended format.

O0001 (Program number)
(Note that the current diameter of the tool is 8 inches)
N005 T0101 (Index to rough face and turn tool)
N010 G96 G00 S600 M03 X1.7 Z0.005 (Start spindle cw at 600 sfm during approach to workpiece)
N015 G01 X0.55 F0.010 (Rough face)
N020 G00 X1.415 Z0.1 (Rapid back to rough-turn diameter)
N025 G01 Z-0.995 (Rough turn)
N030 X1.55 (Feed up face)
N035 G00 G97 S2387 X8.0 Z4.0 (Select rpm for next tool's approach and rapid back to tool-change position)
N040 M01 (Optional stop)
N045 T0202 (Index to rough-boring bar)
N050 G97 S2387 M03 (Start spindle cw at 2387 rpm)
N055 G00 X0.960 Z0.1 (Rapid to rough-bore position)
N060 G96 S600 (Reselect sfm mode)
N065 G01 Z-0.995 F0.009 (Rough bore)
N070 X0.65 (Feed down face)
N075 G00 Z0.1 (Rapid out of hole)
N080 G97 S2376 X8.0 Z4.0 (Rapid back to tool-change position, select rpm for next tool)
N085 M01 (Optional stop)
N090 T0303 (Index to finish-boring bar)
N095 G97 S2376 M03 (Start spindle cw at 2376 rpm)
N100 G00 X1.125 Z0.1 (Rapid to starting position)
N105 G96 S700 (Reselect sfm mode)
N110 G01 Z0 F0.005 (Feed flush to face)
N115 X1.0 Z-0.0625 (Form chamfer)
N120 Z1.0 (Finish bore)
N125 X0.75 (Feed down face)
N130 G00 Z0.1 (Rapid out of hole)
N135 G97 S2139 X8.0 Z4.0 (Rapid back to tool-change position, select rpm for next tool)
N140 M01 (Optional stop)
N145 T0404 (Index to finish-turning tool)
N150 G97 S2139 M03 (Start spindle cw at 2139 rpm)
N155 G00 X1.25 Z0.1 (Rapid to starting position)
N160 G96 S700 (Reselect sfm mode)
N165 G01 Z0 F0.005 (Feed flush to face)

N170 X1.375 Z-0.0625 (Form chamfer)

N175 Z-1.0 (Finish turn)

N180 X1.5 (Feed up face)

N185 G00 X8.0 Z4.0 M05 (Rapid back to tool-changing position, stop spindle)

N190 M30 (End of program)

Notice the subtle changes to the spindle related commands. First in line N010, we start the spindle *during* the approach, making the approach movement internal to the spindle startup. In line N035, we temporarily switch to the rpm mode, selecting the correct rpm for the next tool's approach diameter. This effectively keeps the spindle from decelerating. In fact, it will actually *accelerate* slightly to get the spindle ready for the next tool. Line N050 is only in the program for the purpose of rerunning the rough-boring bar. Since the spindle is already running at this rpm, under normal circumstances, this command will have no impact on program execution time. However, if the rough-boring bar must be rerun, this command is necessary to ensure that the spindle will start. To make rerunning tools more efficient, this command could be easily combined with the approach movement (simply combine lines N050 and N055). These techniques are repeated for the balance of the tools in the program. In line N185, note that we also combine the spindle stop command with the last tool's retraction to its tool-changing position, making the movement time internal to the spindle stopping time.

How much time you save by using these techniques will vary with your spindle's response time and how much the spindle rpm is currently changing between tools. Typically, these techniques save at least four seconds of spindle acceleration and deceleration time per tool. For our example program, that equates to at least 16 seconds of saved program execution time. For a production run of 1000 workpieces, that's over four hours of saved production time.

Changing workholding devices can affect cycle time

Though not commonly known by CNC users, most current CNC controls have *parameters* that affect how quickly a spindle will respond to rpm changes. These parameters protect the spindle drive system from trying to accelerate or decelerate too quickly (possibly causing a fuse or breaker to blow because of excess electricity) and are determined by the size and weight of the workholding device you are using. You can think of this function as being like the governor on a lawn mower engine. Just as the engine's governor protects the engine from trying to run too fast, so do the spindle accel/decel parameters of a turning center control protect the spindle drive motor from trying to accelerate or decelerate too quickly. Also, just as the governor can be adjusted to different speeds based on the engine's application, so can the parameters of the turning center control be adjusted to allow efficient spindle acceleration and deceleration based on the size and weight of the workholding device.

Most machine tool builders are quite conservative when it comes to the factory setting of spindle accel/decel parameters. Rightfully, they adjust these parameters in such a way that the end user will *never* have problems with blown fuses or breakers due to exceeding the electricity limitations of their spindle drive systems. Additionally, they tend to set these parameters for the kind of workholding device that is on the machine when it is originally purchased.

Keep in mind, however, that the same machine tool could be equipped with different workholding devices, to meet changing needs. A very light collet chuck might be used when the company performs bar work. A much heavier 3-jaw chuck might be needed for chucking work. Generally speaking, the light collet chuck can be accelerated and decelerated faster than the heavy 3-jaw chuck (assuming the bar being fed is relatively light and supported in a hydraulic-style bar feeder). Most machine tool builders will set the accel/decel parameters suitable for the chuck that comes with the machine.

If you never change workholding devices on your turning centers, there will never be a need to be concerned with the accel/decel parameters. However, if you change workholding devices according to the kind of work you do, and if there is a great difference in the weight of the workholding devices involved, you will want to learn more about them.

Say, for instance, you purchased a large turning center with a hydraulic 15-inch (381-mm) 3-jaw chuck. Your machine tool builder would have originally set the accel/decel parameters in such a way that the spindle would never try to accelerate this heavy chuck too quickly. Maybe there are times when you run a smaller 8-inch (203.2-mm) air chuck on the same machine. Maybe you have some thin-wall workpieces that might be deformed during clamping. After you change chucks, you will probably notice that the spindle still takes the same amount of time to accelerate and decelerate as it did with the 15-inch chuck. By changing the related spindle parameters, you can make the machine respond more quickly to rpm changes, dramatically improving the program's execution time.

Unfortunately, the documentation from the machine tool builder related to spindle accel/decel parameters is usually rather difficult for end users to interpret correctly. For this reason, we recommend that you contact your turning center manufacturer for help with the setting.

Keep in mind that if you do change these parameters to gain faster program execution time (when going from a large workholding device to a small one), you must remember to change them back when you replace the larger device. Since more and more controls allow parameters to be changed from within the CNC program (usually with a G10 command), it may be wise to include the proper parameter-setting commands at the beginning of *every* program. This way, the spindle-related parameters will always be set in the correct manner for every CNC program being run.

Tooling style can affect program execution time

Most current-model turning centers allow users to use either right-hand or left-hand tooling. In fact, very little (if any) modification must be done to the toolholders of most CNC turning centers in order to have them hold either style. The CNC setup person can load either style with the same ease, making the choice of right- or left-hand tooling almost a matter of personal choice.

Keep in mind, however, that when you switch from using left-hand to right-hand tools within the same CNC program, a change in spindle direction is usually required. For example, if you are using a right-hand rough-turning tool cutting toward the chuck, most turning centers require a clockwise (M03) spindle direction. If this tool is followed by a left-hand finish-turning tool, the spindle direction must be reversed (M04). Spindle reversals require time to complete, meaning lost program execution time.

Many tools used on turning centers are only available in right-hand versions (like drills and reamers). Additionally, certain machining operations require the spindle to be running in the clockwise direction (as when chasing right-hand threads). For this reason, you can eliminate almost all spindle reversals by using only right-hand tools for all machining operations.

The only time we recommend using left-hand tools is when you are performing extremely powerful machining operations. The design of most current-model turning centers is such that the thrust of the machining operation tends to pull right-hand tools away from their direction of support (the turret and bed of the machine). For extremely powerful operations on these machines, it is wiser to utilize left-hand tools to force the thrust of the machining operation into the bed of the machine tool.

Improving the efficiency of bar feeders on turning centers

Ideally, a bar feeder should be designed to handle the same capacities as the turning center being used. If the turning center can handle up to 2-inch-(51-mm-) diameter bars, so should the bar feeder. If the turning center has a maximum spindle speed of 6000 rpm, so should the bar feeder (at the maximum bar diameter and length). Current model bar feeders allow amazing capacities with regard to maximum speed, bar diameter, and length. It is not unusual for a bar feeder manufacturer to boast the ability to rotate at 6000 rpm while holding a 2-inch diameter, 12-foot- (3.66-m-) long steel bar.

While these high-quality bar feeders are available, for cost reasons, many machine tool builders (and CNC users) do not equip them. Additionally, even these high-quality bar feeders do have limitations that may constrain how fast bars can be rotated. For example, most require that bars be perfectly straight when placed into the bar feeder in order to meet their maximum claims.

In reality, many CNC bar-feed turning center users cannot allow their spindles to be rotated up to the machine's maximum (while holding a 12-foot bar) without experiencing vibration in the bar-feed unit. Most are required to limit the maximum rpm in the CNC program in order to eliminate vibration.

One popular turning center control manufacturer uses a G50 for this purpose. Here is an example of the spindle limiting command.

N005 G50 S2000 (Limit spindle speed to 2000 rpm)

This command will tell the control not to let the spindle exceed 2000 rpm during the balance of the CNC program. Once this command is given, even if some machining operation requires it (possibly facing to center under the influence of constant surface speed), the spindle will not exceed 2000 rpm. Many bar-feed turning center users make use of this command often when their bar feeder cannot keep up with the machine's maximum spindle speed.

However, if the CNC program often requires spindle speeds over 2000 rpm, a great deal of program execution time can be lost. Remember that time is inversely proportional to spindle speed. As spindle speed increases, so will the feed rate (if you are programming in the inches per revolution mode). As the feed rate increases, time will decrease proportionally. If a spindle speed of 4000 rpm is required during a given machining operation while the spindle is limited to 2000 rpm, for example, it will take twice as long to complete the machining operation.

If you are faced with spindle-limiting problems when bar feeding, first check into the manufacturer's specifications for your particular bar feeder. If the manufacturer claims your bar feeder should perform better than it does, solicit their help correcting the problem. Possibly the bar feeder is simply out of alignment. Possibly more (or better) spacers are required within the bar feeder. Possibly you can demand that your material supplier provide straighter bars. Possibly you can have your CNC operator perform a bar-straightening operation internal to the machining operation.

If you find that your bar feeder simply cannot keep up with your CNC turning center, your second alternative is to shorten the bar's length. A 6-foot (1.83-m) bar, for example, will allow a much faster maximum rpm than a 12-foot (3.66-m) bar. Unfortunately, this alternative almost defeats the purpose of bar feeding in the first place, which is unattended operation for as long a period of time as possible. Additionally, the added operation of cutting the bars in half may introduce more serious work-handling problems.

A third alternative to handling this maximum rpm problem is to allow the maximum spindle speed of the machine to increase as the bar gets shorter. By using this technique, workpieces being machined late in the bar will be made faster than workpieces machined early in the bar.

To reap the fullest benefits of this technique, there can be no finish consistency requirements on the workpieces being machined. While workpieces run late in the bar will actually have better finishes (they are machined at speeds closer to optimum for the material), there will probably be noticeable differences in workpiece finish. Even this problem can be overcome to some extent if all finishing operations are performed at the same (beginning) speed in rpm mode. At least the time required for all other operations can be minimized.

Programming for this technique can be relatively simple if subprogramming or parametric programming techniques are used. The bar can be broken into segments. Each segment will allow a different maximum spindle rpm. For example, say you have a 12-foot long, 2-inch-diameter (3.66-m × 50.8-mm) bar to machine. Each workpiece is 0.875 inch (22.23 mm) long and the cutoff tool is 0.125 inch (3.18 mm) wide (1 inch [25.4 mm] overall distance from workpiece to workpiece). In this case, each 12-foot-long bar will allow 140 workpieces with a 4-inch (101.6-mm) remnant (144-inch [365.8-cm]) overall bar length. If we divide the bar into four segments, that's 35 workpieces per segment (140 ÷ 4 = 35).

Before running the first bar, the setup person will test for maximum rpm. Say the machine allows a maximum speed of 4000 rpm, yet when the 12-foot bar is rotated, the setup person can only safely get to 2000 rpm before vibration is experienced. At this point the setup person will run the normal CNC program for the workpiece and machine 35 workpieces while limiting spindle speed to 2000 rpm. After the thirty-fifth workpiece, the setup person will stop the cycle and test for maximum rpm a second time. With the shorter bar (three-quarters its original length), it is likely that the maximum speed will be higher than for the bar in its original state. For our example, let's say the setup person is able to safely reach 2500 rpm before vibration occurs. At this point, the setup person will run another 35 workpieces, limiting speed to 2500 rpm, and stop again. With the bar half its original length, the setup person may find that the bar can be rotated at 3200 rpm without vibration. The setup person will then run 35 more workpieces and stop for a final test. With the bar only one-quarter its original length, the operator finds that the bar can be rotated all the way up to the machine's maximum 4000 rpm without vibration. The last 35 workpieces will then be machined from the bar.

For all other bars in the production run, a different program will be run. Using subprogramming techniques, the following example program would run each bar segment with a different maximum rpm.

```
O0001 (Main program)
N005 G50 S2000 (Limit to 2000 rpm for first segment)
N010 M98 P1000 L35 (Run the first 35 workpieces)
N015 G50 S2500 (Limit to 2500 rpm for second segment)
N020 M98 P1000 L35 (Run the second 35 workpieces)
N025 G50 S3200 (Limit to 3200 rpm for third segment)
N030 M98 P1000 L35 (Run the third 35 workpieces)
N035 G40 S4000 (Limit to 4000 rpm for fourth segment)
N040 M98 P1000 L35 (Run the fourth 35 workpieces)
N045 M30 (End of program)
```

This controlling program differs from most bar-feed programs in that it no longer depends on the end-of-bar signal to stop the automatic operation of the bar feeder. Instead, this program will stop in response to the M30 program ending command.

This technique does force the programmer and setup person to do more work for the machining of the first bar in each setup, but if they are running high production quantities, it can dramatically reduce the overall time required to finish the production run. A typical savings from using this technique, when bar feeders limit the maximum rpm to 2000 or less and when some machining operations require 4000 rpm, is *at least* 10 percent of the overall cycle. (Some companies have realized savings of 30 percent with this technique.) For a workpiece with a cycle time of 3 minutes, the 10 percent savings is approximately 18 seconds. For 1000 workpieces, that equates to 5 hours of saved production time.

While the time and effort needed to perform the maximum rpm tests may at first sound prohibitive, remember that the spindle speed limitation characteristics of a given bar diameter and material will remain consistent from one workpiece to the next. For example, given the previous tests, the next time a 2-inch (51-mm) diameter steel bar must be run, the programmer will know that the new bar must be limited to 2000 rpm. A quarter of the way through the bar, the spindle must be limited to 2500 rpm. Half the way through the bar, the spindle must be limited to 3200 rpm. And three-quarters of the way through the bar, the speed must be limited to 4000 rpm. For the same workpiece length, the main program can be easily written without the need for performing more tests. If parametric programming techniques are used, it will be even simpler, since the parametric program can perform the related calculations automatically.

MACHINING CENTER SUGGESTIONS

At this point, we offer cycle time reduction suggestions specifically related to machining center applications.

Multiple identical workpieces

Though today's machining centers boast amazingly short chip-to-chip times, any machining center will be nonproductive during tool changes. One way to minimize tool-changing time is to average it over the machining of two or more workpieces. Here is a realistic example that will illustrate the impact tool changing has on program execution time. For our example, we will use round numbers to make the potential for savings easy to visualize.

We will assume a production run of 1000 workpieces. The workpiece is relatively small and rectangular in shape. A common 4-inch (102-mm) table vise nicely holds the workpiece. Ten tools are used by the program in this operation. The chip-to-chip time for your automatic tool changer is 5 seconds (rather fast by anyone's standards). The overall cycle time needed when running one workpiece in the setup is precisely 5 minutes, meaning the machine is cutting chips for 4 minutes, 10 seconds and changing tools for 50 seconds (10 tools × 5 seconds).

If the setup is changed to machine two workpieces on the table, the tool changing time for one of the workpieces can be eliminated. In this case, the total cycle time will be 2 × 4 minutes, 10 seconds = 8 minutes, 20 seconds, plus the 50 seconds of tool changing time. The time needed to machine two workpieces *should* now be 9 minutes, 10 seconds. However, if running two workpieces per setup, some rapid motions will be required to go from one workpiece to the other within each tool. If the workpieces are positioned approximately 10 inches (254 mm) apart, about 100 inches (254 cm) of rapid motion will be added to the program (10 tools × 10 inches of rapid movement). If the machine rapids at 500 ipm (1250 cm/min) (somewhat slow by today's standards), about 12 seconds of rapid time will be added to the program. The total cycle time will be 9 minutes, 22 seconds.

What was taking 10 minutes to machine two workpieces, machining one workpiece per setup (5 minutes per workpiece), is now taking 9 minutes, 22 seconds. The time per workpiece is now 4 minutes, 41 seconds (a savings of 19 seconds per workpiece). This realistic example stresses that 19 seconds per workpiece can be saved by simply running two workpieces per cycle instead of one. By applying the 1-second rule, the overall production time saved while running 1000 workpieces is 5 hours, 15 minutes (19 seconds × 16.6 minutes per 1000 workpieces). While setup time will increase because the setup person must mount two vises on the table instead of one, the overall time to complete the production run will be dramatically reduced (surely the setup person will be able to mount the second vise in well under 5 hours, 15 minutes).

The more workpieces to be machined per production run, the more tool changing time per workpiece can be saved. If, for example, we increase the number of parts to be run per cycle to three, the total cycle time will be 3 × 4 minutes, 10 seconds + 50 seconds of tool changing time + 24 seconds of rapid time, meaning it will take 13 minutes, 44 seconds to run three workpieces. This equates to 4 minutes, 34 seconds per workpiece. Now the overall savings per workpiece is 26 seconds, and this savings will result in over 7 hours of saved production time in a production run of 1000 workpieces.

While this example should help you easily visualize the effect of running multiple workpieces on cycle time, there is an easier way to calculate the approximate savings per workpiece by the following formula.

Seconds saved per workpiece = (total tool-changing time × number of workpieces) – (total tool-changing time + added rapid-movement time) ÷ number of workpieces.

Given our previous example, if 10 workpieces are run per setup (still 10 inches [254 mm] apart), the time saved per workpiece will be 33 seconds. In a production run of 1000 workpieces, that adds up to a savings of 9 hours, 7 minutes. Knowing this formula is especially useful for horizontal machining centers where it is quite common to use tombstone fixtures to hold many workpieces and machine them during each cycle.

A secondary benefit of applying this technique is that the operator is freed to perform other tasks for a longer period of time. This technique also extends program execution time, which may make it possible to move the on-line task of workpiece loading off line in marginal cases.

Efficient automatic tool-changer programming

Following are three simple suggestions for minimizing tool-changing time that reduce program execution time. While these techniques are commonly taught in basic machining center courses, there are still many CNC programmers who are unaware of them.

Place tools into the machine's magazine in sequential order whenever possible. While most current-model machining centers boast very efficient *random-access* tool-changing systems, some take longer than others to activate, especially when a lengthy magazine rotation is involved. Even double-arm tool changers that exchange the tool in the spindle with a tool in the waiting position require the toolholding magazine to be rotated to the next tool's position before a tool change can occur. While this rotation can be taking place during the previous tool's machining operation, the machine may sit idle waiting for the completion of the magazine's rotation when very short machining cycles are involved. Assume, for example, that the tool in the spindle is center drilling one hole in aluminum. Time for this operation is only one or two seconds. Given the very fast rapid rates with today's machining centers, it is likely that the machine will be back at its tool changing position well before the next tool in the magazine is ready. Placing tools in the magazine sequentially can minimize (if not eliminate) magazine rotation time.

Get the next tool ready. For machines that have double-arm tool changing systems, the tool magazine can be rotating to the next tool while the current tool in the spindle is machining. If you have this kind of machine, be sure to program the next tool station early in the programming commands for each tool in the program. This also will minimize (if not eliminate) magazine rotation time.

Unfortunately, not all machining centers have double-arm tool changers. Some require that the tool currently in the spindle be placed back in its original position *before* the magazine can be rotated. For this kind of machine, about the best you can do to minimize tool-changing time is to place tools in the magazine in sequential order.

Orient the spindle on each tool's return to the tool-change position. Most machining centers require that the spindle be properly oriented before a tool change can occur. While most machining centers will perform the spindle orientation as part of the tool-changing command (commonly M06), program execution time can be saved if the spindle orientation is commanded *during* each tool's positioning movement to the tool-change position. Most machines use an M19 to cause the spindle to orient. Though spindle orientation time varies from one machine to the next, using this simple technique can commonly save from one to four seconds *per tool change*.

Programming efficient spindle range changes

While many (especially smaller) machining centers have only one spindle range, most larger machining centers have at least two. Since spindle range changes are determined by the programmed spindle speed in rpm, they are almost transparent to the CNC programmer. However, the CNC programmer should be well aware of the range-changing cutoff point.

A 30-horsepower vertical machining center may, for example, have two spindle ranges. The low range of this machine may run from 30 to 1500 rpm and the high range from 1501 to 4000 rpm. If a spindle speed of 400 rpm is programmed (with an S400 word), the control will automatically switch to the low spindle range (assuming that the machine is not already in the low range) before the spindle starts. If a speed of 2000 rpm is programmed, the machine will automatically switch to the high spindle range.

As noted during our discussion of turning center spindle range changing, it takes time to change spindle ranges. Most machining centers require that the spindle be stopped before spindle range changes can occur. Additionally, a transmission of some kind must be engaged. Though they vary from one machine tool to another, it is not uncommon for larger machining centers to take as long as three to five seconds to perform range changes.

If programmers are unaware of the cutoff point between low and high spindle ranges, there may be times when they unwittingly cause the machining center to make unnecessary spindle range changes. Assume, for example, the cutoff point between low and high spindle ranges for a given machining center is 1500 rpm, and that a 1.5-inch (38.1-mm) diameter carbide end mill immediately follows a 3/16-inch (4.76-mm) diameter high-speed steel drill. Given a mild steel workpiece, the drill may run at 1528 rpm (75 sfm; 23 m/min), while the end mill may run at 1464 rpm (575 sfm; 175 m/min). Given the proximity of these two tools to the machine's cutoff point between low and high range, spindle-range changing time could be eliminated between tools if both tools are run in the same spindle range. Given the nature of the tools involved in this example, it would be best to compromise the speed for the drill downward (to 1500 rpm), since the end mill is likely to require more horsepower than may be available at the low end of the high range.

Minimizing program execution time for 3-dimensional work

Since more and more CNC machining center users are getting involved in 3-dimensional machining, we now introduce you to some of the program execution time-limiting factors of this relatively new technology. Frankly speaking, until quite recently, CNC machining center controls (we'll call them *traditional CNC controls*) did little or nothing to accommodate high-speed machining for 3-dimensional work.

Admittedly, much of this information is related to understanding the current and future trends of high-speed-machining technology. While we offer little that can aid you with traditional CNC controls, we intend to show how and why high-speed machining controls can dramatically shorten program execution time for 3-dimensional programs. As we proceed through this discussion, keep in mind that many high-speed-machining CNC control manufacturers are willing to update your current CNC machining centers with their high-speed-machining controls. If you perform a great deal of 3-dimensional work, and if you are using traditional CNC controls, you may be amazed at how easy it can be to justify the purchase of a high-speed-machining control for your current CNC machining centers.

Three-dimensional machining defined. Earlier in this chapter we said any time you make the control think, it takes time. Three-dimensional machining is one application that pushes CNC machining center controls to the limits of their calculation capabilities.

By 3-dimensional machining, we mean generating a 3-dimensional shape from a series of tiny 1-, 2-, or 3-axis movements. The smaller the movements, the smoother the finish machined on the workpiece, but the longer the CNC program.

Because of the complex nature of CNC programs used for 3-dimensional machining, this application almost demands the use of a computer-aided manufacturing system. Most 3-dimensional CAM systems import data right from 3-dimensional computer-aided design systems, meaning if the shape can be drawn in the 3-dimensional CAD system, it can be machined by the CAM system.

As with many forms of machining, one or more rough milling cutters are followed by finishing cutters. The milling cutters usually take the form of ball end mills. For roughing operations, the departure distance (increment of movement per command) can be quite large (as much as 0.100 inch [2.54 mm] or more). This minimizes the length of the CNC program segment for roughing.

However, if the workpiece is to have a smooth surface, the finish milling departure distance must be kept quite small. Due to the *current* limitations of traditional CNC controls, many companies perform hand finishing (filing) on the workpiece to minimize the length of the program and speed the program execution time for the finishing process. These companies will use a departure distance of about 0.040 inch (1.02 mm). However, the trend in 3-dimensional machining is to minimize the departure distance for finishing and completely machine the workpiece on the CNC machining center, thus eliminating the need for hand finishing altogether. To accomplish this, the departure distance must be much smaller than 0.040 inch. Many companies use a departure distance of about 0.005 inch (0.127 mm) for finishing. (Note that many CAM systems allow the programmer to specify a kind of *tolerance* to maximize departure distance for the purpose of shortening programs. This tolerance tells the CAM system how far the cutter can stray from the work surface contour before a programmed point will be generated. This means departure distances will not be equally spaced throughout the machining operation.)

The problem of increased program length. Consider how long the typical 3-dimensional machining CNC program will be. Say you are machining a shape in a 5-inch (127-mm) square workpiece. You wish to finish the workpiece with 0.005-inch (0.127-mm) departure distances for each pass through the workpiece (we'll ignore the CAM system's ability to generate larger movements for this example). For a 5-inch-long pass, each pass will require *at least* 1000 commands (5 ÷ 0.005). The more complex the contour of the workpiece per pass, the more commands required. Since you are trying to machine the workpiece without hand finishing, you set your move-over amount between passes (commonly called the *pick amount*) to 0.010 inch (0.254 mm), meaning your program will require 500 passes (5 ÷ 0.01). Since each pass will have to machine some kind of contour, this CNC program will require well in excess of 500,000 commands. If the average command length is 15 characters (not unusual if two axes are moving per command), this equates to 7.5 million characters. That's 7.5 megabytes of CNC program length. And we did not even consider the roughing operation.

As time goes on, the trend will be to continue to decrease departure distances for finishing. When it comes right down to it, even a 0.005-inch departure distance does not allow for a very accurate workpiece (related to finish *and size*). Early NC users faced this very problem when generating circular motions. Before circular interpolation was developed, a circular motion had to be commanded with a series of tiny straight-line movements. The finer the movements, the smoother the circular move but the longer the CNC program. (Sound familiar, old timers?) Most users compromised on the departure distance (making it larger) to keep the program relatively short, but the circular movement was rather crude. When circular interpolation is used, the control (internally) breaks up the circular motion into a series of very tiny single-axis departures. The departure distance used by most current-model CNC controls is under 0.0001 inch.

Imagine how long the program in our previous example would be if making 0.0001-inch (2.5-μm) departures (instead of 0.005 inch). Instead of requiring 1000 commands per pass, we would now require 50,000 commands per pass (5 ÷ 0.0001). Instead of 500,000 total commands, we would now need 25 million. Instead of being at least 7.5 megabytes, the program would be at least 375 megabytes. While a CNC program this long may seem unreasonable by today's standards, the trend in 3-dimensional machining is truly to tighten the resolution (smaller departure distances), requiring vast amounts of data.

The problem of program storage. Most traditional CNC controls have pitifully small program storage capacity, which brings up our first limitation of traditional CNC controls. If the CNC program will not fit into the CNC control's memory, some other means of running the program must be found. For CNC controls that do have large memories (say over four megabytes), some companies break up the program into portions that *will* fit into the control's memory. Each portion of the program will be loaded and run. Then the memory will be cleared for the next program portion. This helps to explain why many CNC users are willing to use large departure distances and hand finish. It can be quite cumbersome to run lengthy programs in this manner.

No traditional CNC controls (with the exception of some PC-based controls) have the ability to store programs in excess of about 16 megabytes, and there are very few that allow this sizable program storage capacity. The vast majority have small memories of well under one megabyte. If 3-dimensional machining is to be done with these controls, another method of running programs must be used.

The limitations of direct numerical control. The most common alternative is to use direct numerical control techniques. With direct numerical control, the program is run from an outside device (like a personal computer). The control is fooled into thinking the program is coming from the machine's tape reader, when in reality, it is coming from the outside device. While DNC may be the only feasible alternative for traditional CNC controls, those DNC systems that use RS-232C serial communications protocol may soon become obsolete for high-speed-machining applications.

The main problem with RS-232C–based DNC systems is data flow. The baud rate (and handshaking) determine how fast data can be transferred. At a baud rate of 9600 bits per second, about 1000 characters can be transferred per second. While this may sound like a sufficient transfer rate, consider the 3-dimensional machining trend to smaller departure distances. Even at the relatively large finishing departure distance of 0.005 inch (0.127 mm) given in our previous example, 1000 characters per second will limit the maximum feed rate to 20 ipm (508 mm/min). (Note: to calculate the required characters per second data transfer rate for a given 3-dimensional application, first divide one by the departure distance. This gives you the number of blocks per inch of motion. Now multiply the result times the number of characters per block to come up with the number of characters per inch of motion. Multiply this result times the desired ipm feed rate to determine how many characters must be transferred per minute. Finally, divide this result by 60 to find the required characters per second transmission rate.)

While 20 ipm may seem like a fast feed rate, keep in mind that many CNC users machine very free machining materials like graphite electrodes used in electrical discharge machining. For them, 20 ipm is very slow indeed. Also keep in mind the trend toward smaller departure distances. For our second example using only 0.0001-inch (0.0025-mm) departure per command, 1000 characters per second will limit the feed rate to 0.4 ipm (10 mm/min).

Note that both of these examples assume there is no handshaking going on during the RS-232C serial transmission. With most RS-232C devices, the buffer of the receiving device will eventually become full. The receiving device will then send a signal to the transmitting device telling it to halt the transmission temporarily. When the receiving device's buffer is emptied, the receiving device sends another signal telling the transmitting device to continue with the transmission. All of this takes time, and will detract from the maximum possible characters-per-second transmission rate.

True high-speed-machining controls have no use for RS-232C type DNC. Instead, either the program resides entirely in the memory or hard drive of the control, or it comes from a computer-based network. Either way, the transmission rate is increased to more than a million characters per second. Even with a 0.0001-inch departure distance, this data flow will allow a feed rate of up to 400 ipm (1016 cm/min). But do not misunderstand. We are *not* saying that all high-speed-machining controls can feed at 400 ipm. We are simply making the point that, with true high-speed-machining controls, there is virtually no limitation in motion rate caused by program transmission.

Though this may be stating the obvious, when the traditional CNC control cannot keep up with the programmed feed rate, program execution time will suffer. The machine will stop at the completion of every command, waiting for the control to figure out what is coming up next. In essence, the traditional CNC control is the *weak link* in the system. While the machine tool could machine faster, the control simply cannot keep up.

This pausing between commands not only causes a longer program execution time, it also opens the door to other (tooling-related) problems. When the end mill pauses in the middle of the cut, for example, it will have the tendency to work-harden the material being machined, and tool life will also suffer.

The impact of the *look-ahead* buffer. The last point we will make related to high-speed-machining controls has to do with their improved look-ahead buffers. Anyone who has been involved with CNC for a long period of time knows that early (NC) machines moved in a strictly point-to-point manner. They would reach the destination point of one command and stop before moving on to the next command. While this form of motion was just fine for hole machining operations, it was very poor for contour milling. If the tool has to come to a stop at the completion of every command, a nasty witness mark will be left at every stopping point throughout the contour. For this reason, control manufacturers developed something called the look-ahead buffer. The look-ahead buffer lets the control see what is coming up in the next few commands for the primary purpose of keeping the machine from having to stop at the completion of the current command (the look-ahead buffer is also required for certain other control functions, like cutter-radius compensation).

While traditional machining center controls have utilized look-ahead buffers since the first CNC controls were manufactured, until recently they did nothing special to handle 3-dimensional applications, which is another reason why traditional CNC controls perform so poorly in 3-dimensional applications.

The high-speed, 3-dimensional machining control must look far enough into the program to determine how fast the tool can feed. If no dramatic change in motion direction is coming up, the tool can feed quite fast (as fast as the programmed feed rate). However, if a change in motion direction is coming up, the control may determine the current motion rate to be too fast for the machine tool to respond. Depending on the acceleration and deceleration limitations of the machine tool's axis-drive systems, the control will automatically slow the feed rate to ensure that the machine tool can accurately make the motions. When you think about it, this function places the limitation of machining speed back in the lap of the machine tool. True high-speed-machining CNC controls are no longer the weak link in 3-dimensional machining applications.

REDUCING TOOL MAINTENANCE TIME

As stated earlier, any period of time that a CNC machine tool is sitting idle during a production run must be included in your overall calculation of cycle time. One classic time the CNC machine sits idle is when an operator performs cutting tool maintenance (indexing carbide inserts, replacing dull tools, etc.). There are many things the typical CNC user can do to minimize this tool maintenance time and, in effect, reduce the production time for the job.

As you know, machining time and tool life are closely related. Generally speaking, as you manipulate cutting conditions for a given cutting tool in order to shorten machining time, you also shorten the length of time the tool will last. The CNC programmer and setup person must ensure during the program's verification and optimizing that cutting conditions perform efficient machining while allowing a feasible period of tool life. Reduced machining time will mean nothing if the CNC operator is constantly having to perform tool maintenance.

For small production quantities, the CNC programmer may elect to use rather conservative cutting conditions to ensure that cutting tools will last through the entire production run. The increase in overall production time caused by applying more conservative cutting conditions will be offset by the time saved by eliminating the need for cutting tool maintenance.

However, as production quantities grow, the more likely it becomes that the CNC operator will have to perform cutting tool maintenance at some time during the production run, regardless of how conservative the cutting conditions are. The larger your production quantities, the more tool maintenance will be required during the production run, and the more you should be willing to do to minimize (if not eliminate) on-line cutting tool maintenance.

PLANNING FOR ON-LINE CUTTING TOOL MAINTENANCE

The ultimate goal will be to move the on-line task of cutting tool maintenance off line. This means that when a cutting tool gets dull and needs maintenance, the tool maintenance is done while the machine is still actually running production. Though current-model CNC machines can be equipped with features that allow this, many CNC machine tools now in production (and especially turning centers) are not well suited to moving tool maintenance completely off line. You may find that, while you can dramatically reduce on-line tool maintenance for your current machines, you cannot completely move it off line. If tool maintenance is an important issue for your company, it should be part of the basic criteria that are used to help select your *next* CNC machine tool. The feature *tool life management*, for example, must be equipped on the machine tool.

In this section, we present a detailed discussion of tool life management and show what it takes to completely move tool maintenance off line, but first we offer some simple suggestions to operators of those CNC machines that require cutting tool maintenance to be performed on line. Most have to do with facilitating the operator's ability to perform tool maintenance tasks.

Grouping on-line cutting tool maintenance tasks

For high production quantities, it is very easy to identify trends in tool maintenance requirements. By tracking how often tool maintenance is required, you can begin to minimize on-line tool maintenance time while reducing machining time.

Remember that CNC operators can perform tool maintenance on several tools together faster than they can perform the same tasks on tools independently. This is because several of the tasks they perform during tool maintenance must be duplicated every time tool maintenance is required.

Say we have a turning center job using a rough-turning tool, a carbide drill, a rough-boring bar, a finish-boring bar, and a finish-turning tool. There are 10,000 workpieces to be machined. After the program is verified, the operator begins running workpieces. After 100 workpieces, the rough-turning tool becomes dull. The operator stops the cycle, opens the door, indexes the turret so the insert can be changed, and changes (or indexes) the carbide insert. The operator closes the door and activates the program to continue running production. After 20 more workpieces, the rough-boring bar needs maintenance. The tool maintenance procedure must be repeated. After 30 more workpieces, the carbide insert drill needs maintenance. Fifty more workpieces and the finishing tools require maintenance.

Performing cutting tool maintenance in this manner can be very wasteful. Every time tool maintenance is required, for example, the operator must break out of production to gather the hand tools needed to change inserts, open the door to the turning center, place the machine in the manual mode, and index the turret to the proper tool station. Once finished with tool maintenance, the operator must close the door, place the machine back in the automatic mode, reactivate the cycle, and put away the hand tools used for tool maintenance. If tool maintenance can be performed on two or more tools during the production stoppage, these tasks need only be done once.

Yet, if tool maintenance is not well planned, the operator must be close by the machine at all times, since tool maintenance will be required often. This reduces the opportunities to have the CNC operator doing other things during the production run.

While every production run will be different, and production quantities must be high enough to justify the effort involved, the programmer and setup person can work together to find a way to group cutting tool maintenance. To do this, it will be necessary for all tools (or at least groups of tools) to require maintenance at the same time. For tools that require more frequent maintenance, the cutting conditions can be reduced (adding to machining time). For tools that require less frequent maintenance, cutting conditions can be made more aggressive (reducing machining time). By compromising from both directions, your people can usually minimize tool maintenance without adversely affecting cycle time.

Turning center suggestions

While more and more CNC turning centers are being equipped with externally mounted automatic tool changers that resemble those used on CNC machining centers, most CNC turning centers do not yet allow access to cutting tools while the machine is in operation. Most require the cycle to be stopped and the door to the work area opened. In fact, many turning centers have a *door interlock* that will place the machine into the feed hold mode if the door is opened during a machining cycle. For this kind of machine, there must *always* be some on-line time taken for tool maintenance.

This dramatically limits your ability to reduce on-line cutting tool maintenance time. About all you can do is facilitate the CNC operator's ability to perform tool maintenance efficiently. Truly, anything you can do to help the CNC operator perform tool maintenance will effectively reduce on-line tool maintenance time. It should go without saying that every CNC operator in your shop must have a personal set of hand tools to be used for tool maintenance procedures. These tools must be kept close to the machine (some companies even mount a small tool box *inside the work area* to provide quick access to cutting tool maintenance hand tools).

Helping the CNC operator replace twist drills. Many cutting tools used on turning centers utilize inserts made from carbide or ceramic materials. While these tool inserts do take time to be indexed or replaced, at least the operator will be able to ensure that the cutting edges of a new carbide insert will be in the same location as the previous insert. While a tool offset may have to be changed in order to compensate for the previous tool's wear, the program zero setting values for the tool will not have to be changed.

Unfortunately not all cutting tools used on turning centers utilize inserts. For those that do not, the entire cutting tool must be replaced when the tool gets dull. Additionally, not all tools held in the turret of a turning center can be qualified (without presetting), meaning the program zero value in Z will change for cutting tools not placed in exactly the same position as the previous tool. Straight-shank internal tools, like twist drills and reamers, present the biggest problems in this regard. Measuring the program zero value in Z repeatedly during a production run can be very time consuming, and should be avoided. If at all possible, a way should be found to precisely replace straight-shank internal tools.

To minimize the amount of time it takes to precisely replace dull straight-shank internal tools, some companies utilize a gage made to the length the tool must protrude from the face of the turret or internal toolholder. While this technique works well, a different setup gage must be made for each tool length required.

A more flexible (even programmable) way to handle the problem is to incorporate optional block skip techniques to allow the CNC operator to control whether a new drill is being positioned. For most of the time, optional block skip will be left on, causing the control to skip the commands related to setting a new drill (or other straight-shank internal tool).

When the tool needs to be replaced and positioned, the operator first loads the drill into the internal toolholder in such a manner that the drill is well short of its required length. The operator snugs down the clamping device holding the drill, turns off optional block skip, and runs the program. When it comes time for the drilling operation (probably right after the rough-facing operation), the commands for drill positioning will be run since the optional block skip switch is now off. The internal tool will rapid to within the specified distance from the qualified face of the workpiece (or other fixed Z position) and stop.

We recommend keeping this approach position precisely 0.100 inch (2.54 mm) farther away from the surface than required. This way, the operator uses a 0.100-inch gage block when positioning the internal tool. The operator holds the gage block flush with the face of the workpiece, releases the clamp holding the drill, and pulls the drill out until it touches the gage block. Then the drill is securely clamped. The operator turns the optional block skip switch back on and activates the cycle. (The 0.100-inch gage block keeps the drill from rapiding into the next workpiece if the operator forgets to turn on the optional block skip switch.) Here are the commands related to doing this in the format for one popular CNC turning center control:

O0001 (Program number)
(Rough face workpiece leaving 0.005 inch stock for finishing)
.
.
.
N085 T0202 (Index to drill station)
/N090 G00 X3.0 Z0.105 (Stay 0.100 inch away from face)
/N095 M00 (Program stop for drill positioning)
/N100 G00 Z1.0 (Rapid away for clearance)
N105 G97 S500 M03 (Program for drill continues)
N110 G00 X0.0 Z0.1 (Rapid to approach position)
.
.
.

In line N090, the X3.0 indicates a position in X that is below the diameter of the workpiece just faced. The Z0.105 value reflects the fact that 0.005 inch (0.127 mm) of stock is being left for finishing. At the program stop in line N095, the operator holds the 0.100-inch gage block flush with the face of the workpiece and pulls the drill to touch it. The drill is then clamped.

How quick-change tooling can reduce cutting tool maintenance time. During our discussion of setup time reduction, we said that quick-change tooling can dramatically shorten the time it takes to load and adjust cutting tools. The reason we gave was that quick-change tooling manufacturers make these tools with amazing repeatability. A programmer can truly predict the position of the cutting tool tip *before* the setup is made. For this same reason, quick-change tooling will allow turning center users to move most of the tasks related to tool maintenance off line.

For each tool in the machining cycle, two (or more) identical tools can be assembled. While the machine is running production with one tool, the operator performs maintenance on the additional tools. When a cutting tool in production eventually needs maintenance, the operator simply replaces it with a fresh one. The on-line tool maintenance time will be equal to the time it takes to swap tools, and with quick-change tooling, will be well under 10 seconds per tool.

What about tool offsets? As you know, the physical task of changing cutting tools and inserts is only part of the operator's responsibility in cutting tool maintenance. As single-point turning tools (for turning, boring, threading, and grooving) continue to machine workpieces, they are prone to wear. This tool wear will cause variations in workpiece size. Depending on the machining operation, the CNC operator may have to make several adjustments to tool offsets during a given cutting tool's life (especially for finishing tools). Eventually the tool must be replaced (usually by indexing or changing the insert) and the offset must be set back to its original value.

The task of adjusting tool offsets to adjust for tool wear should *always* be an off-line nonproductive task. All current-model CNC turning center controls allow offsets to be changed while the machine is running production. This means the CNC operator, after measuring a workpiece and determining which offset must be adjusted and by how much, can easily change the offset during the machining of the next workpiece.

Though this is an off-line task and will not detract from cycle time, many companies expect their operators to perform other tasks during the machining cycle. As stated, many companies expect their CNC operators to inspect workpieces, perform secondary operations, and/or even run two or more CNC machine tools. The more you expect your CNC operator to do internal to the turning center machining operation, the more you should do to simplify the task of adjusting offsets due to tool wear.

For relatively small production quantities, there may be little you can do to help with tool-wear offsetting. Fortunately, when you are running small production quantities, there shouldn't be much tool maintenance required during the production run. However, as production quantities grow, you may be able to completely automate tool-wear offset adjustments.

CNC operators who run CNC turning centers for large production quantities will eventually memorize what has to be done with tool offsets during the production run. For example, the operator for a given job may know that after every 50 workpieces, the offset for a critical finish turning tool, say offset number 2, must be reduced by 0.0001 inch (0.0025 mm). After a total offset of 0.001 inch (0.0254 mm) (10 adjustments), the tool is dull and needs replacing. After changing or indexing the finish-turning tool's insert, the operator will need to increase the current value of offset number 2 by 0.001 inch to make the new insert edge cut properly. The same offset-related changes will eventually be memorized for every tool used in the program. Once this happens, the changing of tool offsets to compensate for tool wear becomes a rather mindless task, one that can be eliminated on most turning center controls.

If you confirm through experience how offsets must be adjusted during a tool's life, why waste the CNC operator's time (and open the door to potential mistakes) when offset changes can be programmed? This question is especially poignant for bar-feed turning centers, where the goal is unattended operation through the machining of (at least) one entire 12-foot- (3.66-m-) long bar. Almost all turning center controls made today allow offsets to be changed through programming commands. One popular control uses a G10 for this purpose. For this particular control, a U word within the G10 command specifies an incremental change to the X value in the offset while a W word specifies an incremental change to the Z value of the offset. A P word in the G10 command tells the control which wear offset is being changed. Given our previous example requiring the offset for a finish-turning tool to be changed by 0.0001 inch after every 50 workpieces, here is a simple program that uses subprogramming techniques to automate the changes required for offset 2.

```
O0001 (Control program for tool-wear offsetting)
N005 M98 P0002 L50 (Run 50 workpieces)
N010 G10 P2 U-0.0001 (Reduce offset 2 X by 0.0001 inch)
N015 M98 P0002 L50 (Run 50 workpieces)
N020 G10 P2 U-0.0001 (Reduce offset 2 X by 0.0001 inch)
N025 M98 P0002 L50 (Run 50 workpieces)
N030 G10 P2 U-0.0001 (Reduce offset 2 X by 0.0001 inch)
N035 M98 P0002 L50 (Run 50 workpieces)
N040 G10 P2 U-0.0001 (Reduce offset 2 X by 0.0001 inch)
N045 M98 P0002 L50 (Run 50 workpieces)
N050 G10 P2 U-0.0001 (Reduce offset 2 X by 0.0001 inch)
N055 M98 P0002 L50 (Run 50 workpieces)
N060 G10 P2 U-0.0001 (Reduce offset 2 X by 0.0001 inch)
N065 M98 P0002 L50 (Run 50 workpieces)
N070 G10 P2 U-0.0001 (Reduce offset 2 X by 0.0001 inch)
N075 M98 P0002 L50 (Run 50 workpieces)
N080 G10 P2 U-0.0001 (Reduce offset 2 X by 0.0001 inch)
N085 M98 P0002 L50 (Run 50 workpieces)
N090 G10 P2 U-0.0001 (Reduce offset 2 X by 0.0001 inch)
```

N095 M98 P0002 L50 (Run 50 workpieces)
N100 G10 P2 U-0.0001 (Reduce offset 2 *X* by 0.0001 inch)
N105 M98 P0002 L50 (Run 50 workpieces)
N110 G10 P2 U0.001 (Increase offset 2 *X* by 0.001 inch for new insert)
N115 M30 (End of program)

While this is a very good technique to use with proven tool-wear offset changes, and it actually borders on what can be done with a true tool life management system, there are two problems with this technique in production. Since we are using simple subprogramming techniques, if the cycle must be stopped for some reason, it will be impossible to restart the machine and have the control know the exact status of the tool offset. Additionally, a program stop command (M00) will now have to be used in the subprogram (O00002) to stop at the end of each workpiece. The operator may not recognize the difference between the M00 in the subprogram and the M30 end-of-program command in the main program. Therefore, the operator may not recognize the end of the cycle and will simply continue machining workpieces.

These two problems can be overcome if parametric programming techniques are used. Here is a revised example, using the most popular version of parametric programming, Custom Macro version B. Program number O0001 is the machining program that calls the custom macro (O1000).

O0001 (Main program)
(Machine entire workpiece)
.
.
.
N545 G65 P1000 (Call custom macro for tool offsetting)
N550 M30 (End of program)

O1000 (Custom macro for tool wear)
#500 = #500 + 1
IF [#500 LT 50] GOTO 99
G10 P2 U-0.0001
#500 = 0
#501 = #501 + 1
IF [#501 LT 10] GOTO 99
#501 = 0
G10 P2 U0.001
#3000 = 100 (INSERT MUST BE CHANGED)
N99 M99

This program is still somewhat limiting. Custom macro techniques are available that can make this application much more powerful and easier to use. Variables could be passed telling the custom macro which tool offset is involved, how often to adjust the offset, and the maximum offset change. We could even incorporate more than one tool in the offset adjustment procedure with one custom macro. This limited custom macro, if invoked just before the end of the machining program, will perform the tool-wear-related offset changes discussed in our example.

The custom macro uses permanent common variable #500 (originally set to zero by the operator before the production run) to count up to 50 workpieces. Then the X value of offset 2 is reduced by 0.0001 inch (0.0025mm). The custom macro uses permanent common variable #501 to count to 10 (overall offset change of 0.001 inch [0.025 mm]). Then it resets the offset and generates an alarm (caused by the #3000 statement) which places the message "INSERT MUST BE CHANGED" on the display screen, making it very clear to the operator that the insert must be indexed or changed. Since #500 and #501 are retained even after the power is turned off, the control will keep track of the current offset status even if the program has to be stopped and restarted.

While you may not thoroughly understand what is going on in the custom macro, the main point is that your programmers probably have the ability to automate the offset changes on your CNC turning center controls by one means or another. For high production quantities when offset changes are predictable, you can eliminate the need to have your CNC operators make offset changes and free them to perform other off-line tasks.

Machining center suggestions

The general nature of machining centers makes it easier (than on turning centers) to perform most cutting tool maintenance tasks off line. While the machine is using one tool (in the spindle) to machine the workpiece, all other tools are exposed outside the work area in some form of magazine. The operator can easily see the cutting tools on most machining centers during the machining cycle. This makes it possible to visually check for tool condition.

Can your operators remove tools during the machining cycle? Unfortunately, most machining center manufacturers do not allow the CNC operator to actually rotate the tool magazine during the machining cycle. This means the operator will not be able to remove or replace cutting tools while the machine is in production, limiting what can be done in the way of off-line tool maintenance.

This manual interlock of the tool magazine rotation is done for safety reasons. If the operator is rotating the magazine when a tool change command is given by the program, the operator could be injured. Or if the operator leaves the tool magazine out of position after the manual operation, the magazine may not have the correct tool in the *next tool* position, causing serious problems after the next tool change.

These safety-related problems can be resolved with a relatively simple *standby* switch. This is the same kind of switch used for machines with automatic pallet changers to keep a pallet change from occurring while an operator is still working on a pallet. Prior to performing a manual rotation of the magazine, the standby switch can be turned on. When turned on, the machine will memorize the current tool in the waiting position. If during manual operation a magazine rotation is commanded by the program, the machine will wait (standing by) until the operator finishes with the manual use of the magazine and turns off the standby switch. When the switch is turned off, the magazine will automatically rotate back to its memorized position, eliminating the possibility for the magazine to be misaligned with the CNC program.

While very few machine tool builders equip their machining centers with this kind of manual intervention capability, the operator must be able to exchange cutting tools in the magazine while the machine is running production if tool maintenance is to be moved *completely* off line. If your machine does not currently have this ability, contact your machine tool builder. A builder who understands what you are trying to do should be willing and able to help interface the standby switch. If not, there are any number of aftermarket subcontractors and consultants available whose primary purpose is to help CNC users interface new features to their CNC machine tools.

Here is an example of how the ability to remove and replace cutting tools during the machining cycle allows on-line tool maintenance to be moved off line. In a 15-minute machining cycle, say the operator notices that one of the high-speed steel drills is dull and needs to be sharpened (or replaced). The operator goes to the tool magazine, turns on the standby switch, rotates the magazine to allow removal of the drill, and removes the drill.

Depending on how much time will pass before the drill is required in the program and how long the current tool in the spindle will be machining, the operator may or may not immediately turn off the standby switch at this point. If the switch is turned off, the control will continue with other tools in the program. However, if the operator cannot finish the tool maintenance on the drill before the drill is needed in the cycle, the cycle will have to be (manually) stopped. If the standby switch is left on, the machine will halt at the next magazine rotation command. This problem can, of course, be easily overcome if the operator maintains a duplicate drill. Instead of having to perform tool maintenance during the machining cycle, the new (sharp) drill can simply be loaded after removing the dull drill.

Once the tool maintenance on the drill is performed, the new tool length determined, and the new tool-length compensation value entered into the offset table, the operator goes back to the tool magazine, turns on the standby switch (if it was turned off), rotates the magazine, replaces the drill, and turns off the standby switch.

If the standby switch is left on during tool maintenance and another tool is soon commanded, at worst this will save but a small portion of the cycle execution time. At best, the operator will be able to complete the tool maintenance on the drill without affecting the machining cycle.

Determining tool offset values off line. During our setup time reduction discussions, we presented several techniques aimed at minimizing the on-line setup tasks related to setting up cutting tools and determining tool offset values. Though we will not repeat those discussions here, we wish to remind you that the same cutting tool techniques used by setup people during setup can be used by CNC operators when tool maintenance must be performed.

Remember that if the tool's length is used as the tool-length compensation value, for example, the CNC operator can easily measure the tool-length compensation value for each tool off line. A simple height gage can be used. Also keep in mind that the tool-length compensation values for taper tools like taper taps and taper reamers can be easily determined with the use of a special setup gage.

We also give other special techniques in the discussion of setup time reduction for sizing workpieces when new tools are used. These same techniques can be used after tool maintenance is performed. By using optional block skip in conjunction with trial machining, for instance, you can ensure that critical finishing operations will be to size on the very first workpiece machined by a new tool.

TOOL LIFE MANAGEMENT SYSTEMS

Tool life management systems are usually developed by CNC control manufacturers. Unfortunately, this means they vary dramatically from one control manufacturer to another (some control manufacturers even vary the version of tool life management used for each control *model* within their own product line). In this regard, you can think of tool life management systems as being like any form of personal computer-based software product. No two word processors, for example, work exactly the same. While all allow documents to be created and printed, the actual techniques involved vary substantially from one to another. This is one reason for the confusion about what can be done with tool life management. While one version of tool life management may be powerful and easy to use, another may be rather crude and difficult to use.

Additionally, just as with computer software products, tool life management systems have come a long way since they were originally developed. Many companies have shied away from tool life management systems after having tried to use early versions. Current versions of tool life management from most CNC control manufacturers are quite powerful and easy to use. If you run high production quantities and wish to move tool maintenance completely off line, you will need to evaluate the tool life management capabilities of future CNC machines you intend to purchase.

It is not within the scope of this text to cover the details of how all tool life management systems work. Since they vary so dramatically, this would be an impossible task. In this section, we limit our presentations to showing what you should look for in tool life management systems and how tool life management systems can help move cutting tool maintenance off line.

Applications of tool life management

Unless the tool life management system can be applied over multiple production runs (which is not commonly feasible), the applications of tool life management are limited to higher production quantities. Since tool maintenance is minimal during low-quantity production runs, there will be no need for tool life management. The exception to this statement is if you run many jobs that use the same cutting tools. In this case, tool life management techniques can be applied over the course of several production runs, still minimizing on-line tool maintenance. We begin by discussing how tool life management systems are applied to larger production quantities.

Tool life management systems are used for two basic purposes. Though these purposes are similar, they can sometimes conflict, leading to misunderstandings about how tool life management systems should be applied. First, they are used to extend the length of time between tool maintenance procedures, which is important in any form of unattended machining system. If your company owns bar-feed turning centers, for example, and especially if you use bar feeders with automatic bar loaders, the cutting tools in the machine must last for the entire length of the bars if the machine is to run completely unattended.

A tool life management system will allow the bar-feed turning center programmer to program tool *groups* instead of individual tool stations. For those tools that need maintenance most often, like roughing tools, duplicate tools can be placed in the machine's turret. On the basis of criteria set by the user during setup, the tool life management system will determine when a tool requires maintenance (is dull). It will then stop using the dull tool and begin using the next tool in the group.

Note that in this kind of application, tool maintenance is still being done on line. Eventually all tools in all groups will require maintenance. While the machine may be able to run unattended for a very long period of time (possibly overnight), the company is simply postponing the inevitable need for eventual on-line tool maintenance.

The second and more efficient purpose for tool life management systems is to completely move the on-line task of tool maintenance off line. By programming tool *groups* instead of individual tool stations, the CNC user is assured that at least one active tool in the group is capable of machining in the current cycle at all times. As long as a tool can be removed for maintenance without having to stop the machining cycle, *all* tool maintenance can be performed off line. Even if the machine must be stopped to remove exhausted tools, it will only be for the period of time required to remove and replace exhausted tools. On-line tool maintenance time will be very short.

For turning center applications, the only way to fully take advantage of tool life management (moving all tool maintenance off line) is to equip the machine with an external automatic tool changer. Even then, the machine tool builder must allow tools to be removed from the tool changer while the machine is in cycle. In similar fashion, machining centers must allow tools to be removed from the magazine during the machining cycle to fully take advantage of all that tool life management systems can do. The same kind of standby switch discussed earlier in this chapter can be used to allow safe removal and replacement of cutting tools to and from the tool magazine while production is running.

Determining *when* tool maintenance is required is just a *part* of what must be done by the tool life management system. Wear offsets must commonly be changed during a tool's life to hold size, especially in turning center applications. Of additional concern to CNC users considering tool life management systems will be how many tools can be included in each group, as well as how the tool's life is specified (usually in time and/or number of workpieces being machined). Also, remember that tool life management data will likely change from one production run to the next, unless cutting tools, cutting conditions, and workpiece material remain exactly the same. If you will be using tool life management systems for several different workpieces, the tool life management criteria (tool groups, tool life, offset changing data, etc.) must be *programmable* to free the setup person from changing the criteria from one production run to the next. While all of this means that tool life management systems can become rather sophisticated, the best ones allow all of the functions discussed in this section and are still relatively easy to work with.

PREVENTIVE AND CORRECTIVE MAINTENANCE

Though Chapter Twelve is devoted to machine maintenance issues, we wish to mention the impact of machine maintenance on production. Any time a machine is down, the time it takes to complete the production run increases. This includes any time spent repairing or maintaining CNC machine tools. There are two types of machine maintenance.

Preventive maintenance is any task that helps prolong the machine's life. Changing fluids, cleaning wipers, and replacing filters are among the many procedures machine tool builders recommend to prolong machine life. *Corrective maintenance* is any task done to repair the machine after a breakdown.

Preventive maintenance can be scheduled. Your company management can select times when the machine is not in production (during weekends or off shift) to perform preventive maintenance procedures. This means preventive maintenance should not affect production schedules.

On the other hand, corrective maintenance cannot be scheduled. When it is required, the machine is down, waiting to be repaired. Remember that there are only three reasons why corrective maintenance is required.

1. *Lack of preventive maintenance.* If preventive maintenance is not performed, the machine tool will break down more frequently. Many preventive maintenance procedures, like changing fluids and filters, are aimed at prolonging the life of critical machine components. Additionally, during preventive maintenance, the maintenance person will notice those machine components that show excessive wear. These components can be replaced *before* they fail.

2. *Machine or control component failure.* Even the best preventive maintenance programs cannot protect against the eventual breakdown of *every* component. About the best a company can do to minimize this kind of corrective maintenance is to maintain a comprehensive supply of those parts that are most prone to failure. Most machine tool builders can supply a list of such parts when requested to do so. Additionally, during the initial purchase of each CNC machine, it is wise to confirm that the machine tool builder and control manufacturer stock the components that are most prone to failure.

3. *Human error.* There are many mistakes an operator or setup person can make that will cause damage to the machine tool. The best protection you can have against human error is a good training program, which is the topic of Part II of this text. Repeated mistakes should be taken as a signal that your company needs to review its training program.

PART IV

ADDRESSING OTHER IMPORTANT CNC ISSUES

There are three more areas of importance to help you improve CNC machine tool utilization. In Chapter Ten, you will see how wasted time and duplication of effort can be minimized or eliminated through documentation. In Chapter Eleven, we will discuss how wise choices in the selection of devices for program preparation, storage, transfer, and verification can enhance optimum utilization. In Chapter Twelve, we show how the implementation of a preventive maintenance program can ensure that your CNC machines remain in condition to be utilized at their highest levels.

Chapter Ten

Documentation Issues

More companies are trending toward a *paperless* environment. What has been traditionally done with paper, pencil, and file cabinets, is now being done with computers. This trend filters through all areas of a company. In design engineering, blueprints are now being created with computer-aided design systems instead of paper and pencil. In process engineering, instead of writing down the process for each workpiece, the process is engineered and entered with the help of a computer. In production control, work orders for production workpieces are issued by computer instead of manually.

The CNC environment is no exception to this trend toward a paperless environment. At the very least, CNC-related documentation can be stored in a computer's hard drive (just as any other computer file can be stored). When a job is run, the related documentation can be printed and sent out to the shop with other paperwork for the job. This eliminates the need for the traditional file cabinet approach to storing master copies of documentation. What in the past might have required an entire roomful of file cabinets now requires just one personal computer.

With recent falling prices for personal computers, and with the ease of networking many computers to one central *serving* computer, more companies are taking the paperless environment a step further. By placing several networked personal computers (or workstations) in key locations throughout the shop, CNC people have access to all CNC documentation (as well as many other manufacturing-related functions) from the shop floor. Some companies even go so far as to place one networked computer next to every CNC machine tool, making it possible for the CNC setup person and operator to access important information while working on the CNC machine tool. This eliminates the conflicts that can occur when two or more people simultaneously require access to one computer.

Though computers offer numerous advantages over traditional documentation techniques, the method by which you document functions in the CNC environment is not nearly as important as *what* you document. In this chapter, we emphasize what *should* be documented. All techniques we offer can be applied to computer-based documentation as well as more traditional documentation methods.

WHY DOCUMENTATION IS SO IMPORTANT

One of the reasons commonly given for *not* documenting is that it takes time. The programmer, setup person, and others in the CNC environment may claim they are so busy they do not have time to document what they do. While this is a typically human response, in reality the opposite is usually true. Documentation *saves* time.

Keep in mind that *documentation is an off-line task*. It is done while production is running. Any documentation that facilitates the setup person's ability to make setups effectively reduces *on-line* setup time. Any documentation that helps the CNC operator understand how production should be run enhances the CNC cycle and may even reduce cycle time. Since the shop rate of any CNC machine tool is commonly much more than the hourly wage of *anyone* in the CNC environment, and since value is only being added to your products while machines are in production, time-saving documentation related to CNC setups and production runs should be considered a mandatory part of *every* CNC environment.

Documentation is especially important for companies that perform a great deal of repeat work. In a product-producing company, for example, a given workpiece may be run once every six weeks. The more times per year the job is run, the more potential savings emerge through good documentation. Any uncertainty caused by poor (or no) documentation that leads to wasted production time must be considered very wasteful indeed.

The importance of documentation also has a great deal to do with the size of your CNC environment. The larger your CNC environment, the more important the documentation. If your company has several CNC setup people and CNC operators, it is quite likely that different people will be involved with setup and production each time a job is run. Only good documentation can ensure that *everyone* understands how the job must be set up and run.

We compare good documentation in the CNC environment to the documentation included with any product that requires assembly after purchase. Consider, for example, the assembly of a 10-speed bicycle. *Good* documentation will assume very little knowledge on the part of the person performing the assembly. The documentation may begin by showing the components of the bicycle required for assembly, acquainting the assembler with each part, and allowing an inventory to be taken. There will likely be a section listing the tools required to perform the assembly, ensuring that once started, the assembler can complete the assembly without having to find more tools. The written instructions will be easy to read and given step by step. Sentences will be kept short and will avoid words the assembler may not understand. A box will be provided next to each step to allow the assembler to check off completed steps, thus preventing missing or duplicating any step. There will be a series of diagrams or photographs illustrating what is to be done during each step of the assembly procedure, making it very easy for the assembler to visualize the assembly process. An exploded view of the entire bicycle may be given to let the assembler see the relationship of all component parts at any time.

With good documentation of this kind, the assembler can quickly and easily assemble the bicycle without frustration. However, consider the assembly of the same bicycle with poor (or no) documentation. Say, for instance, the instructions for assembly have been lost. While the person performing the assembly may (or may not) eventually figure out how to assemble the bicycle, the time required to complete the assembly will be much longer than for the person with good documentation (assuming the people involved have similar assembly skills). And once completed, the assembler without instructions may not be confident that the method used to complete the assembly was correct.

Just as the person assembling the bicycle with good documentation will outperform the person with poor documentation, so will people with good CNC documentation be able to outperform those without it. By facilitating each CNC task, good documentation can save countless hours of CNC production time over the life of your CNC equipment.

Also, just as the person assembling the bicycle with good documentation will be less likely to make mistakes, so will CNC people be less likely to make mistakes in the CNC environment. One of the most common causes of mistakes is confusion about what *should* be done. Good documentation will eliminate this confusion.

SUGGESTIONS FOR DOCUMENTATION

As you read our suggestions for CNC-related documentation, keep in mind that there are many ways to effectively communicate information. Your most important goal should be to eliminate human memorization from the CNC environment. The less a person has to commit to memory, the better the documentation.

Another goal should be to lower the skill level required to perform CNC-related tasks (especially for CNC setup people and CNC operators). The less people must figure out for themselves when making setups and operating CNC machine tools, the less skill they must possess, and the easier and faster tasks will be accomplished.

Good documentation should make tasks in the CNC environment as simple as following a set of instructions. While certain simple and redundant tasks may eventually become memorized, good documentation will remove all doubt from the mind of any person reading the documentation.

A NOTE ABOUT UNWRITTEN RULES

In almost all companies, there are certain procedures that are done repeatedly— so repeatedly they eventually become second nature and almost taken for granted. Many companies minimize documentation for these procedures by making the *assumption* that everyone understands what must be done. In essence, the company simply expects everyone in the CNC environment to understand unwritten rules about how given problems should be handled.

One example of this kind of unwritten rule has to do with tool offsets. Most companies that have turning centers (logically) make the tool offset number for a given tool the same as the tool station number. Tool 1 uses offset 1. Tool 2 uses offset 2. And so on. While it may make perfect sense to assign tool offsets in this manner, beginners in the CNC environment cannot be expected to automatically know that this is how you handle tool offsets. By one means or another, they must be made aware of this unwritten rule.

While unwritten rules do minimize documentation for tasks that are constantly performed in a consistent manner, care must be taken to ensure that *everyone* understands them. Any procedure that is not expressly documented with each job must be presented and practiced during the training of new personnel. Also keep in mind that as the number of unwritten rules you incorporate grows, so does the potential that beginners will forget them.

PROGRAM DOCUMENTATION

Most CNC programmers need to prepare for writing CNC programs. This preparation allows the programmer to break up the complex task of developing a CNC program into smaller and easier-to-handle segments. Generally speaking, this makes the overall task of programming simpler. The more complex the job, the more important the preparation.

Sequence-of-operations planning form

The CNC program may require a complex process using 20 tools. Since the machining order (process) will almost always impact the quality of the workpiece, the wise CNC programmer will develop a sequence of machining operations *before* the program is written. While documenting each machining operation, the programmer can also come up with other critical machining information, including spindle speed and feed rate for each tool (some computer-aided manufacturing systems will determine speeds and feeds automatically). Since tooling is an important part of the process, the programmer can also document the tool type and tool station number used to hold the tool with the sequence of operations. Figure 10-1 shows a sequence-of-operations planning form that can be used to help the CNC programmer prepare the machining process.

Though the primary purpose for this kind of documentation is to make programming simpler, think of how easy this sequence of operations planning form will make it for any *other* programmer (or even the original programmer after a long period of time) to understand exactly what the CNC program will do. In essence, this sequence of operations planning form is the English version of the CNC program. For this reason, many companies require that their CNC programmers document the program's process in this manner to minimize future confusion about what each program will do.

Marking up the blueprint

Many programmers use a working copy of the blueprint just for programming purposes. This blueprint is kept with other documentation for the program. The programmer will document the blueprint with anything that might be helpful during the actual programming process.

Sequence of Operations Planning Form

Part No. A-2355-2C		Rev: F		Date: 1/16/95		By: MLL
Seq.	Description	Tool	Station	Speed	Feed Rate	Restart Block
1	Rough face and turn	80 degree tool	1	400 SFM	0.014 IPR	N005
2	Drill 1.5 diameter hole	1.5 Carbide insert drill	2	1,215 RPM	0.006 IPR	N085
3	Rough bore	1.25 Boring bar	3	400 SFM	0.012 IPR	N145
4	Finish bore	1.25 Boring bar	4	550 SFM	0.007 IPR	N220
5	Finish face and turn	55 degree tool	5	550 SFM	0.007 IPR	N275
6	Neck 0.125 wide groove	0.125 wide groove tool	6	450 SFM	0.005 IPR	N345
7	Chase 3/4 - 10 thread	60 degree thread tool	7	2,011 RPM	0.100 IPR	N420
8						
9						
10						
11						
12						
13						
14						
15						
16						
17						
18						
19						
20						
21						
22						
23						
24						

Figure 10-1. A sequence-of-operations planning form used to help CNC programmers plan the process.

Since the location of program zero determines the position from which all co-ordinates needed for programming are taken, the programmer documents this position on the blueprint. Since it is possible that the programmer may forget about clamps and other obstructions while programming, he or she marks their locations on the blueprint. Depending on the complexity of the blueprint and how much machining is to be done on the workpiece, the programmer may even document just exactly what surfaces must be machined by the CNC program. Depending on how cluttered the blueprint is getting, manual programmers may also wish to document all coordinates needed for the program directly on the blueprint. If the blueprint is too cluttered for this, the programmer may elect to use a separate coordinate sheet to document coordinates needed for the program.

Again, all of this documentation is done principally to help the programmer develop the CNC program. However, this documentation also makes it easy for anyone to determine exactly what the CNC program is doing if the program must be modified at some future date.

DOCUMENTATION WITHIN THE CNC PROGRAM

All current-model CNC controls allow the programmer to include messages within the CNC program. Most use parentheses to do this. Any information the control sees within parentheses will be shown on the display screen of the control but ignored during the execution of the CNC program. This gives the CNC programmer an excellent way to document information for the setup person and operator.

General information about the job

A well-documented CNC program begins with a series of messages that remove any doubt about the program's use. The following sample program beginning includes sufficient documentation for this purpose. (Note that many controls allow only upper case characters for messages.)

```
O0001
(        MACHINE:  MORI SEIKI SL4)
(  PART NUMBER:  A-2355-2C)
(     PART NAME:  BEARING FLANGE)
(       REVISION:  F)
(      CUSTOMER:  ABC COMPANY)
(     OPERATION:  20, MACHINE BORED END OF PART)
(  PROGRAMMER:  MLL)
( DATE FIRST RUN:  4/11/93)
(  LAST REVISION:  6/30/94 BY CRD)
(            RUN TIME:  00:05:25)
N005 T0101 M41
N010 G96 S400 M03
```

N015 G00 X3 Z.1 M08
.
.
.

Notice that anyone viewing this header information can easily tell which CNC machine the program is for, what workpiece and operation the program is machining, who wrote the program, who last changed it, and important dates. Though this kind of information may seem quite basic, remember that many companies eventually accumulate *thousands* of CNC programs. Without this basic documentation in each program, it can be very difficult to keep track of which programs are used for a given job.

Of special importance in this sample header is the current revision for the workpiece. Remember that the designs for production workpieces are commonly changed, which means that the CNC programs are also changed. These changes can wreak havoc with the organization and maintenance of CNC programs. In the event that a revision may not be permanent (the design engineer may delete the revision), many companies maintain a master copy of each CNC program for *every revision*, meaning a given operation for one workpiece may eventually have several CNC programs. The setup person and/or CNC operator must be very careful to confirm that the CNC program about to be run will machine the workpiece to its most recent revision. Documenting and maintaining the revision information at the beginning of every CNC program makes this checking easy.

Also notice the specification of run time. When the program has run once, it can be helpful to document its run time in the CNC program. Anyone looking at the program in the future (while the program is not currently running) can easily determine how long the program takes to run.

Tool information

A CNC program, especially a long one, can be quite difficult to read, even for experienced CNC setup people and operators. Since the CNC operator will have to rerun tools often, the CNC programmer can make it much easier to find critical restart positions in the program by documenting the beginning of every tool. By having this documentation available in the program, CNC operators can easily confirm that they have found the correct position within the CNC program from which to start a given tool. The following turning center program illustrates this technique.

O00001
(MACHINE: MORI SEIKI SL4)
(PART NUMBER: A-2355-2C)
(PART NAME: BEARING FLANGE)
(REVISION: F)
(CUSTOMER: ABC COMPANY)

(OPERATION: 20, MACHINE BORED END OF PART)
(PROGRAMMER: MLL)
(DATE FIRST RUN: 4/11/93)
(LAST REVISION: 6/30/94 BY CRD)
(RUN TIME: 00:05:25)
N005 T0101 M41 (ROUGH TURNING TOOL)
N010 G96 S400 M03
N015 G00 X3.040 Z0.1
N020 G01 Z-1.995 F0.017
N025 X3.25
N030 G00 X6.0 Z5.0
N035 M01
N040 T0303 M41 (2-INCH DRILL)
N045 G97 S300 M03
N050 G00 X0 Z0.1
N055 G01 Z-2.6 F.009
N060 G00 Z0.1
N065 G00 X6.0 Z5.0
N070 M01
N075 T0404 M41 (1.5-INCH ROUGH BORING BAR)
N080 G96 S400 M03
N085 G00 X2.085 Z0.1
N090 G01 Z-1.995 F0.010
N095 X2.0
N100 G00 Z0.1
N105 X6.0 Z5.0
N110 M01
N115 T0505 M42 (1.5-INCH FINISH BORING BAR)
N120 G96 S600 M03
N125 G00 X1.125 Z0.1
N130 G01 Z-2.0 F0.006
N135 X2.0
N140 G00 Z0.1
N145 G00 X6.0 Z5.0
N150 T0202 M42 (FINISH TURNING TOOL)
N155 G96 S600 M03
N160 G00 X3. Z0.1
N165 G01 Z-2.0 F0.006
N170 X3.25
N175 G00 X6.0 Z5.0
N180 M01
N185 M30

Notice how in lines N005, N040, N075, N115, and N150, the message makes clear which tool is being used. An operator wishing to rerun the finish boring bar, for example, could easily confirm that line N115 is the correct block from which to start.

At *every* program stop

Depending on the machining operations being performed, there may be times when the operator must perform a special (manual) task during the CNC program's execution at a program stop, commonly commanded by an M00. For example, a vertical machining center program may be performing several tapping operations. Because chips build up inside the holes, the CNC programmer may include a program stop just before the tapping operations. During this program stop, the operator is expected to brush away the chips and add tapping compound to the holes. To confirm that the CNC operator knows what is supposed to be done, the CNC programmer should place a message in the program close to the program stop command.

.
.
.

N095 M00 (CLEAR CHIPS AND ADD TAPPING COMPOUND)

.
.
.

Other occasions when the CNC operator may be expected to perform manual operations that should be well documented include breaking certain clamps loose on machining centers for finishing operations, reducing chuck pressure on turning centers for finishing operations, and turning a workpiece around in the chuck of a turning center to machine the opposite end of the workpiece. Truly, any time a programmer includes a program stop in the program, a message should be included close by to tell the operator what to do.

Messages used for this purpose assume the operator is monitoring the program page of the control's display screen. In this case, when the machine stops due to the M00, the message will be visible. However, if the operator is monitoring another page of the display screen, the position page for example, the message will not be visible. The operator can, of course, simply activate the program page to see the message. However, this assumes the operator knows that the machine stopped due to an M00.

Some controls allow a more fail-safe method of displaying messages during a program stop. At the program stop, the display screen of the control will automatically switch to a special message page and display the message, regardless of what display screen page the operator happens to be monitoring. The example command we show is given in the format used with Custom Macro version B.

#3006 = 101 (TURN PART AROUND)

In Custom Macro version B language, #3006 is a message-generating system variable. When this command is read the display screen of the control will automatically switch to the message page. Message number 101 that states TURN PART AROUND will be shown. After the part is turned around in the chuck and the operator reactivates the cycle, the message will disappear.

To document anything out of the ordinary

It only makes sense to strive to do things in a consistent and logical manner. CNC programmers should strive to keep their programming techniques as consistent as possible to ensure that everyone is comfortable with their methods of programming. However, there may be special considerations that force a CNC programmer to do things not commonly done. Whenever this happens, a message in the program can clarify the reason *why* common practices are not being followed and give instructions about how to proceed.

For example, most CNC turning center programs require only one tool offset per tool. For this reason, most CNC programmers make the tool offset number for each tool the same as the tool station number. In this way, the CNC operator can easily associate the tool offset with the tool. Tool 1 will use offset 1. Tool 2 uses offset 2, and so on.

There may be times, however, when more than one offset is required for a given tool (or set of tools) within a CNC program. For example, say a grooving tool is necking two grooves. One groove is being machined in an area of the workpiece that has good support. The other groove is being machined in a different area of the part that has weak support. If we say this difference in support is causing one groove (the one in the unsupported area of the workpiece) to come out larger than the other, this may be a time the programmer elects to use two offsets for one tool. One offset controls how grooving will be done in the weak area. The other controls the groove in the well-supported area. Since this technique is rarely used (by this particular company), the programmer would be wise to include messages in the program to alert the setup person and operator to what is being done. This message should be included at the beginning of the program's header to ensure that the setup person and operator see it. Here is an example.

```
O0010
(SPECIAL NOTE! TWO OFFSETS ARE USED FOR GROOVING TOOL!)
(USE OFFSET 5 FOR 3.25-DIAMETER GROOVE AND OFFSET 25 FOR)
(4.00-DIAMETER GROOVE.)
(        MACHINE:  MORI SEIKI SL4)
(   PART NUMBER:  A=3566-2D)
(     PART NAME:  CAP)
(      REVISION:  C)
(      CUSTOMER:  ABC COMPANY)
(     OPERATION:  20, MACHINE GROOVED END OF PART)
(   PROGRAMMER:  MLL)
```

```
( DATE FIRST RUN:  3/30/93)
( LAST REVISION:  6/25/94 BY CRD)
(       RUN TIME:  00:05:25)
N005 T0101 M41 (ROUGH TURNING TOOL)
.
.
.
```

For changes made after a dispute

There are times when a CNC programmer is asked to make changes to the CNC program with which he or she does not agree. A change that improves machining time may (in the programmer's opinion) open the door to safety-related problems. Whenever a CNC program change is made after any dispute, it is a good idea to document the circumstances that lead to the change. The most obvious place to document such a change is within the CNC program.

For simple setup instructions

Most CNC machine setups require more documentation than can be easily included in a CNC program. Many require, for instance, a drawing or picture to clarify any instructions in writing. However, there are simple setups that can be explained with nothing more than verbal instructions. Here is an example that shows the beginning of a program including setup instructions. Note that general information about the job is also included at the beginning of this program.

```
O0001
(        MACHINE:  MORI SEIKI MV-40)
(  PART NUMBER:  A-3325-2B)
(    PART NAME:  TOP PLATE)
(     REVISION:  C)
(     CUSTOMER:  DEF COMPANY)
(    OPERATION:  25, MACHINE HOLES IN TOP OF WORKPIECE)
(  PROGRAMMER:  CRD)
( DATE FIRST RUN:  6/11/94)
( LAST REVISION:  6/30/94 BY MLL)
(       RUN TIME:  00:05:25)
(  )
(          SETUP INSTRUCTIONS              )
(TOOLS REQUIRED:)
(STATION 1, 1/2-INCH-DIAMETER SPOT DRILL)
(STATION 2, 13/64-INCH DRILL)
(STATION 3, 1/4-20 TAP)
()
(WORKHOLDING SETUP:)
(USE COMMON 4-INCH VISE MOUNTED TO TABLE IN TABLE SLOTS
```

ONE AND FOUR. USE BUMP STOP FLUSH WITH LEFT SIDE OF VISE.)
(USE 1-INCH PARALLELS UNDER WORKPIECE. PROGRAM ZERO IS)
(LOWER LEFT CORNER OF PART IN X/Y AND THE TOP OF THE PART)
(IN Z. MEASURE THIS POSITION AND STORE THE PROGRAM ZERO)
(NUMBERS IN FIXTURE OFFSET 1.)
N005 G91 G28 Z0 M19 (Program begins)
N010 T01 M06
N015 G54 G90 S600 M03 T02
.
.
.

As you can see, though this technique limits how elaborate the programmer can be with setup instructions, it can be very effective for reducing paperwork in simple setups, especially when making similar setups on a regular basis.

SETUP DOCUMENTATION

Though almost all companies tend to develop at least some unwritten rules about how setups are to be made, nothing beats the clarity of good documentation. This is especially true as a new setup person is becoming familiar with your company's CNC environment. The clearer the setup documentation, the easier (and faster) it will be to make the setup; good setup documentation can effectively minimize setup time.

Universal setup sheets

Most companies utilize some form of setup sheet to specify how setups are to be made. If a standard form is used as the setup sheet, it will soon become familiar to the setup people, making it easier to read and follow. Just as any standard form, if used often enough, it will soon become familiar to its user.

For relatively simple work, many companies try to minimize paperwork by including all setup information on one page. Most setup sheets of this nature include general information, instructions related to how to make the workholding setup (including room for a setup sketch or drawing), information about the cutting tools, and any special information about how the machine tool must be adjusted. Figure 10-2 shows an example of a machining center setup sheet.

While this kind of universal setup sheet can concisely relate a great deal of information in a very small space, it also limits the amount of information that *can* be related to the setup person and almost forces certain assumptions to be made. In the general information given at the top of the setup sheet shown in Figure 10-2, you may be able to think of other information of importance to your setup people. In the cutting tool section, notice that only the actual cutting tools themselves are specified. This assumes that the setup person will be able to locate and assemble the other components that each tool comprises (toolholder, extension,

Universal Machining Center Setup Sheet

Part No. A-3455-3C	Rev: D	Programmed by: CRD	Work holding: 4" vise
Part Name: Top Plate	Date: 3/24/93	Machine: MV-40	CNC Program File: F:\MILLS\MV40\O3255.CNC

CUTTING TOOLS:

WORK HOLDING INSTRUCTIONS:

Station	Tool	Tool Notes
1	3" face mill	
2	#4 Center drill	Minimum 6" long
3	13/64 drill	
4	1/4-20 tap	
5	27/64 drill	Minimum 6" long
6	1/2-13 tap	Minimum 6" long
7	31/32 drill	
8	1.0001 reamer	
9	3/4 roughing end mill	Hogging style
10	3/4 finishing end mill	4 flute
11	1.25 end mill	4 flute
12	2.5000 boring bar	Cartridge type
13		
14		
15		
16		
17		
18		
19		

Locate the vise in table slots one and four. Use bump stop flush with the left side of the vise. Use 1" parallels under workpiece. Program zero is the lower left corner of the workpiece in X/Y and the top surface of the part in Z. Use fixture offset number one to store program zero setting values.

SETUP DRAWING:

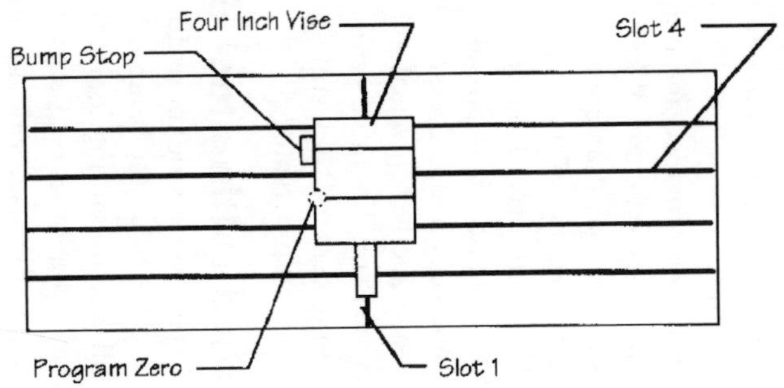

Bump Stop — Four Inch Vise — Slot 4 — Slot 1 — Program Zero

Offsets: Make length offset numbers same as tool station numbers. For radius offset numbers, add 30.

Figure 10-2. An example of a universal-style single-page setup sheet for machining centers.

collet, etc.). Also notice that only 19 tools can be included on this setup sheet. Many complicated machining center setups require more. In the workholding instructions, the limited space available will minimize the amount of information a programmer can relate. The programmer will not be able to offer any kind of step-by-step instructions. In the setup drawing section, space is also constrained for all but the simplest setups.

The less information provided about making setups, the higher the skill level that is required of the setup person. While relatively simple changes in this universal setup sheet's design may dramatically improve its usability, all universal setup sheets will have constraints of the nature discussed, meaning they generally require a rather high skill level from your setup people. Since many companies *expect* their CNC setup people to possess a high level of CNC skill, the universal setup sheet is commonly and successfully used.

Figure 10-3 shows another universal style setup sheet for turning centers. Though turning centers tend to require more similar setups from one production run to the next than machining centers, they still constrain the amount of information that can be given to the setup person.

More detailed cutting tool documentation

Companies that wish to provide more documentation than the typical universal setup sheet tend to break up the setup information into three or more segments. One segment will contain cutting tool information. Another will contain written setup instructions. Yet another will contain graphic images including drawings, photographs, and possibly even videos to illustrate how the setup should be made.

Cutting tool lists. One way to improve the cutting tool segment of the universal setup sheet is to list every component of each tool. This improved documentation can still be provided in list form. Figure 10-4 shows an example of this kind of cutting tool information. As you can see, the setup person can easily determine exactly what components are required to assemble each tool.

When it comes to the identification of each component, companies vary with regard to how components are specified. Some simply use the tool manufacturer's designation number. This is especially true with turning center cutting tools, since there is a high degree of standardization in component specification, even among different cutting tool manufacturers. While this type of designation makes it easy to specify each tool component, it does nothing to help the setup person *locate* the needed components. For this reason, highly organized companies will establish their own cutting tool component designations. These designations will correspond to the storage bins located in the tool crib, making it easy to find the cutting tool components required for a given setup.

For example, an easy way to designate cutting tool components is to use a kind of acronym for the tool designation name. Once everyone understands the nature of the acronym, anyone can easily determine exactly what each tool component is just from its name. For example, the first letter or two may specify the type of

Universal Turning Center Setup Sheet

Part No. A-3457-4A	Rev: C	Programmed by: CRD	Work holding: 8" hydraulic three-jaw chuck
Part Name: Hub		Date: 2/14/92	Machine: SL4	CNC Program File: F:\LATHE\SL4\O3323.CNC

CUTTING TOOLS:

Station	Tool	Tool Notes
1	80 deg rough turn tool	1/32 tnr
2	2" Carbide drill	
3	1.25 Rough boring bar	1/32 tnr
4	1.25 Finish boring bar	1/64 tnr
5	55 deg finish turn tool	1/32 tnr
6		
7		
8		
9		
10		
11		
12		

Tailstock pressure setting: None required

Chuck pressure valve setting: 14

Coolant pressure valve setting: 5

Offsets: Make offset numbers the same as tool station numbers

Other notes:

WORK HOLDING INSTRUCTIONS:

Use standard hard jaws mounted in the 7th seration of master jaw as shown to grip stock diameter. Program zero is right end of finished workpiece in Z. Store distance from program zero in Z to chuck face in shift offset.

SETUP DRAWING:

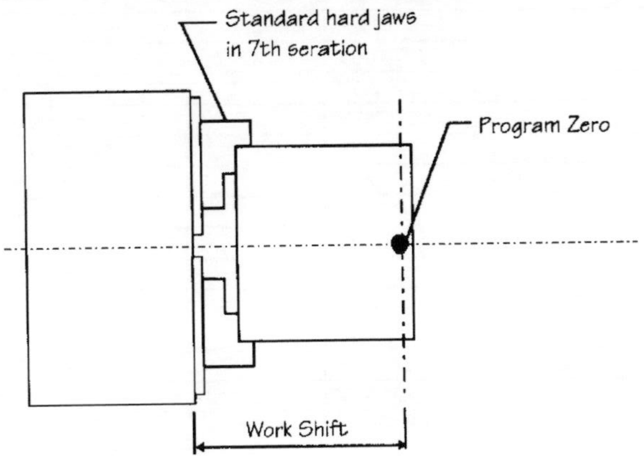

Standard hard jaws in 7th seration

Program Zero

Work Shift

Figure 10-3. A universal-style single-page setup sheet for turning centers.

Tool Assembly Information

Part No. A-3455-3C		Rev: D	Programmed By: CRD	Sheet: 1 of 1			
Part Name: Top Plate		Date: 3/24/93	Machine: MV-40				
Station	**Tool Name**	**Tool Number**	**Cutting Tool**	**Insert**	**Adaptor**	**Compon. 1**	**Compon. 2**
1	3" face mill	FM3000CI-1	3" Carbide insert face mill	CNMG-432	AFM1250		
2	#4 center drill	CD4HS-1	#4 HSS center drill		AC1000	E1000-3	CO3125
3	13/64 drill	D0203HS-2	13/64 high helix drill		AC1000	E1000-3	CO203
4	1/4-20 tap	TO250HS-1	1/4-20 high helix tap		AC1000	TE1000-3	TCO250
5	27/64 drill	D0421HS-5	27/64 HSS twist drill		AC1250	E1250-3	CO421
6	1/2-13 tap	TO500HS-1	1/2-13 HSS tap		AC1250	TE1250-3	TCO500
7	31/32 drill	D0968HS-2	31/32 HSS drill		AC1250	E1250-3	CO968
8	1.0001 reamer	R1001HS-1	1.0001 HSS reamer		AC1250	E1500-4	CO750
9	3/4 roughing end mill	EMRO750HS-2	3/4 HSS 4 flute hog mill		AEMO750		
10	3/4 finishing end mill	EMFO750HS-1	3/4 HSS 4 flute end mill		AEMO750		
11	1.25 end mill	EMF1250C-1	1.25 4 flute carbide end mill		AEM1250-E		
12	1.5000 boring bar	BB2500CI-2	1.5000 carbide boring bar	Cartridge #2	ABB1500		
13							
14							
15							
16							
17							
18							
19							
20							
21							
22							
23							

Figure 10-4. A more elaborate method of specifying all components of each cutting tool.

component (C for collet, TC for tap collet, E for extension, A for adapter, and so on). The next series of numbers may designate the size of the component. If more criteria must be designated, more letters or numbers can be used to make up the acronym specification. If the component is supplied by several tool manufacturers, for example, a designation within the component's specification can be used for this information as well. With this component naming system, the designation C1250 might specify a 1.250-inch- (31.75-mm-) diameter collet. E0750 could specify a 0.750-inch- (19.05-mm-) diameter extension. AEM1000 may be a 1.000-inch- (25.4-mm-) diameter end mill adapter.

As long as the tool crib is organized by tool component type, finding tool components will also be easy. Any tool component beginning with the letter *A* will be easily found with tool adapters. Any component beginning with a *C* will be found in the collets section, and so on. Additionally, if components are organized by size from smallest to largest, it will be easy to locate a specific tool component within each section.

Cutting tool assembly drawings. The tool list method of documenting cutting tools is quite flexible, allowing virtually any combination of components to be designated. While this flexibility allows the programmer to specify a given cutting tool exactly as desired for each program, this method does lead to many similar, but slightly different, tool assemblies. Depending on the variety of different workpieces being machined, this flexibility may lead to a great deal of duplication of effort.

Say, for instance, your company produces a product and has 50 different workpieces requiring 1/4-20-tapped holes to be machined. Each of these 50 workpieces may be run as many as six or eight times a year. As the programmer (or group of programmers) writes the individual programs for each of the 50 workpieces and develops the cutting tool assemblies, he or she should do everything possible to minimize the number of different 1/4-20 tap drill and 1/4-20 tap assemblies required for the 50-workpiece production run. While it may not be possible to come up with one tap drill and one tap assembly that will work for all 50 workpieces (due to differences in workpiece material and configuration), the smaller the number of different drill and tap assemblies, the more potential there will be for being able to keep a given drill and tap in use from one setup to the next, eliminating the need for tool assembly.

We discuss standardizing cutting tools in much greater detail during our presentation of setup time reduction techniques in Chapter Eight. Here we limit our presentation to showing one method of *documenting* the standard tools your company uses.

Another limitation of the cutting tool list method is that it does not allow the setup person to visualize the assembled tool. The easier it is to visualize, the faster the tool can be assembled, especially by new setup people.

Tool assembly drawings can be used to help with both problems. The programmers will be able to easily track previously used tools for the purpose of minimizing duplication of effort, and the setup person will be able to easily visualize the tool's construction.

If documentation of this nature is done in a paperless environment, all tool drawings could be stored within the computer network and organized with simple database techniques. The tool list on the setup sheet will simply specify the *tool assembly number* for each cutting tool. The setup person will simply call up the computer file drawing of each tool to see how it is assembled.

If documentation is to be done with more traditional methods, we recommend using one size A drawing for each tool. The tool drawings can then be placed in a standard 3-ring binder and organized by *tool assembly number* for easy location. Figure 10-5 shows an example tool assembly drawing.

Documenting cutting tool material. Cutting tools are available in various materials to suit the different workpieces they machine. Drills, for example, can be made from high-speed steel, cobalt, or carbide, and can even have tool-life-prolonging coatings applied. In similar fashion, many cutting tool holders are designed to accept a variety of different *inserts*. The inserts can be made from different materials and may also have special coatings.

If your company uses identical cutting tools or inserts made from different materials, your cutting tool documentation must make clear which of the tool materials you expect the setup person to use. In the tool assembly drawing shown in Figure 10-5, for example, it is made very clear that a high-speed steel twist drill is to be used. In the drill's tool assembly number, the letter H in D0312H-1 also designates that a high-speed steel drill be used.

If a titanium-coated drill must be used, the tool assembly number will be D0312T-1. For many tools, like twist drills, taps, reamers, and common end mills, the tool shank is made of the same material as the cutting edge. Since there will be a limited number of different tool materials available, it will be very easy for the setup person to determine which tool material is required by a simple 1-letter abbreviation (H for high-speed steel, T for titanium-coated, C for carbide, etc.). However, carbide inserts come in *many* different grades. For this kind of tool, it may be necessary to include a legend on the tool assembly drawing that shows which insert is related to the 1-letter abbreviation. Figure 10-6 shows an example.

While we have shown a few simple techniques that can get you started with your cutting tool designations, keep in mind that every company will have its own special needs in this regard. Be sure the method you come up with will allow your people to designate all of the cutting tools your company uses.

More detailed workholding instructions

As stated, the easier the setup person finds it to follow the setup instructions, the faster a setup can be made. The more complicated the setup, the more important it is to provide good workholding setup documentation.

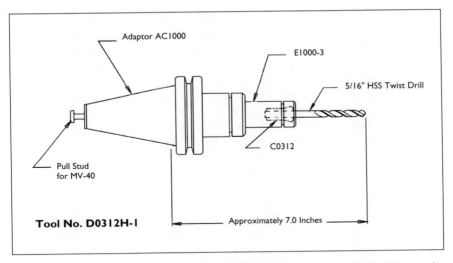

Figure 10-5. Cutting tool assembly drawing clearly illustrates how the tool should be used.

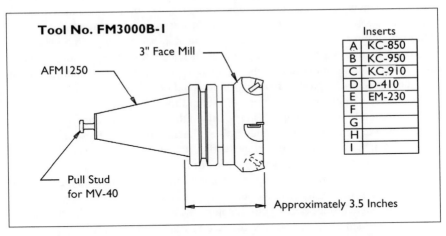

Figure 10-6. The insert grade can be easily determined from a 1-letter abbreviation by using the legend on the tool assembly drawing.

Written documentation. While experienced CNC setup people should eventually become adept at making setups (especially if your company runs the same jobs over and over), beginners may struggle to become proficient. And even experienced setup people may have occasional questions about how a new setup should be made. In many companies, when a setup person is having problems making a setup, the person who planned the setup (commonly the programmer) is called in to help.

Keep in mind that the machine will be down during this period of confusion. Depending on the availability of the setup planner and how quickly he or she can explain the solution to the setup person's problem, a great deal of production time can be wasted while basic questions are answered. If the instructions provided with the setup documentation are in the form of a step-by-step procedure, *all* steps must be included. *Any* questions the setup person has about the setup should be answered by the documentation.

Consider once again the step-by-step procedure for assembling a 10-speed bicycle discussed earlier in this chapter. Many CNC machine setups can be dramatically simplified by using the same techniques.

For example, a list of all components required to make the workholding setup can be provided to allow the setup person to gather everything required *while the machine is still running production*. In similar fashion, a list of all hand tools and gaging tools required to make the setup will further help the setup person to get ready. If everything is readily available to the setup person before the setup is actually made, no production time will be wasted while the setup person frantically searches the shop for a given component, tool, or gage.

The step-by-step procedure can be as elaborate as required. If your setup people are inexperienced, even the minor steps of making the setup can be described. If your setup people are experienced, only major steps can be included. We recommend developing procedures based on your most *inexperienced* setup person.

Graphic documentation. Remember the saying "A picture is worth a thousand words?" While some setup people may have problems understanding written instructions, almost anyone can easily understand drawings and photographs. At the very least, a drawing of the completed setup should be included with the setup documentation to let the setup persons confirm their understanding of individual steps along the way (similar to a bicycle's exploded view in the bicycle assembly instructions). With complicated setups, difficult steps can be more easily related with drawings and photographs than with words.

If you know a complicated setup is going to be made often, one way to enhance your setup person's ability to make future setups is to take a photograph after the completion of each step. Better yet, videotape the entire setup. This kind of documentation will surely eliminate any future questions during setup.

If your company is trending toward a paperless environment, keep in mind that many current distributive numerical control systems (most commonly used for distributing CNC programs) now have the ability to provide much more manufacturing information, including setup documentation. Additionally, more than simple written setup instructions can be distributed for setup purposes. More and more DNC systems allow photographs and even video clips to be easily brought into the DNC system for viewing during setup.

Setup instructions must be available in time to allow the setup person to begin visualizing how the setup should be made *before* beginning the setup. As long as the documentation is given to the setup person in time for study, setup people can easily familiarize themselves with how the setup is to be made. Just as reading the entire set of 10-speed bicycle instructions prior to starting will speed the assembly process and minimize the potential for mistakes, so will having your setup people study instructions prior to making setups minimize downtime and setup mistakes.

Other important setup documentation

Remember that one of the primary reasons for documenting in the first place is to eliminate human memory from the tasks in the CNC environment. Anything that may be forgotten or that will take time to find should be well documented. Here we list some of the many things we recomment you include in your own setup documentation.

Offset documentation. Offsets are used for a variety of purposes with CNC equipment. The three most common machining center uses are with tool length compensation (to specify tool length data), cutter radius compensation (to specify cutter radius or diameter data), and fixture offsets (to specify the position of program zero points used within the program). The three most common turning center uses are dimensional offsets (to allow sizing of the workpiece), tool-nose radius compensation (to allow the specification of the tool type and radius), and geometry offsets (to allow the specification of the program zero point for each tool).

The actual total number of tool offsets available will vary from one CNC control to another. Most controls will come with an ample supply, allowing at least twice the number of offsets as tools the machine can hold. Since tool offsets are extremely important to the success of the setup, documentation for setups should, at the very least, specify which offsets are to be used with each tool.

Most companies set up unwritten rules about how tool offset numbers are to be determined. Many companies using machining centers, for example, will make the tool-length compensation offset number the same as the tool station number. If the cutter-radius compensation value cannot be stored with the tool length, many companies will add a constant number (that is greater than the number of tools the machine can hold) to the tool station number to determine a cutter-radius compensation offset number. For instance, if the machine can hold 25 tools, they may pick 30 as the constant number. The tool-length compensation value for tool station 1 will be stored in offset 1. The cutter-radius compensation value for tool station 1 will be stored in offset 31. Similar techniques are used on turning centers to determine primary and secondary offset numbers.

While this is a simple and effective way to handle the determination of offset numbers, many companies rely entirely on word of mouth to relate unwritten rules. In some cases, it is almost assumed that new setup people and CNC operators already know the company's practice before they begin.

Since major problems are caused if the wrong offset is entered or changed, and since there may be times when a programmer will not adhere to the standard method of determining an offset number, we recommend that offset specifications be included with setup documentation. A simple standard note can be used on the setup sheet, like the one shown at the bottom of the cutting tool list in Figure 10-2.

CNC program loading documentation. While the actual act of loading a CNC program into a control is likely a skill you expect your setup person to possess, you will want to ensure that the setup person knows *which* CNC programs are related to the setup and *where to find them.* If your company is trending toward a paperless environment, you have probably already eliminated paper tapes as a method of saving and retrieving CNC programs. However, many companies still use portable program transfer devices like floppy disk drive systems and laptop or notebook computers. If this is the case, this device should be made readily available to the setup person (with the associated program storage disks) at the time of setup. At the very least, the setup sheet should specify the location where the program storage disks are kept.

More and more companies are utilizing one central computer that acts as the serving computer for all CNC machines in the shop. With this kind of system, the setup sheet should specify the directory path and names of the files containing the CNC programs. The general information sections of the setup sheets shown in Figures 10-2 and 10-3 illustrate this method of documentation. With this information, the setup person can easily call up the programs needed for the current setup being made.

Keep in mind that many CNC controls have a feature called *background edit,* which allows a program to be loaded or edited while the machine is running another program. With this feature, the setup person can be loading the programs for the next job while the machine is running the current job.

PRODUCTION-RUN DOCUMENTATION

Once a job is up and running, many companies do little, if anything, to document what the operator must do during the running of workpieces. At first glance, it may seem that the CNC operator would not need the kind of help documentation can provide. Admittedly, most CNC operators have learned enough about their CNC machines not to need any help. However, depending on the skill level of the new CNC operator you hire, production run documentation can eliminate a number of mistakes that result in scrapped workpieces.

Workpiece loading documentation

While this may be considered a duplication of what is included within the setup documentation, a simplified set of workpiece loading instructions and a related drawing (or photograph) can be used to show the CNC *operator* just how to load workpieces, and can minimize loading mistakes with complicated jobs.

For example, if several workpieces are machined on a machining center, it may be possible that the order of loading (and clamping) will affect how well the workpieces are supported during machining. Or maybe fixtures are being used that can hold more than one type of workpiece. Special locators may have to be properly positioned to allow proper workpiece loading.

Keep in mind that more than one operator may be involved in the production run, complicating the task of informing each as to how to load workpieces when changing over from one job to the next. A workholding setup may, for example, be made during the first shift. The first-shift operator begins running production. When the second-shift operator comes in, many companies simply expect the first-shift operator to relate the pertinent details of the job to the second-shift operator. However, if for some reason the first-shift operator is unavailable, the second-shift operator may be left alone to determine how to run the job. The same goes, of course, for the third shift. This confusion due to lack of production run documentation can lead to (at best) wasted time, or worse, damage-causing mistakes.

Offset setting documentation

In almost all CNC operations (with the possible exception of automation systems and other unattended systems), the CNC operator is expected to maintain offsets during the production run. Offsets are commonly changed on turning centers, for instance, to allow for tool wear. Many companies expect their CNC operators to be able to determine the relationship of offsets used by the program to the surfaces being machined. While this may not be a very difficult task, production run documentation can make it easy to determine which offsets are related to the various surfaces being machined, and eliminate the possibility of mistakes.

One easy way to document the offsets used during a production run is to provide a marked-up print or workpiece sketch for the operators to use during the production run. Using a set of colored pens, the programmer can easily mark each surface that is machined by a different tool with a different color. Each color will represent an offset number that controls the sizing for the surface. For example, the color red may correspond to offset 1. The color blue may be used for offset 2, orange for offset 3, and so on. If the same color coding is used for every job, operators will eventually become familiar with the code and will easily recognize which offset must be changed by simply seeing the related color. This color-coding method of documenting offsets is especially helpful to new CNC operators who have little previous CNC experience.

Other production-run documentation

There are a number of things you expect your CNC operators to do during production runs that require documentation. Many companies, for example, expect their CNC operators to perform measurements on the workpieces they machine. They may also have to enter statistical process control data based on the measurements they take for quality control tracking purposes. Additionally, some CNC operators are expected to perform secondary operations on the workpieces they machine. Just as documentation can clarify what the CNC operator is to do on the CNC machine tool itself, so can it clarify other functions the CNC operator must perform during the CNC operation.

Chapter Eleven

Program Preparation, Transfer, Verification, and Storage Issues

T
hough all CNC machines require programs to operate, the need to create, transfer, verify, and store CNC programs varies dramatically from one company to the next. This leads to a great deal of confusion about which devices are best for a given company. Additionally, countless devices are available to help with the preparation, transfer, verification, and storage of CNC programs. Since the sales people commonly make their devices appear to fit everyone's needs, it can be difficult to get an objective opinion about which devices best suit *your* needs. In this chapter, we discuss a company's need for CNC-program-related devices based on the four most important usage factors: the number and variety of different CNC machine types the company uses, the complexity of the work to be done, typical production quantities, and the amount of repeat business involved.

PROGRAM PREPARATION AND STORAGE DEVICE ISSUES

There are four basic types of program preparation devices. They are introduced here in order of complexity. The first is the *CNC control* itself. While the CNC control does not make a very efficient program preparation device, all current controls do allow CNC programs to be entered and modified at G-code level through the use of the keyboard and display screen. Basic functions like alter, insert, and delete CNC words, as well as more global editing (find and replace, cut and paste, etc.) are among the features commonly available for CNC program editing.

The second program preparation device is the *CNC text editor*. Most commonly a software application running on a personal computer, CNC text editors allow (manually written) CNC programs to be entered through the keyboard of the personal computer and stored on the computer's hard drive. Most have the same global editing capabilities found in word processors. Additionally, most CNC text editors include a series of special functions unique to CNC (automatic sequence numbering, trigonometric functions, decimal point format conversion, etc.). Most CNC text editors have the capability of transferring the CNC program through the serial port of the computer to the CNC control once the program is created.

The third program preparation device is the *conversational CNC control*. Conversational controls enable the CNC user to develop programs right at the CNC control using a method commonly called *shop-floor programming*. Instead of working with cumbersome G-code level manual programs, the conversational programmer works at a much higher level. Program entry is usually quite graphic, allowing visual checking at every step along the way. Once the conversational program is complete, many conversational controls even translate the conversational program into a true G-code level program, which is run to machine workpieces. Others actually run directly from the conversational program.

The fourth type of program preparation device is the *computer-aided manufacturing system*. While CAM systems vary dramatically, most are personal computer-based. All allow CNC programmers to work at a much higher level, freeing them from having to write programs at G-code level (manually). Most use a graphic interface that allows the programmer to actually draw the workpiece on the display screen of the computer. Once drawn, machining operations can be defined. Each step along the way to creating a CNC program, the CAM system will display a visual confirmation that entries are correctly made. When finished, the CAM system will create a true CNC program from which workpieces will be run.

USING THE CNC CONTROL TO ENTER PROGRAMS

Almost all CNC users should reserve the program editing functions of CNC controls for simply making corrections to the CNC program during verification and optimizing of the program. Unless the CNC control has a feature called *background edit*, programs cannot be entered by using the keyboard and display screen of the control while the machine is running production. Even with controls that do have background edit, it is rather cumbersome to enter programs during production since the CNC operator will often need to use the keyboard and display screen for other purposes (setting offsets, for example). This means a great deal of setup time will be wasted if programs are entered through the keyboard and display screen of the control. Stated another way, CNC controls used to enter CNC programs make very expensive typewriters.

Given the availability and affordability of personal computer-based CNC text editors, there is virtually no reason for entering programs through the keyboard and display screen of the CNC control. About the only exception might be if a company is running extremely high production quantities and dedicates its CNC machines to running only one simple job. In this case, once the program for the job is loaded into the control (by any means), there will be little need to *ever* enter another program. The only program-related device this company will need is a very simple program transfer device used to *back up* the program in the control's memory.

USING CNC TEXT EDITORS TO CREATE PROGRAMS

Regardless of how their CNC programs are originally created, almost all CNC users have the need for a CNC text editor. Again, given their affordability, there is no reason *not* to have one. Aside from the inexpensive nature of CNC text editors, many companies simply have little need for anything more elaborate when it comes to CNC program preparation. For them, the CNC text editor may be the only form of program preparation device required.

For predominantly simple work requiring little in the way of math calculations for CNC programming, an experienced manual programmer can commonly out-perform even a CAM system programmer. Examples of *simple* work include hole machining operations on machining centers and straightforward turning and bor-ing operations on turning centers. This kind of work is especially simple when programming for CNC controls that have special and helpful canned cycles. At rare times when more complex math is required, the manual programmer can work with the design engineer to determine coordinates going into the (manually written) program as long as the company uses a computer-aided design system. As the frequency and difficulty of performing complex calculations increase, how-ever, the manual programmer will find it more difficult to compete with a good CAM system programmer. With highly complex work, like 3-dimensional work, it may actually be impossible to prepare programs manually.

CNC text editors are also commonly the only form of program preparation device needed by companies that have only a few CNC machines, as long as the company performs relatively simple work. However, as the number of machines increases, and especially when there are many different types of CNC machines, one manual programmer may not be able to easily keep track of the different techniques required for programming all types, even if the company does pre-dominantly simple work.

The use of CNC text editors as the sole program preparation device is also common in companies that run very high production quantities, since there will be plenty of time to prepare the CNC program during production runs. In similar fashion, companies that see a great deal of repeat business may require nothing more than a CNC text editor. While the company may need to make changes to some CNC programs based on revision changes (done with the CNC text editor), the original creation of a CNC program is required only once.

Taking repeat business one step further, many product-producing companies run a finite number of different workpieces. While new products require new CNC programs, once CNC programs *are* created, the company will run the same jobs over and over again. Depending on the complexity of the workpieces, it is quite likely that a CNC text editor will suffice as the only program preparation device required.

USING CONVERSATIONAL CONTROLS TO CREATE PROGRAMS

Conversational controls enjoy a rather narrow band of popularity. A great deal of confusion and controversy surrounds the question of whether it is wise to utilize conversational controls to create CNC programs. Companies that use them exclusively tend to swear by them, while companies that do not tend to downplay their effectiveness. As we contrast the use of conversational controls with the four factors contributing to the best choice of CNC program preparation device, the reasons for the controversy will become obvious.

When it comes to number of CNC machines, conversational controls tend to be most effective in companies that have only a few machines. In fact, some of the best applications for conversational controls and shop-floor programming are in companies that have one or two CNC machine tools. These companies tend to expect their CNC people to perform *all* CNC-related tasks, including processing, programming, setup, program verification, and running production. In this kind of company, programming time is an *on-line task* performed during setup, meaning the machine tool is commonly down while a program is being created. Anything that can be done to facilitate the CNC person's ability to develop the program will effectively reduce setup time. Conversational controls offer many features aimed at minimizing programming time at the CNC machine tool.

Unfortunately, as the number of CNC machines a company owns grows, it becomes increasingly difficult to justify the use of conversational controls for program preparation, for three reasons:

1. A company can dramatically shorten setup time if programs are prepared *off line* while the machine is currently running production. The more CNC machines involved, the more setup time can be saved by off-line programming.

2. The skill level required to program and operate a conversational control is much higher than that required for operating a conventional CNC control. The more conversationally controlled CNC machines a company uses, the more highly skilled operators it needs. A company can limit the number of highly skilled personnel required in its CNC environment if conventional CNC equipment is used.

3. Conversational controls commonly cost much more than conventional CNC controls. One popular CNC control manufacturer, for example, charges about $5000 more for a given conversational control in its product line than for a comparable conventional CNC control. At this rate, it can be quite expensive to equip the entire shop with conversationally controlled machines, especially in light of the falling prices of CAM systems. One CAM system can usually handle the programming of *all* machines in the shop.

Workpiece complexity is also a factor in deciding whether to employ conversational CNC controls. Compared to manual programming, most conversational controls dramatically simplify programming, especially for programs requiring many math calculations. Generally speaking, current-model conversational controls can easily handle the bulk of the most common machining operations. However, compared to today's most powerful CAM systems, conversational controls have several limitations.

With one rather weak conversational machining center control, for example, very basic operations like hole machining, round and square pocket milling, and face milling can be commanded with relative ease. However, with this particular control, even relatively simple 2-axis contour milling presents problems. Whenever contour milling must be done, a rather cryptic series of commands must be given. There are many milling scenarios that this particular control cannot handle, and if a mistake is made, the user must go back and redefine the entire contour.

Companies that use conversational controls extensively commonly have a range of workpiece complexity that easily falls within the capabilities of their particular controls. As you evaluate conversational controls for your own CNC environment, you will need to confirm the effectiveness on your own workpieces of the conversational control you intend to purchase. Be sure you see a wide range of your own workpieces programmed on the conversational control you intend to buy.

When it comes to workpiece quantities and repeat business, conversational CNC controls also apply very well in companies that run small production quantities and see very little repeat business. In fact, some of the best applications for conversational controls are in companies having typical quantities of fewer than 10 workpieces. With these small quantities and short cycle times (especially if there is no repeat business), it is likely that no single CNC programmer could keep up with more than one CNC machine tool, regardless of how programs are prepared off line. In this kind of company, the more machines the company owns, the worse the problem gets, and the more the company will need the ability to create CNC programs right at the machine tools. Companies utilizing conversational controls for low production quantities include mold and die shops and prototype shops, as well as toolrooms and research and development departments in product-producing companies.

Keep in mind that many conversational controls allow the best of *both* programming worlds, meaning they can be programmed conventionally (by manual programming techniques or with a CAM system), as well as conversationally. This makes the conversational control an excellent choice for companies having fluctuating needs. If a company is running high production quantities or a great deal of repeat business, programs can be prepared using conventional techniques. If production quantities drop to a point where the CNC programmer cannot keep up, at least an experienced CNC operator has the ability to prepare programs at the machine.

USING CAM SYSTEMS TO CREATE PROGRAMS

The revolution in low-cost, high-powered personal computers, combined with the falling cost of software products, have made CAM systems increasingly popular in recent years. Not long ago, even relatively simple CAM systems were priced in the tens of thousands of dollars. Today, a wide range of excellent CAM systems is available at well under $5000, some of which even include high-level features like 3-dimensional milling.

Given the large number of CAM systems currently available, it is impossible to address the specific pros and cons of each one. Instead, we base our discussion on the four important factors mentioned earlier that determine the kind of program preparation system that best suits your needs.

One of the best applications for a CAM system is preparing programs for a variety of different types of CNC machine tools. Though a single manual programmer may find it quite difficult to keep track of the programming differences from one machine to another, the typical CAM system can do so easily. However, even CAM systems vary with regard to how easily they go from programming one type of CNC machine tool to another.

If streamlining the programming of multiple machines is the principal reason for your needing a CAM system, there are two important issues you must consider. If your people will be programming multiple *similar* machines, you may often need to run a given program on the first available machine tool. This means your CAM system must be able to quickly create the CNC program for a specific machine, regardless of any minor programming differences among your machines. For example, if you have five similar 20-horsepower turning centers, each made by a different machine tool builder and capable of running a given workpiece, the machine that comes free first will run the next job. However, there may be substantial differences as to how the CNC program must be formatted for each individual machine tool, especially if the machines have controls made by different control manufacturers. Make sure your CAM system allows you to take one CAM system program (commonly called the source program) and quickly create the CNC program for *any* of your CNC machine tools.

If your people must program machines that have little in common, your CAM system must make it equally simple to program each machine type. If, for example, your company owns machining centers, turning centers, wire electric discharge machining machines, and CNC turret punch presses, your CAM system must include at least *four modules*, one for each machine type.

While many CAM systems boast of allowing multiple machine applications, some make it easier to work with each of the modules. Some of the best CAM systems share the same geometry definition functions from one module to another in order to minimize the learning curve required to draw workpieces. With this kind of CAM system, the programmer can draw and trim workpieces in essentially the same manner, regardless of which machine tool is being programmed. In similar fashion, the tooling specifications and cutting condition specifications (speeds, feeds, depths of cut, etc.) should remain consistent from one module to the next. Some of the best systems share tool and material files among their machining modules. Machining functions of the CAM system should also remain similar among the different machining modules. While there are substantial differences in the machining operations required among different CNC machine tools, a good CAM system will make it very easy to learn another machining module once one is mastered.

Unless the CAM system is originally developed to incorporate the wide range of CNC machine tools being used, one of the machining modules (usually the first one developed) will commonly be easy to use and quite powerful, while others may be quite cumbersome and weak. Prior to acquiring anything, make sure you see demonstrations of your intended CAM system with all forms of machines you will be working with.

If your company owns several CNC machines, you will also need to confirm the cost of tailoring your new CAM system to the machines your company currently owns, as well as setting your company up for growth. Most CAM systems are sold by the module, meaning your company will only need to purchase those modules required for your company's machine tools. However, you may wish to determine what other modules are available to allow for growth.

Additionally, some CAM systems charge extra for supplying each *postprocessor.* The postprocessor fine tunes the CNC program output to a given CNC machine tool, and one is commonly needed for *each* machine your company owns. If you own a large number of machines, postprocessor costs can equal or even exceed the cost of the CAM system itself. The best CAM systems allow end users to tailor their CNC programs without the help of the CAM system supplier. In essence, they allow users to create their own postprocessors, resulting in substantial savings.

As to workpiece complexity, the more complicated the workpiece, the more a CAM system can help with programming. In fact, numerous examples exist of workpieces being manufactured that cannot be feasibly programmed by any other means, making the CAM system an essential part of the CNC environment.

If your main reason for using a CAM system is to handle the programming of highly complex workpieces, keep in mind that many CAM system suppliers specialize in *niche* markets. Generally speaking, the niche market suppliers tend to handle the special problems of workpiece programming in their niche better than general-purpose CAM systems purveyors trying to be all things to all users. For example, a CAM system that specializes in creating 3-dimensional milling programs will tend to be more powerful and will outperform the 3-dimensional milling capabilities of general purpose CAM systems that happen to have a 3-dimensional milling module. Other examples of niche market CAM systems used for very complex programming include turbine milling, cam milling, and engraving.

ISSUES RELATED TO PROGRAM STORAGE

How important is it to efficiently store and retrieve CNC programs? This depends mostly on the amount of repeat business. Generally speaking, the more times a company must run a given program, the more important it is to be able to store and retrieve the program and related documentation as quickly as possible. Conversely, a company that never does repeat business will be relatively unconcerned with storing programs for future use.

The media on which CNC programs are stored have changed dramatically in recent years. There are still CNC (or NC) machines in production that require programs to be loaded by 1-inch-wide paper tape. With paper tape, cumbersome storage cabinets are required for tape filing. All storage and retrieval is done manually and can be time-consuming, requiring a person to find tapes as needed and replace them once finished. Also, since tapes are prone to wear and tear, they require constant maintenance. A great deal of production time can be wasted in finding, retrieving, replacing, and maintaining tapes.

While companies that use machines requiring tape must still address program storage issues, the bulk of the company's effort should be directed toward achieving a tapeless environment. As older machines are replaced, the day will come when tape is completely obsolete. All current-model CNC controls allow programs to be loaded by standard serial communications, just as all kinds of data can be transferred over the phone lines. This allows CNC programs to be easily placed on computer storage devices, like floppy diskettes and hard disks.

Computer storage devices also lend themselves to easy program organization. Even with no special database software, a CNC user can easily use the directory and file structure of computer storage devices to organize program storage in a logical manner. All CNC programs, for example, can be stored under the hard disk's subdirectory PROGRAMS. Under the subdirectory PROGRAMS, the end user can organize yet further, making subdirectories based on machine type or customer name. This makes it quite easy to find all programs related to a given customer or machine tool.

Tracking other important CNC program information will require a little more effort. Some companies keep a logbook that itemizes all important information about each program. Entries in the logbook can include the program's file name, the date it was created, the workpiece name and part number, the programmer's name, the CNC machine the program is for, and any other important data.

While a manual logbook works just fine, this kind of information can be enhanced by using a computer database. Very basic program-related information, setup instructions, cutting-tool component locations, fixture locations, and any other important data about the program can be included in the program log database. If this information is made available to everyone in the shop (through a network of computer terminals), everyone can easily locate needed programs and related documentation quickly and efficiently. Once programs are located, they can be transferred by using the same network.

Not only is this paperless form of program storage easier to organize, the amount of space required to store programs is dramatically reduced. One 3.5-inch high-density floppy diskette (1.44 megabytes) can store 12,000 feet of tape. This means that what at one time filled an entire room of tape storage cabinets with paper tape now easily fits on the hard drive of a single personal computer.

Keeping backup copies of CNC programs is also easy with computer-based program storage. Countless backup systems are available for individual personal computers as well as computer networks. Program security is much enhanced over paper tape. With paper tape, a master tape is commonly kept in an office environment while a working tape is sent out to the shop. In many companies, the entire cumbersome tape storage system is duplicated.

PROGRAM TRANSFER DEVICE ISSUES

Most CNC machines cannot run production while CNC programs are being saved or retrieved (unless the machine has a feature called *background edit*). This means program loading becomes an on-line setup-related task. Anything you can do to reduce program loading time effectively reduces setup time by the same amount. For this reason, one way to dramatically enhance machine utilization is to ensure that programs are being loaded and saved as efficiently as possible.

In this section, we first introduce the basic types of program transfer devices. We then discuss a company's best choice of devices based on the same four factors used for determining the best program preparation device: the number of machines the company owns, the complexity of the work, production quantities, and the amount of repeat business.

DISTRIBUTIVE NUMERICAL CONTROL DEVICES

There are two categories of program transfer devices. Unfortunately, they both go by the acronym DNC which tends to add to the confusion surrounding these devices. *Distributive* numerical control systems are used to transfer the entire program from the serving device to the CNC control (and vice versa). When the CNC program resides in its entirety in the CNC control, it is available for running production. With distributive numerical control systems, the CNC program must be small enough to fit in the control's memory.

Many types of distributive numerical control devices are available. For the purpose of this text, we classify any device that can transfer programs in their entirety to and from the CNC control as a distributive numerical control device. Examples of *manually* activated devices include portable floppy drive systems, portable tape reader/punch devices, portable random-access memory (RAM) devices, and notebook, laptop, and desktop computers. The most common form of *automatically* activated distributive numerical devices is the personal computer or workstation.

Any manual distributive numerical control device applies nicely to companies that require a limited number of program transfers. A company having only one CNC machine tool, for example, will not have many program transfers to make. A simple portable manual distributive numerical control device can be placed near the machine for program transfers. In similar fashion, a company that dedicates its CNC equipment to running ultrahigh production quantities may be interested only in being able to back up their CNC programs in case of catastrophe.

However, as the number of program transfers increases, so will the need to seek out a more efficient way of transferring CNC programs. To illustrate, suppose a company uses a portable program transfer device (perhaps a portable floppy drive system) for its five CNC machine tools, which all have different CNC controls. Whenever a program transfer is required, the setup person must first find the portable program transfer device and adjust it to the particular machine needing the program transfer. Only then can the program transfer be made. Meanwhile, the machine will sit idle during the period of time it takes the setup person to search for and connect the program transfer device. It is not uncommon for this procedure to take at least 10 minutes (if all goes well), but more likely will take 20 to 30 minutes *per program transfer*. If your company is currently using this kind of device, watch your setup people make program transfers. You will likely be unpleasantly surprised with how long they take.

An *automatic* distributive numerical control system can virtually eliminate program transfer time. With today's automatic systems, an operator or setup person can command the program to be saved or retrieved *from the keyboard and display screen of the CNC control*. Program transfer time is commonly less than 30 seconds. Additionally, many automatic devices allow more than simple CNC programs to be transferred. Many (more expensive) systems allow other information about each job to be transferred to a computer terminal close to the CNC machine.

Automatic distributive numerical control systems find best application in companies that do a great deal of repeat business. Once a program is in the system, it can be called by any machine operator at any time. If the company runs a finite number of different workpieces, no more maintenance need be performed on the automatic distributive numerical control system once all programs are in the system. The more program transfers you make, the more justification you have to purchase an automatic distributive numerical control system.

DIRECT NUMERICAL CONTROL DEVICES

Note that there are times when a CNC program may be too large to fit in a CNC control's memory. One example is when performing very complex 3-dimensional milling operations. Three-dimensional programs tend to be extremely long, made up of many very tiny movements. *Direct* numerical control systems are used to feed only small portions of the CNC program to the control at a time. As the control runs a small program portion and is able to accept another portion, it requests more of the program. This is commonly called *spoon feeding* the CNC program to the control.

As with distributive systems, direct numerical control systems can be manual as well as automatic. Additionally, they can be stationary as well as portable. As long as the CNC control has the ability to execute the CNC program from the RS-232C port (as many current models do), almost any distributive numerical control device can double as a direct numerical control device. However, some older controls will not allow programs to be executed through the RS-232C port. They require the direct numerical control system to be connected through the machine's tape reader. This kind of direct numerical control system is commonly called a *behind-the-reader* system.

Additionally, some high-end automatic distributive numerical control systems (commonly workstation- or mainframe-based) allow both forms of DNC. The automatic DNC system can be spoon feeding one or more machines with direct numerical control techniques at the same time other controls are requesting entire programs by distributive numerical control techniques.

A new style of CNC control is currently evolving that overcomes the problems attending direct numerical control. This control type is called a *high-speed machining control*. As companies strive to get better and better finishes directly from the CNC machine to eliminate hand polishing, each movement within the CNC program must get smaller and smaller. As each movement gets smaller, the speed of data transmission (called the *baud rate*) must increase if a given feed rate must be maintained. The maximum transmission rate allowed through standard RS-232C serial communications protocol is fast becoming the weak link in 3-dimensional milling operations. An example stressing the limitations of RS-232C communications in high-speed machining is given on pages 300-301 in Chapter Nine.

True high-speed-machining controls eliminate RS-232C communications altogether. By using a form of computer network, transmission rates of about 1 million characters per second are allowed. This effectively overcomes the transmission rate problems of direct numerical control. If your company does a great deal of 3-dimensional milling work, you will want to learn more about high-speed-machining controls.

PROGRAM VERIFICATION DEVICE ISSUES

All new CNC programs must be cautiously verified. However, many companies do not even begin the program verification procedure until the program is loaded into the CNC control. Though some CNC controls do have graphic capabilities to help with program verification, if program verification is completely done at the CNC machine tool, a great deal of setup time can be wasted. Remember, if a problem is found with the CNC program during on-line program verification, correcting the program becomes an on-line setup task. However, many excellent computer-based off-line program verification devices are available to help precheck CNC programs.

All systems discussed in this section make it very easy to visualize your program's motions. However, there are three types of mistakes that can be made during the CNC machine setup that will cause even a perfectly written CNC program to fail. Since no program verification system can show these mistakes, your setup people will still have to be quite careful in these areas. The first setup mistake has to do with how program zero is assigned. If the setup person makes a mistake during the measurement and/or entry of the program zero points, it will cause mispositioning during every motion command in your CNC program. The second setup mistake is with tool offsets. If the setup person makes a mistake measuring or entering a tool offset value (tool-length offset value, cutter-radius offset value, etc.), the program's motions will be incorrect. The third mistake has to do with cutting conditions. No program verification system to date allows for checking or optimizing cutting conditions (feeds, speeds, depths of cut, etc.).

TOOL-PATH PLOTTERS

Tool-path plotters are available in two styles. The first and most popular simply shows the tool path generated by a CNC program on the display screen of a computer. Each tool is commonly displayed in a different color, and cutting motions are displayed by solid lines, while rapid motions are displayed by dotted lines. CNC machine panel functions such as single-block, dry-run, and feed-hold are commonly simulated with this kind of tool-path plotter.

The second style of tool-path plotter actually works with a separate plotting device. A set of colored pens held by the plotter are used to draw the tool paths on a piece of paper. This kind of plotter can easily provide a hard copy of the tool path, which can be included with the documentation for a given job. However, since this form of tool-path plotter is costly, many companies opt for a computer-screen tool-path plotter.

While many tool-path plotters are included as part of CAM systems, most work only with CAM system programs. This means they cannot show what will happen if the actual CNC program is changed. The best tool-path plotters actually work from the CNC program at G-code level and will plot CAM-system-generated CNC programs as well as manually written CNC programs.

ANIMATION SYSTEMS

Keep in mind that most tool-path plotters do not work in *real time*. Instead, they simply display the tool path as quickly as possible. While this is usually good enough, certain machine types may require more.

Imagine, for example, a 4-axis, single-spindle turning center. With this kind of machine, two tools commonly work on the workpiece at the same time. The timing of the two tools is extremely critical. While a simple tool-path plotter will work nicely to show the tool path generated by each tool (one at a time), the timing of the program cannot be verified unless *both* tools are shown together in real time.

Though it generally involves much more work to set up than simple tool-path plotters, a true animation system will demonstrate the CNC program running in real time. Most animation systems will also show much more than a simple line representing a tool path. Instead, the entire tool holder, insert, workpiece, and workholding device will be shown as part of the animation.

Though this statement may change with advancements in software technology, true animation systems are commonly considered overkill for most simple CNC applications (you can normally see everything you need to see with a simple tool-path plotter). Animation systems that show real-time motions are most helpful when verifying any automated device that performs articulated motions, such as robotic systems.

Chapter Twelve

Service and Maintenance Issues

As we have been stressing throughout this text, CNC machine time is expensive. Downtime of any kind causes a drop in productivity, leading to underutilization of the CNC machine tools. The two types of downtime are: *Planned downtime*, which occurs on the company's terms. While it can still be costly, at least everyone knows the machine will be down for a period of time and can get prepared. Setup time is one example of planned downtime. When a CNC machine is close to finishing a production run, the manager of the department can organize other things for the operator to do while the setup is being made. If the CNC machine is feeding other machines with workpieces, planning can ensure that operators of other machines in setup can also stay busy. While the CNC machine itself is not productive during planned downtime, at least people in the CNC environment can be.

Unplanned downtime is much more costly. This kind of downtime takes everyone in the CNC environment by surprise and there is almost nothing that can be done in preparation to keep people busy. The most common cause of unplanned downtime is machine failure. Something goes wrong with the machine that keeps it from being able to run.

Unplanned downtime can wreak havoc with your company's productivity and must be considered part of the current production run's cycle time. Since all CNC-using companies depend heavily (if not solely) on their CNC machines to keep production flowing, when a CNC machine goes down unexpectedly, almost all areas of the company will suffer. Any machines performing secondary operations on workpieces supplied by the disabled CNC machine also go down. All workers involved with secondary operations will have nothing to do. In product-producing companies, assembly of component workpieces into the company's product is also halted. This can be devastating, especially if your company employs just-in-time techniques.

We have shown many techniques to minimize planned downtime throughout this text. Indeed, one of the keys to improving CNC machine utilization is reducing the planned downtime caused by setup, which was the primary topic of Chapters Seven and Eight. In this chapter, we discuss the only way to minimize unplanned downtime: through preventive maintenance.

PREVENTIVE VERSUS CORRECTIVE MAINTENANCE

An automobile provides an excellent analogy for the two kinds of maintenance required for CNC machine tools. Preventive maintenance (PM) tasks for automobiles, like changing the oil, keeping all fluid reservoirs full, checking tire pressures often, and changing belts, are recommended by all automobile manufacturers. In every new car manual, you will find a chart that lists the intervals (usually in time or miles) for each preventive maintenance procedure. These procedures are always done at the convenience of the owner, when the automobile is not needed.

The preventive maintenance procedures you see in an automobile manual are the minimum requirements to help ensure that *corrective* maintenance will not be necessary. Corrective maintenance is done to repair some kind of problem—usually a serious problem. If a fan belt is not checked often and replaced when worn, it will eventually break. When it will break is unpredictable, and it will probably leave the driver stranded. Though no amount of preventive maintenance can ensure that there will never be a need for corrective maintenance, one way to guarantee that corrective maintenance *will* be required is to ignore the preventive maintenance procedures.

Though it should almost go without saying, corrective maintenance *must* be performed as soon as it is required. There can be no postponing it. Ignoring the need for corrective maintenance will always lead to much more serious problems. In the automobile analogy, if the fan belt breaks and the owner ignores the need to have it replaced, serious damage to the engine will occur within a few miles of continued driving.

While the automobile analogy may seem somewhat simplistic, I am amazed by the number of companies I see that do little, if any, preventive maintenance on their CNC machines. Instead, they foolishly run their machines *until* they break, and then worry about what they must do to repair them. Given the high cost of CNC machines and their importance to the company's production, it should be obvious that CNC machines deserve a very high degree of concern and care.

All machine tool builders include a chart in their operation or maintenance manuals listing recommended PM procedures. Most are very similar to the scheduled maintenance performed on automobiles. Examples of machine tool builders' recommendations include checking and adding way oil, changing hydraulic oil, changing spindle lubrication oil and air filters, and replenishing other fluid reservoirs.

Keep in mind that, just as with an automobile, preventive maintenance can be planned for nonproductive times. It may be possible to perform quick and easy procedures, like changing air filters, during lunchtime and breaks. More time-consuming procedures, like hydraulic oil changes and cleaning way wipers, can be scheduled during times when the machine is normally down (weekends, evenings, before shifts, etc.).

This cannot be overstressed. *You can schedule and plan for preventive mainte-nance.* Preventive maintenance minimizes the need for corrective maintenance, which you cannot plan or schedule. Perform the PM procedures as your machine tool builder recommends.

IMPLEMENTING A PM PROGRAM

The amount of effort required to implement a PM program is based primarily on how many machines must be maintained. If your company owns only one or two CNC machines, it may be simple enough to manually track your maintenance procedures. Using the chart recommended by the machine tool builder as your guide, you can easily make a list of daily procedures your maintenance person (possibly the CNC operator) must perform. The maintenance person can initial the items as they are done, documenting the fact that scheduled maintenance has been done.

For less frequent and especially more complicated PM procedures, your chart can highlight those (nonworking) times when preventive maintenance can be done. Again, the person performing the maintenance can initial the chart as each main-tenance procedure is completed. Figure 12-1 depicts a maintenance chart for a typical turning center.

As the number of machines increases, manually tracking and scheduling PM becomes more difficult. There are several very good personal computer-based software products available for helping with your PM program.

Depending on your company's maintenance department, there may be PM pro-cedures that require you to seek the help of your machine tool builder. Replacing spindle bearings is an example of a maintenance procedure that most CNC users do not wish to perform on their own. Most machine tool builders will teach their users how to perform even the most difficult procedures. However, unless you have a sufficient number of CNC machines to justify a highly qualified mainte-nance technician, you may be better off soliciting the help of the machine tool builder for very complicated maintenance tasks.

MACHINE CRASHES

While ignoring preventive maintenance can lead to corrective maintenance, most CNC people will agree that, by far, human error is the single largest cause of corrective maintenance. The frustrating fact is that there is no excuse for human-error crashing of a CNC machine tool. If your company has well-trained setup people, no mishaps will occur during changeover. If your company has well-trained programmers, mistakes will be minimized during programming. Any program-ming mistake that will cause a mishap will be found by the well-trained setup person during the program's verification. If your company has well-trained opera-tors, no mishaps will occur during production runs.

Preventive Maintenance Procedures   Okuma Howa Act 35 - September, 1994

Daily Procedures	1	2	3	4	5	6	7	8	9	10	11	12	13	14	15	16	17	18	19	20	21	22	23	24	25	26	27	28	29
Check hydraulic oil level	CL	CL			JK	CL	CL	JK	JK			CL	CL	CL	JK	JK			CL	CL	CL	CL	CL			JK	JK	JK	JK
Check slideway lubricant level	CL	CL			JK	CL	CL	JK	JK			CL	CL	CL	JK	JK			CL	CL	CL	CL	CL			JK	JK	JK	JK
Check coolant level	CL	CL			JK	CL	CL	JK	JK			CL	CL	CL	JK	JK			CL	CL	CL	CL	CL			JK	JK	JK	JK
Check air system lubricant level	CL	CL			JK	CL	CL	JK	JK			CL	CL	CL	JK	JK			CL	CL	CL	CL	CL			JK	JK	JK	JK
Check hydraulic pressure level	CL	CL			JK	CL	CL	JK	JK			CL	CL	CL	JK	JK			CL	CL	CL	CL	CL			JK	JK	JK	JK
Check chuck pressure level	CL	CL			JK	CL	CL	JK	JK			CL	CL	CL	JK	JK			CL	CL	CL	CL	CL			JK	JK	JK	JK
Check slideway oiler pressure	CL	CL			JK	CL	CL	JK	JK			CL	CL	CL	JK	JK			CL	CL	CL	CL	CL			JK	JK	JK	JK
Check oil flow of spindle cooler	CL	CL			JK	CL	CL	JK	JK			CL	CL	CL	JK	JK			CL	CL	CL	CL	CL			JK	JK	JK	JK
Clean taper part of spindle bore	CL	CL			JK	CL	CL	JK	JK			CL	CL	CL	JK	JK			CL	CL	CL	CL	CL			JK	JK	JK	JK
Other Procedures																													
Clean oil filter of spindle cooler			CR																										
Change hydraulic sys oil filter			CR																										
Change hydraulic sys oil			CR																										
Change air filter in control			CR																										
Clean way wipers			CR																										
Adjust spindle belt tension																													
Adjust spindle timing belt																													
Check for X axis backlash																													
Check for Z axis backlash																													
Adjust ball screw starting torque																													
Clean axis limit switches																													
Clean electric system panels																													

Figure 12-1. This monthly schedule of maintenance procedures makes it very clear when maintenance must be done and provides an excellent way to document maintenance procedures.

The key to eliminating crashes caused by human error is training. If your company is experiencing what you consider to be excessive corrective maintenance caused by human error, it should be taken as a signal that your people need more training. During training it should be stressed that every human-error-related crash can be prevented through cautious programming, setup, and operation techniques.

While some crashes are caused by CNC control failure, the days are gone when faulty electronic circuitry would cause CNC machines to go haywire on a regular basis. Most current-model CNC machine tools are extremely reliable, and mishaps caused by faulty control components are rare. (Most current-day machine tool builders and control manufacturers boast uptimes of well over 95 percent.) Though this is the case, many CNC people are quick to point to the CNC control as the cause of a mishap.

Knowing that crashes are rather commonplace, most machine tool builders do their best to design machines with crash prevention in mind. For example, shear pins that can be quickly replaced are commonly used in a turning center's turret to hold the turret in place firmly, but not so firmly that severe damage to the turret will occur in case of a crash. Most linear-axis drive systems incorporate a kind of brake to hold the ball screw firmly in relation to the axis drive motor, but not so firmly that the brake will not slip in case of a crash. The headstock of most turning centers is not pinned into place. Instead, it is simply held down with bolts. If the turret crashes into the chuck, the headstock will slip along the machine bed, minimizing the possibility of more severe headstock damage.

While no CNC machine is totally crashproof, machine tool builders have set procedures to check and repair any damage caused by a crash. Depending on the severity of the crash, most can tell your maintenance people exactly what to do (over the phone) to repair the problem. Also, crash recovery is commonly taught during the machine tool builder's maintenance training. As your maintenance personnel learn how to repair specific problems caused by the crashes your people have, it is very important that you have them document precisely what must be done during repair. This will minimize the time it will take to repair similar future crashes. It is also important to have on hand at all times the spare parts required to repair a damaged machine. Most machine tool builders can supply a spare parts list that includes those parts most prone to breakage during a crash.

In all but the most advanced CNC environments, unfortunately, the first thing that happens after a crash is a cover-up. The person responsible for the crash is commonly afraid of reprisals. If damage caused by the crash is not obvious, the person may not report the crash and may attempt to continue running production as usual. As you know, continuing to run a damaged machine can be very dangerous. At the very least, workpieces will probably not come out correctly. More likely, the machine will sustain more damage with continued running. And in extreme cases, the machine could actually be dangerous to run.

Your people must not be afraid to report crashes. In fact, crashes should be documented along with your company's other service and maintenance records.

Instead of placing blame, instruct the person about what went wrong and caused the crash. Be sure the person understands how to avoid the same situation in the future. Again, if you are experiencing excessive crashes caused by human error, it should be taken as a signal that your people have not received the proper amount of training.

HOW MUCH SHOULD MAINTENANCE PEOPLE KNOW?

Every company has a person who just seems to have a knack for fixing things. This is the person people call on when their cars will not start on a cold winter afternoon. Generally speaking, this kind of mechanical aptitude is about all it takes to perform the PM tasks recommended by the machine tool builder. However, as corrective maintenance is required to repair damaged machines, your maintenance personnel will need more training.

Knowing that many of their customers do not have highly skilled service technicians, most machine tool builders and control manufacturers include easy-to-use diagnostic systems within their CNC machines and controls. In many cases, a person unskilled with electronics and machine service (possibly even the CNC machine operator) can act as the hands of the machine tool builder's service technician while talking on the telephone. Given a description of a problem, the machine tool builder's service technician will ask the CNC operator to perform tests and other diagnostics on the machine tool step by step. Without needing to completely understand what is going on, the operator can relay the results of such tests, giving the machine tool builder's service technician a way to diagnose the problem. Many machine failures are diagnosed in this manner and, if required, repair parts can be shipped by overnight express for replacement by the CNC user's own personnel. In worst-case scenarios that require service by a factory technician, at least the technician will know exactly what is wrong with the machine, and bring along any parts required to repair it.

Some machine tool builders even incorporate a modem in their CNC machines for communication during a machine problem. By connecting to the machine through the modem, the machine tool builder's service technician can see many of the machine functions in the same way the operator will. Depending on the severity of the problem, the repair may even be effected through the phone line.

As long as your machine tool builders and control manufacturers have knowledgeable and helpful service departments, and as long as they are readily available for help, your own maintenance people do not need to be expert technicians. However, some machine tool builders are better at servicing their customers than others. Additionally, as machines and controls get older, machine tool builders and control manufacturers tend to minimize (if not eliminate) the service they provide. For these reasons, you may need a more competent service and maintenance staff, especially if your company owns many CNC machine tools.

Most machine tool builders provide excellent service and maintenance courses. These courses are best taken while your machine is new. Again, machine tool builders will eventually minimize the training they provide for older equipment. Course content commonly includes all facets of machine maintenance and service, including mechanical and electronic repairs.

SPECIFIC SERVICE ISSUES

During my 10 years working as an applications engineer for a machine tool builder, I received many phone calls from end users who suspected a service problem, but wanted to confirm that they were not doing something wrong that would cause the problem they were having. There are many application-related mistakes that can appear to be machine or service problems. Conversely, many machine or service problems can appear to be application mistakes.

PROBLEMS RELATED TO BACKLASH COMPENSATION

Almost all current-model CNC machines have a feature called *backlash compensation*. Any mechanical linkage, including each axis drive system of CNC equipment, is prone to wear. When the axis drive systems of a CNC machine wear, backlash will develop. Backlash is the amount of nonmotion during an axis motion-direction reversal. For example, after moving in the plus *X*-axis direction, a minus motion of precisely 1 inch (25.4 mm) is commanded. If there is any backlash in the *X* axis, the table or turret will not start moving in the minus direction the instant the ball screw direction reverses. This results in a minus *X*-axis motion of something less than 1 inch. The amount the machine moves less than 1 inch is the amount of backlash.

Backlash usually shows up on machined workpieces in the form of unpredictable dimensions, circles that are not truly round, and unacceptable witness marks. If the *X* (diameter-controlling) axis of a turning center develops backlash, for example, a noticeable witness mark will be left on the workpiece whenever a cutting tool reverses direction in the *X* axis (as would be the case in machining a spherical shape, like a trailer hitch ball).

Backlash should be checked on a regular basis. The check is simple. A common dial indicator can be mounted on the machine in such a way that turning the handwheel in one direction causes the axis to move against the stylus of the dial indicator. Once this happens, the handwheel can be turned in the other direction. When the indicator starts moving, the amount of backlash can be determined.

In all cases, the best way to eliminate backlash is to do so mechanically, if possible, by tightening gibs or making other adjustments. However, backlash compensation is a control feature that will automatically add the amount of backlash to *every* axis reversal.

WHEN TO USE THE CURRENT COMMAND PAGE

Many functions of programming can, if misused, cause problems that make it look like machine misfunction. In some cases, a programmer, setup person, or operator may accidentally invoke a command that places the machine in an unexpected mode. A CNC person who does not know that the command has been given may suspect that something is wrong with the machine.

For example, G90 specifies the absolute mode for most machining centers, and is often commanded in a CNC program. Suppose, for example, the programmer transposes numbers and specifies G09 by mistake. G09 on one popular control specifies an *exact stop check*, which causes the machine to stop at the end of each motion command. If the G09 command is unfamiliar to the CNC operator, the motions the machine makes while in the exact stop mode will appear very unusual. The CNC person will probably suspect that something is wrong with the machine.

In similar fashion, M98 and M99 on many turning center controls specify subprogramming commands. But similar G commands for many turning centers specify feed-per-minute (G98) and feed-per-revolution (G99) modes. If beginning turning center programmers mistake G98 for M98 when trying to command a subprogram, they will unwittingly place the machine in feed-per-minute mode. Since most turning centers are programmed predominantly in feed-per-revolution mode, *all* subsequent cutting commands will move very slowly. Again, the beginning CNC person will probably suspect a machine problem.

These are but two examples of times when CNC programming commands are unknowingly given that cause the machine to behave in an unusual manner. Other examples include inch versus metric mode commands, mirror image, scaling, and forgetting to cancel cutter-radius compensation.

It is important to know about a special display screen page that is available on almost all CNC controls that will show the current commands at any time. Commonly called the *program check page*, this function will allow the CNC operator to determine all currently invoked CNC commands. In many cases, this will help the operator spot the reason why a CNC machine is behaving in an unusual way.

UNDERSTANDING THE IMPORTANCE OF PARAMETERS

CNC controls that can be applied to a number of different machine tools use *parameters* to inform the control about certain machine functions for the particular machine being used. One very popular machining center control model can be applied to vertical and horizontal machining centers made by any number of different machine tool builders. Parameters are used to inform this machining center control model about specific details regarding the actual machine tool being controlled. Things like maximum spindle speed, rapid rate, maximum feed rate, and axis travels are among the many hundreds of functions parameters control.

All CNC users should maintain records of parameter settings for all CNC machines they use. According to one popular control manufacturer, the largest control-related cause of downtime is related to parameters. When a CNC control malfunctions, many times a service engineer must *initialize* the control. Initializing involves clearing all memory of the control, including CNC programs, offsets, *and parameters.* Once the control is repaired (possibly within minutes), parameters must be reloaded. If the CNC user has not kept an updated copy of parameter settings, the machine could be down for days while a service engineer uses trial-and-error techniques to get the machine up and running again.

While some machine tool builders or distributors will keep parameter records for you, do not depend on them to do so. As the CNC user, you *must* do so for yourself. Almost all current-model controls allow you to save and retrieve parameters in much the same manner as CNC programs, meaning you can probably use your current distributive numerical control system to transfer parameters to a personal computer's hard disk. If you do not have a distributive numerical control system, write down all of the parameters within your control on a piece of paper. The time you spend doing so is nothing compared to the time you will waste trying to determine and reload parameters the first time your machine requires initializing.

The importance of parameters cannot be overstressed. A very close analogy is backing up the hard drive of a personal computer. If you do not back up your data, when the hard drive of your computer fails and must be reformatted, a great deal of information will be lost.

Parameter types

Though the parameter documentation that accompanies most CNC controls is rather difficult to understand, you will want your CNC people to become well versed about the things parameters control. Many parameters, for example, affect the way programs are executed. Knowing that there is a parameter controlling a certain programmable function may sometimes allow your people to do things previously thought impossible. For example, a parameter on one popular turning center control determines whether the work-shift function is enabled or not. If this parameter is set such that the work-shift function is turned off, the display screen will not even show the work-shift page, meaning work-shift will be impossible.

This is but one example of the kind of thing program-related parameters can control. There will be many more within your particular controls. Other parameters of interest to end users include those:

- For protecting important CNC programs,
- For setting a maximum offset value,
- Related to how compensation types behave,
- Related to editing functions,
- Related to program transfers,
- Related to initialized states of the machine,
- Related to canned cycles.

Learn more about parameters. At the very least, read the parameter section of your control's maintenance manual. Question your machine tool builder about those you do not understand.

THE PC LADDER

Until relatively recently, the CNC control manufacturer handled only the electronic components related to the CNC control and axis drive systems of most machine tools. The machine tool builder was responsible for all other electronic components, including those related to miscellaneous devices, automatic tool changers, spindle drive motor, and other important machine functions. The connection of mechanical and electronic devices to the CNC control is commonly known as the *interface* between the machine tool and control.

Today, CNC control manufacturers are taking on more of the responsibility of controlling miscellaneous machine tool builder-supplied devices. They do so with *programmable controllers*. Though our explanation is somewhat simplistic, almost all functions related to CNC equipment are basic timing functions. Imagine, for example, the activation of an automatic tool changer on a vertical machining center. All that really happens during a tool change is that a series of solenoid valves and hydraulic or pneumatic cylinders are activated in sequential order. The cylinders activate the mechanical devices that actually make the tool change.

The machine's programmable controller (PC) is what controls the *sequence* of what happens during any mechanical device's activation. Indeed, almost every electrical and mechanical device supplied by the machine tool builder is controlled through the use of the programmable controller.

Each function of the programmable controller is commonly represented as a rung of a ladder on the schematic diagrams that accompany every machine (hence the name PC ladder diagram). Many CNC controls can even display the PC ladder diagram on the control's display screen, and some even allow programming of the programmable controller using the keyboard and display screen of the CNC control itself.

Though most end users will never have to actually program their machines' programmable controllers, your maintenance people should at least be able to display the desired PC ladder screens should they need the help of a service technician from the machine tool builder.

Index